# PLANT POPULATION
## ECOLOGY

# PLANT POPULATION ECOLOGY

THE 28TH SYMPOSIUM

OF THE BRITISH ECOLOGICAL SOCIETY

SUSSEX 1987

EDITED BY

## A. J. DAVY
School of Biological Sciences
University of East Anglia

## M. J. HUTCHINGS
School of Biological Sciences
University of Sussex

## A. R. WATKINSON
School of Biological Sciences
University of East Anglia

BLACKWELL SCIENTIFIC PUBLICATIONS

OXFORD LONDON EDINBURGH

BOSTON MELBOURNE

First published 1988
Typeset in Great Britain by William Clowes
Limited, Beccles and London
Printed and bound in
Great Britain by
Alden Press and Green Street Bindery

DISTRIBUTORS

USA
 Publishers' Business Services
 PO Box 447
 Brookline Village
 Massachusetts 02147
 (*Orders:* Tel. (617) 524-7678)

Canada
 Oxford University Press
 70 Wynford Drive
 Don Mills
 Ontario M3C 1J9
 (*Orders:* Tel. (416) 441-2941)

Australia
 Blackwell Scientific Publications
 (Australia) Pty Ltd
 107 Barry Street
 Carlton, Victoria 3053
 (*Orders:* Tel. (03) 347-0300)

British Library
Cataloguing in Publication Data

Plant population ecology.
 1. Plants. Population   I. Davy, A. J.
 II. Hutchings, M. J.
 III. Watkinson, A. R. 581.5′248

Library of Congress
Cataloging-in-Publication Data

British Ecological Society. *Symposium (28th:
1987: Sussex)*
 Plant population ecology: the 28th
 Symposium of the British Ecological
 Society, Sussex, 1987 / edited by A. J.
 Davy, M. J. Hutchings, A. R. Watkinson.
 1. Plant populations—Congresses.
 2. Vegetation dynamics—Congresses.
 3. Botany—Ecology—Congresses.
 I. Davy, A. J.  II. Hutchings, M. J.  III.
 Watkinson, A. R.   IV. Title. QK910.B75
 1987 581.5′248—dc19

ISBN 0-632-02349-X

# CONTENTS

# PREFACE

This book comprises twenty-one invited papers given at an international symposium that was held at the University of Sussex from 7–9 April 1987.

Interest in the population ecology of plants has burgeoned over the last 20 years. Much of the initial impetus came from one very influential Presidential Address to the British Ecological Society given by Professor Harper in 1967. Subsequently, the challenge has been taken up worldwide; British Ecological Society Journals, particularly *The Journal of Ecology*, have provided an outlet for much of the work. Textbooks and several important symposia in various parts of the world have followed. Paradoxically, no symposium meeting in Britain has ever been devoted to a synthesis and review of progress in plant population ecology. We perceive this as an omission, which we have attempted to rectify by organizing this symposium.

The last time that the symposia of the British Ecological Society specifically addressed population biology, it was to consider population dynamics of organisms in general. Hence, we may reasonably be asked why the current meeting (and volume) should be restricted to plants; surely the same general principles apply to all populations? We admit to positive discrimination in favour of what has been formerly a minority interest. But there is practical justification for this: a more broadly based meeting that did justice to the body of work on plants would have been unmanageably large and would have blurred our focus. It must also be recognized that populations of many higher plants, by virtue of their indeterminate, metameric growth, relative lack of mobility, diversity of breeding system and autotrophic nutrition, do have properties that are not typical of animal populations. Plant population biology is now a recognizable discipline in its own right, whose techniques range from classical demography to those of molecular genetics.

This symposium volume is arranged as follows. The introductory group of four papers (Chapters 1–4) takes an evolutionary perspective; it is concerned with life-history variation and the genetic structure of populations. The following three contributions (Chapters 5–7) examine aspects of variation in individual behaviour within plant populations. Physiological, demographic and genetical influences on the distribution and abundance of plants are considered in Chapters 8–10, whilst Chapters 11–15 explore the dynamics of plant populations with a range of contrasting life-histories in both the seed and vegetative phases of the life-cycle. Chapters 13–15 also examine the dynamics of individual species' populations in a community context and lead to a discussion (Chapters 16–20) of how species' interactions affect the population ecology of plants. The final chapter provides an overview of the volume—an apophasis of plant population biology.

We are indebted to numerous referees for their help in the preparation of this book. Two of us (A. J. Davy and A. R. Watkinson) are grateful to Dr M. J. Hutchings for the large part he played in making the meeting such a success, through his excellent local arrangements at the University of Sussex. He is in turn grateful to Christine Robinson at the Conference Office and Catering Manager at the University of Sussex for ensuring that everything ran smoothly, and to Andrew Slade for valuable secretarial assistance.

A. J. DAVY
M. J. HUTCHINGS
A. R. WATKINSON

# 1. LIFE-HISTORY VARIATION AND ENVIRONMENT

## A. J. DAVY[1] AND H. SMITH[2]

[1]*School of Biological Sciences, University of East Anglia, Norwich NR4 7TJ, UK and*
[2]*Department of Zoology, University of Oxford, South Parks Road, Oxford OX1 3PS, UK*

## SUMMARY

**1** Variations in plant life-histories arise in response to variations in the environment. Plastic changes imposed directly by the environment may tend to reinforce or oppose the evolutionary changes that result from selection by the environment. Life-history theory frames the selection pressures in tantalizingly abstract terms: they are difficult to quantify or test rigorously. Syndromes of co-adapted traits, referred to as 'life-history strategies', can be correlated with particular environments but the interpretation of such correlations as adaptations is dangerously susceptible to circular argument and 'adaptive story-telling'.

**2** The life-history characteristics of a plant (dispersal, dormancy and age-specific survival, growth and fecundity) are components of its evolutionary fitness, defined as its relative, life-time capacity for leaving offspring. Reciprocal transplantation of related populations between different environments allows precise estimation of selection coefficients at each stage in the life-cycle. It is then possible to test whether selective forces predicted *a priori* materialize as measurable features of the environment at the appropriate stage.

**3** It is important that the transplants reflect the natural range of population density. Where there is a density-dependent element in population regulation, as in many annual plants, it operates through particular life-history characteristics (survival or fecundity); the finite rate of population increase and the effects of selection on it may also be density-dependent. Equilibrium population sizes may change as a result of transplantation.

**4** Coastal salt marshes provide a natural field laboratory for the investigation of infra-specific evolution in life-histories. The spatial heterogeneity and temporal predictability of their environments suggest, *a priori*, both the development of genetic differentiation and the nature of many of the selective forces. We illustrate extensive reciprocal transplants with an analysis of two populations of *Salicornia europaea* agg., an inbreeding annual with non-overlapping generations. The probability of survival to establishment, growth phase survival, fecundity and the finite rate of population increase all showed negative density-dependence.

**5** Selection was generally against the alien populations but was variable in time and space; it varied considerably with life-history stage, often asymmetrically between the two populations at a particular stage, and it varied between replicate plots. Whole life-cycle selection coefficients, based on finite rates of population increase, were decisively against the alien transplants. Transplantation into a matrix of perennial vegetation induced stronger selection than that into cleared plots. Selection varied between growing seasons, depending on the weather; between-season differences in selection at one stage in the life-cycle may be compensated by differential selection at a later stage.

**6** Even when selection by whole environments has been measured rigorously, it can be difficult to isolate specific selective agents, because of these complex interactions. We need a fuller understanding of the genetic basis of life-history differentiation to know the extent to which it is the product of past, or intermittent, selective forces rather than strictly contemporary ones.

## INTRODUCTION

It is axiomatic that different environments support distinctive assemblages of species and, perhaps, distinctive assemblages of genotypes within species. The assemblage of species within a particular environment generally displays a spectrum of variation in life-history characteristics (dispersal, dormancy and age-specific survival, growth and fecundity); but a suite of co-adapted characteristics that represents a consensus, or 'strategy', may be presumed to be an evolutionary response to the selection pressures of the environment (Southwood 1976; Stearns 1976, 1977; Grime 1979). A very similar argument applies to the range of genotypes within a species. Genotypes whose life-histories allow them, on average, to produce more offspring will displace less fit genotypes over successive generations (Cole 1954; Schaffer 1974; Antonovics 1976; Schaffer & Rosenzweig 1977, Charlesworth 1980).

The intuitive attractiveness of such arguments can, unfortunately, camouflage the serious difficulties encountered in attempting to test the specific predictions of life-history theory. One approach involves large-scale comparisons of unrelated plant species in particular, often extreme, types of environment. Few would dispute that the convergent life-histories of mangroves represent adaptive, evolutionary responses to the tropical intertidal environment (Tomlinson 1986), or that those of desert ephemerals are evolutionary responses to periods of prolonged drought punctuated by unpredictable rainfall; the very distinctive life-history traits of taxonomically divergent weeds must be the results of selection by the agricultural environment (Baker 1974). Yet in many ecosystems there is apparently a diversity of solutions to common evolutionary problems. Annuals coexist with perennials, semelparous with iteroparous plants, and herbs with trees. The forces that structure such communities are themselves imperfectly understood but there can be no doubt that the environment of any individual is determined partially by its neighbours, of the same and different species. The environment thus comprises numerous pressures, possibly conflicting or acting differently at different stages in the life-history. Recruitment, for instance, may be favoured by circumstances very different from those associated with mature individuals (Grubb 1977). It may not be helpful at all to generalize about the life-history consequences of particular 'environmental factors'.

An alternative approach is to compare closely related species (Harper *et al.* 1961) in contrasting environments (Abrahamson & Gadgil 1973; Wilbur 1976; Hickman 1977). Unless the pressures exerted by the environment are predictable *a priori*, however, there is a real danger of circular argument and 'adaptive story-telling' (Gould & Lewontin 1979) in the interpretation of life-histories. Ideally the relevant features of the environment should be measured. Even where this has been done, much of the evidence for plants arises from correlation rather than experiment and therefore it is difficult to establish causation. The environmental measurement may itself be correlated with the selective variable and selection may not be acting on the measured character but on a developmentally or genetically linked character. Many studies have examined correlations between combinations of life-history traits and different types of environmental variability (reviewed by Stearns 1976). Correlations have been sought that support theoretical predictions based on assumptions about optimal adaptive strategies (Stearns 1977). Most studies fail to take account of the possibility of non-adaptive differences or of genetic or developmental constraints on theoretical optimality (Gould & Lewontin 1979; Stearns 1980).

The value of experimental manipulation of natural environments by transplantation has been appreciated since the work of Bonnier and Kerner von Marilaun in the 1880s (see Briggs & Walters 1984). Only recently has such manipulation been applied to investigation of the adaptiveness of life-history traits (Antonovics & Primack 1982; Schemske 1984; Davy & Smith 1985). We too often seek insight into life-histories from the environments where plants are successful when it may be more instructive to investigate how they fail elsewhere. Small-scale variation in the environment provides the potential for development of genetically differentiated populations within species. Differentiation will occur wherever the disruptive selection pressures are sufficient to overcome the effects of gene flow (Jain & Bradshaw 1966; McNeilly & Bradshaw 1968; Snaydon 1970; Bradshaw 1972; Snaydon & Davies 1972). Plant species also of course may adjust to heterogeneous environments through phenotypic plasticity, which is itself a capacity that is selectable by the environment, and the plasticity of life-history traits is no exception (Bradshaw 1965; Schlichting 1986). Both plasticity and the strength of selection in the field can most appropriately be measured by experiments involving reciprocal transplantation between the phases of heterogeneous environments.

Examination of the infraspecific life-history responses of plant populations to variation in the environment clearly requires that we distinguish between life-histories that are imposed directly on the plastic phenotype by the environment from those arising through genetic differentiation. Where there are genotypic effects, the interpretation of their significance will depend on variations in selection pressure in time and space. Variations in time include both those within the span of a single life-cycle and those affecting successive generations. Variation in space raises the question of how the distributions of other species affect selection: is it possible to separate components of selection attributable to biotic and physical aspects of the environment or are they always inextricably interwoven?

## PHENOTYPIC PLASTICITY AND LIFE-HISTORY

Plastic responses to the environment can probably be found in all plant species; they may or may not be accompanied by genetic differentiation. Where both types of response occur together, plasticity may reinforce the variation associated with genetic differentiation and thus be considered adaptive. Alternatively it could play a neutral role or even oppose the effects of differentiation. The distinction emerges from a reciprocal transplant using the seedling progeny of tetraploid ($2n = 56$) birch trees of the *Betula pendula*/ *B. pubescens* complex from adjacent bog and heath habitats (Davy & Gill 1984). Eight different quantitative characters that together describe size and shape of leaf found in the saplings were measured 3 years after transplantation. These were summarized using principal components analysis (Fig. 1.1). In each habitat there is a clear, consistent division between the domains of variation occupied by the progenies of bog and heath mothers respectively: this reflects genetic differentiation. There is also an equally consistent displacement between the domains of the progenies of particular mothers when planted in bog and heath habitats respectively: this is phenotypic plasticity. The two types of displacement are of similar magnitude but in

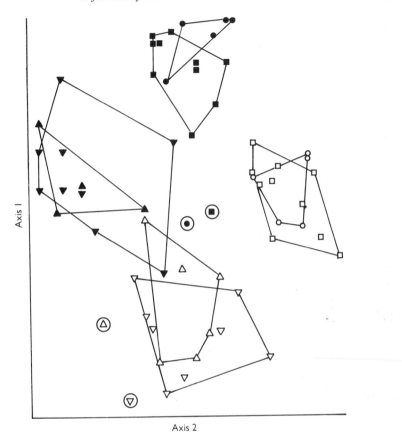

FIG. 1.1 Principal components analysis ordination of a matrix of eight leaf characters × sixty-one *Betula pubescens/pendula* trees transplanted reciprocally between bog and heath at Dersingham, Norfolk (fifty-seven seedling transplants and four parents). Closed symbols, trees growing on bog; open symbols, trees growing on heath. Heath provenance: circles, tree 3 and its progeny; squares, tree 4 and its progeny. Bog provenance: triangles, tree 14 and its progeny; inverted triangles, tree 15 and its progeny. The four parents are encircled and the domains of the eight groups of progeny are shown as polygons. Redrawn from Davy & Gill (1984).

quite different directions, which suggests that the plasticity is not reinforcing the effects of differentiation in this case.

The relationship between these changes in leaf growth and particular life-history traits, and their possible significance for overall fitness are not obvious; the longevity of a tree deters investigation. There are, however, many reports of plasticity in life-history characters, especially in relation to the components of yield. Different traits may respond differently to an environmental stimulus, changing the phenotypic correlation between them dramatically (see Schlichting 1986).

The direct influence of environments in determining life-histories should not be underestimated. Antonovics & Primack (1982) made a detailed demographic study of field reciprocal transplants of *Plantago lanceolata* between a range of contrasting sites. *P. lanceolata* has an iteroparous, perennial life-history and in their study showed generally concave (Deevey type III) survival curves. They concluded that environmental effects on life-history were large relative to genetic differences. Populations showed no significant life-history differences when averaged over all sites; within a site there was no evidence of population differences in survival, even in the early stages of growth when numbers were large. They also reported that the large differences between populations in the proportion of biomass devoted to reproduction were largely phenotypic in origin, a circumstance likely to retard the evolution of life-history differences (Primack & Antonovics 1982).

*Bromus tectorum*, another weedy species, is an example of a semelparous annual or ephemeral whose life-history is apparently dominated by the environment. Mack & Pyke (1983) examined its recruitment, survival and fecundity in moist, mesic and dry habitats over three consecutive generations. Even without transplant experiments, the demographic responses to year-to-year variations in weather and predator activity clearly showed that the environment overrode the intrinsic differences between the habitat types. Different cohorts showed a variety of survival curves, depending for example on drought or snow cover, but the commonest was convex (Deevey type I).

Plants with fundamentally very different life-histories can be susceptible to these substantial plastic modifications to their life-histories resulting from environmental variation. Relatively unpredictable environments are more likely to impose life-histories on their species and are less likely to permit genetic differentiation within them.

## GENETIC DIFFERENTIATION AND LIFE-HISTORY

Life-history characteristics have frequently been used as measures of performance, since these may be expected to influence fitness directly. For example, Law, Bradshaw & Putwain (1977) reported that samples from populations of *Poa annua* experiencing predominantly density-independent regulation had shorter pre-reproductive periods, higher seed output early in life and shorter lives than those from populations experiencing density-dependence, when they were grown in a common environment. Selection has been measured on the basis of survival at different stages in the life-cycle (Cavers & Harper 1967; Lindauer & Quinn 1972; Clegg & Allard 1973;

Davies & Snaydon 1976), of growth (McNeilly & Bradshaw 1968; Baker 1978), and of seed production (Clegg & Allard 1973). The presence of demonstrable selection at any one stage in the life-cycle, however, does not necessarily predict overall fitness. More recent studies have attempted to measure selection at all stages in the life-cycle (e.g. Clegg *et al.* 1978; McGraw & Antonovics 1983) or by the finite rate of population increase over a whole generation (Davy & Smith 1985; Schmidt & Levin 1985; Watkinson & Gibson 1985).

Coastal salt marshes provide a natural field laboratory for the investigation of infraspecific evolution in life-histories. Their properties are such that we would predict, *a priori*, both the development of genetic differentiation and the nature of many of the selective forces. Salt marshes are essentially heterogeneous in space: the overall gradient from sea to land has superimposed upon it a mosaic of smaller scale topographic variation associated with the anastomozing, sinuous creeks with their naturally raised banks ('levees'), and the pools, depressions and salt-pans that punctuate the 'interfluve' areas between them. In marshes of complex physiography, such as those of the north Norfolk coast of England, there may also be physical barriers between areas of upper and lower marsh in the form of ancient shingle ridges (Jefferies, Davy & Rudmik 1979). Such a barrier creates separate drainage systems and greatly inhibits gene flow, as floating seeds, between marshes. In contrast coastal marshes are extremely predictable in time, in the sense that the dominant environmental feature, tidal inundation, is cyclic; the cycle from solstice to equinox and the lunar monthly cycle modulate the twice-daily circadian cycle in a complex fashion which is nonetheless consistent from year to year.

Sediment salinity depends on the frequency of tidal inundation and the net difference between precipitation and evapotranspiration. Consequently it is affected by the more or less deterministic tidal cycle and by the weather, which has a stochastic element. At Stiffkey Marsh, north Norfolk, there is a clear distinction between upper and lower marsh habitats. The lower marsh is flooded by most of the twice-daily high tides of the year. Sediment salinity fluctuates around that of sea water with a low amplitude. The upper marsh is covered completely only by spring tides, and is not even covered by these in the cycles immediately before and after the solstice. The upper marsh can thus be exposed continuously for up to 8 weeks. In winter precipitation exceeds evapotranspiration and so the sediments become hyposaline. In summer the contrary is normally true; salinities up to three or four times higher than sea water (2 mol dm$^{-3}$ with respect to Na$^+$) and concomitantly negative water potentials are not exceptional (Jefferies *et al.* 1979; Smith 1985), depending on the weather.

Short-lived plants have obvious advantages for life-history studies. Annual plants in the genus *Salicornia* occur typically throughout the marsh systems of north Norfolk; their populations may be large, extensive and variable in density. Genetic differentiation in *Salicornia* would be predicted *a priori* because it is believed to be substantially inbreeding, through cleistogamy (Ball & Brown 1970; Jefferies & Gottlieb 1982). Another genetic barrier arises from the existence of diploid ($2n = 18$) and tetraploid ($2n = 36$) forms. There is certainly variation in phenotype, especially in growth-form, size and pigmentation, but many of the differences between local populations are small and statistical. The complex has defied coherent taxonomic treatment, although a number of microspecies have been recognized whose fertile phenotypes appear to correspond with populations that inhabit particular microhabitats on the marsh. For instance plants with the single-flowered cymes and disarticulating branches of *S. pusilla* are found predominantly on the highest, best-drained areas, whereas the robust tetraploid with strongly basal branching, *S. dolichostachya*, is characteristic of the sand flats at the seaward edge of the marsh. We have experimented with several types but here we wish to present a detailed analysis of only the typical, diploid interfluve forms (*S. europaea* complex) of the lower and upper marsh. The lower marsh and upper marsh populations correspond nearly with the microspecies *S. europaea sensu stricto* and *S. ramosissima* respectively.

One of the striking differences between these two populations is in the phenology of their growth. Lower marsh plants increase in dry mass fairly steadily from May to September; plants on the upper marsh, however, show relatively slow growth in the early part of the growing season, up to July, before making a spurt of growth that is associated with flowering. The period of slow growth includes the period when hypersaline sediments are likely to develop. When plots on the high marsh were irrigated regularly with sea water to relieve hypersalinity the same delayed pattern of growth was observed; even the addition of nitrogen only stimulated growth appreciably in the later part of the season (Jefferies, Davy & Rudmik 1981). A reciprocal transplant of spaced seedlings revealed essentially the same determinate pattern of growth in upper marsh plants irrespective of whether they were grown on upper or lower marshes (Fig. 1.2). It seems probable, therefore, that it is a genetic response to a period when conditions very adverse to growth can occur. Although this experiment takes no account of mortality, a selective advantage associated with delayed growth on the upper marsh leads to a prediction of increased average risk of mortality for low marsh migrants to the upper marsh.

The life-history of *Salicornia* comprises annual non-overlapping genera-

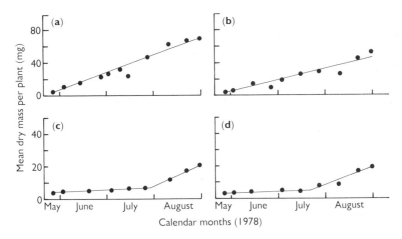

FIG. 1.2    Growth of seedling reciprocal transplants between populations of *Salicornia europaea* agg. from upper and lower marshes, respectively at Stiffkey, Norfolk. (a) Lower marsh population transplanted to lower marsh; (b) lower marsh population transplanted to upper marsh; (c) upper marsh population transplanted to lower marsh; (d) upper marsh population transplanted to upper marsh. Redrawn from Jefferies, Davy & Rudmik (1981).

tions, with no persistent seed bank. The lack of a persistent seed bank in an annual is consistent with the predictability of the salt marsh environment. The upper marsh population is representative of the general pattern (Fig. 1.3), as revealed by random cores of sediment sampled between November 1980 and May 1982. Mature plants decomposed and shed the seeds embedded in their succulent segments into the sediment from late September to December (Fig. 1.3a). Initially, a very high proportion of seeds was innately dormant (Fig. 1.3b). This dormancy was lost after a few weeks of exposure to winter temperatures and there was virtually no dormancy from January to April. Germination started in January but most of it took place continuously from February to April and so did not produce discrete cohorts (Fig. 1.3c). The small proportion of seed that had not germinated by May became dormant again. This summer dormant seed germinated readily when removed from the cores and placed in distilled water, which suggests that dormancy was enforced by salinity. By August virtually the entire population was present as growing plants; the very few seeds remaining by the following September had lost their viability, as judged by tetrazolium staining. As its generation time corresponds with the annual cycle of tidal influence, *Salicornia* is eminently suitable for an examination of temporal variations in selection pressure within and between generations.

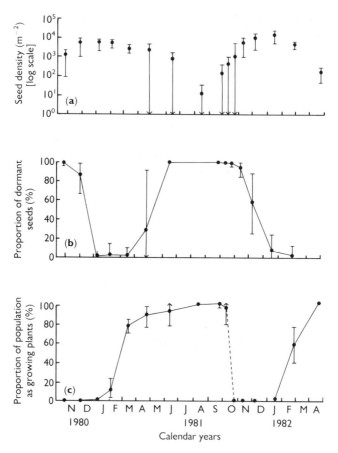

FIG. 1.3 Characteristics of a representative *Salicornia europaea* agg. population on the upper marsh at Stiffkey, Norfolk during the annual life-cycle. (a) Total viable seed bank; (b) proportion of dormant seeds in the seedbank; (c) proportion of the total population present as vegetative plants rather than ungerminated seed. Vertical bars represent 95% confidence intervals, $n = 20$.

### Variation in selection at different stages of the life-cycle

The most detailed evidence comes from reciprocal transplants of seeds made between interfluve areas of the upper and lower marshes in November 1981. At each site four replicate 2 × 2 m plots had been cleared of their indigenous *Salicornia* populations prior to flowering and seed set. Each population was planted at each plot in a randomized series of nine quadrats (100 × 100 mm) representing the range of density normally found in the field. We designed the transplanting procedure to mimic the natural process as closely as possible, because individual dry seeds dissected from the maternal tissues have an abnormal risk of being removed by tidal scour. The seeds were sown as precalibrated mature, fertile segment lengths of the parent populations, which were held in place on the surface of the sediment with 100 × 100 mm squares of 4 mm aperture 'Netlon' mesh ('turves'), themselves anchored at the edges with hooked pins. The segments were allowed to decompose naturally *in situ*. We monitored the fates of these transplants over the following year to reproductive maturity in October 1982. We present coefficients of selection at each stage that are relative to the best performing population at each site rather than necessarily to the home population (McGraw & Antonovics 1983);

$$\text{Selection coefficient} = 1 - \frac{\text{Performance of population}}{\text{Performance of best-performing population}}$$

### Survival to establishment

Seedling populations developed to a maximum in April, with net recruitment to the lower marsh population tending to lag behind that of the upper marsh one at both sites (e.g. Fig. 1.4). Effective survival after mortality and emigration of seeds and germinating seedlings ranged from 27–96%, and it

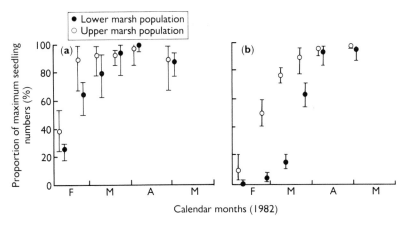

FIG. 1.4 Number of seedlings of *Salicornia europaea* agg. during the establishment phase, expressed as a proportion of the maximum number of seedlings recorded, in representative plots of the reciprocal transplant experiment on (a) the lower marsh and (b) the upper marsh at Stiffkey, Norfolk. Each point represents the back-transformed mean of the arcsin-transformed percentages for all turves in each set on which more than 100 seeds had been sown. Vertical bars represent 95% confidence intervals.

varied considerably between the four replicate plots in every case. The upper marsh population survived better consistently on all of the plots. This is reflected in the selection coefficients (Table 1.1). On the lower marsh there was selection against the home form of 0.12–0.25, although the coefficients do not achieve statistical significance. On the upper marsh the selection against the alien population of 0.27–0.47 was highly significant. The selective forces exerted by the two environments appear to be asymmetrical at this stage.

TABLE 1.1. Selection coefficients for survival during the period between sowing and the end of the seedling establishment phase in *Salicornia europaea* agg. populations transplanted reciprocally between upper and lower marsh sites at Stiffkey, Norfolk.

| Transplant site | Selection coefficient | F | Significance |
|---|---|---|---|
| *Lower marsh* | | | |
| Plot 1 | 0.20 H | 2.32 | NS |
| Plot 2 | 0.10 H | 1.03 | NS |
| Plot 3 | 0.25 H | 3.55 | NS |
| Plot 4 | 0.12 H | 0.83 | NS |
| *Upper marsh* | | | |
| Plot 1 | 0.43 | 91.6 | *** |
| Plot 2 | 0.36 | 14.8 | ** |
| Plot 3 | 0.47 | 13.3 | ** |
| Plot 4 | 0.27 | 11.0 | ** |

Coefficients are calculated from the back-transformed means of arcsin transformed survival percentages in all turves sown with more than 100 seeds, after adjustment for variation in sowing density using one-way analysis of covariance. H indicates selection *against* the home population; otherwise selection is against the alien population. (significance: ***$P <$ 0.001; **$P < 0.01$; NS: not significant.)

*Survival during the growth phase*

Net numbers of plants declined in all plots during the growing season (Fig. 1.5). The small numbers of late recruits (after April) mostly survived for a very short period and will have resulted in a slight overestimate of seedling mortality and a slight underestimate of growth phase mortality. Overall survival was generally lower than at the establishment phase, although the survival rate was actually higher. Survival was rather better on the higher than the lower marsh but the most striking result is that the alien populations fared very poorly indeed on both marshes. The selection coefficients against alien populations (0·66–0·89) were all highly significant (Table 1.2).

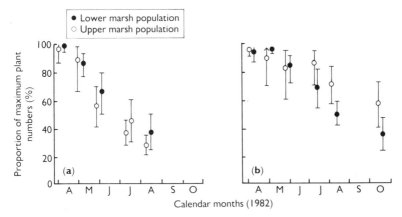

FIG. 1.5 Number of *Salicornia europaea* agg. plants during the growth phase, expressed as a proportion of the maximum number of seedlings recorded, in representative plots of the reciprocal transplant experiment on (a) the lower marsh and (b) the upper marsh at Stiffkey, Norfolk. Each point represents the back-transformed mean of the arcsin-transformed percentages for all turves in each set on which more than 100 seeds had been sown. Vertical bars represent 95% confidence intervals.

## Fecundity

The fecundity of the survivors in the transplant showed strong negative density-dependence (e.g. Fig. 1.6), a plastic response found in other *Salicornia* populations, both natural (Jefferies *et al.* 1981) and transplanted (Davy & Smith 1985). This is a phenomenon too important to be ignored in designing

TABLE 1.2. Selection coefficients for survival during the growth phase, in *Salicornia europaea* agg. populations transplanted reciprocally between upper and lower marsh sites at Stiffkey, Norfolk.

| Transplant site | Selection coefficient | F | Significance |
|---|---|---|---|
| *Lower marsh* | | | |
| Plot 2 | 0.72 | 53.42 | *** |
| Plot 3 | 0.89 | 26.59 | *** |
| Plot 4 | 0.87 | 40.74 | *** |
| *Upper marsh* | | | |
| Plot 1 | 0.77 | 32.03 | *** |
| Plot 2 | 0.77 | 39.10 | *** |
| Plot 4 | 0.66 | 27.00 | *** |

Coefficients are calculated from the back-transformed means of arcsin transformed survival percentages in all turves sown with more than 100 seeds, after adjustment for variation in the density of established seedlings using one-way analysis of covariance. Selection is against the alien population (significance: ***$P < 0.001$).

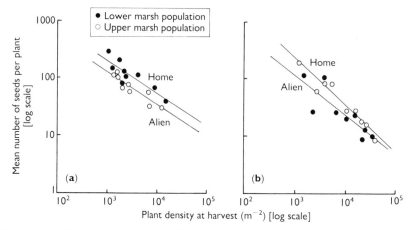

FIG. 1.6   Density-dependent fecundity of *Salicornia europaea* agg. plants in representative plots of the reciprocal transplant experiment on (a) the lower marsh and (b) the upper marsh at Stiffkey, Norfolk. All regressions are significant at $P < 0.01$.

transplant experiments because it appears to be the basis of population regulation in many annual plants (Watkinson & Davy 1985). Yet rigorous comparisons of the performance of the upper and lower marsh populations must take account of the density effect.

It is a moot point as to whether fecundity would be influenced most by density at seed maturity, density at the beginning of reproductive growth, or even density at the establishment of the population of rooted plants; additional reproductive growth in response to declining density requires time. We examined analyses of covariance that adjust for variation in plant density at three dates representative of these phases. Although there are significant differences at earlier dates, the harvest density is probably the most appropriate, unless there is severe mortality shortly before maturity. Selection coefficients on this basis (Table 1.3) suggest strong, significant selection against upper marsh plants at the lower marsh site but much weaker and statistically insignificant selection against lower marsh plants on the upper marsh. Again there was asymmetrical selection, now in the opposite direction to that at the establishment stage.

*Finite rate of population increase*

Undoubtedly the best overall measures of fitness and selection are derived from the capacity of members of one generation to contribute offspring to the next. For a population with discrete generations this is represented by

TABLE 1.3.  Selection coefficients for mean fecundity in *Salicornia europaea* agg. populations transplanted reciprocally between upper and lower marsh sites at Stiffkey, Norfolk.

| Transplant site | Selection coefficient | Significance |
|---|---|---|
| *Lower marsh* | | |
| Plot 2 | 0.43 | 12.42** |
| *Upper marsh* | | |
| Plot 1 | 0.22 | slopes sig. different |
| Plot 4 | 0.13 | 0.58 NS |

Significance levels are based on the log-transformed means, adjusted for variation in harvest density by one-way analysis of covariance (significance: **$P < 0.01$; NS: not significant).

the finite rate of population growth, or in this context, the number of seeds produced on average for each seed sown one generation previously. Mean finite rates of population growth in the *Salicornia* transplant (e.g. Fig. 1.7) showed significant negative density-dependence. Generally, the equilibrium densities were near the upper limit of sowing density: transplantation typically resulted in much lower equilibrium population densities in the alien populations than in the home ones. Selection coefficients derived from these rates of population increase (Table 1.4) are presented for representative plots with sufficiently complete life-history data.

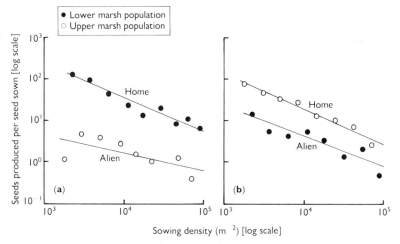

FIG. 1.7  Finite rate of population increase of *Salicornia europaea* agg. in relation to sowing density, in representative plots of the reciprocal transplant experiment on (a) the lower marsh and (b) the upper marsh at Stiffkey, Norfolk.

TABLE 1.4. Selection coefficients for mean finite rate of population increase in *Salicornia europaea* agg. populations transplanted reciprocally between upper and lower marsh sites at Stiffkey, Norfolk.

| Site | Home population | Alien population | Selection coefficient |
|---|---|---|---|
| *Lower marsh* | | | |
| Plot 1 | 24.9 | 1.39 | 0.94*** |
| *Upper marsh* | | | |
| Plot 1 | 33.0 | 23.3 | 0.29* |
| Plot 2 | 3.9 | 0.8 | 0.80** |

Significance levels are based on the log-transformed means, adjusted for variation in harvest density by one-way analysis of covariance (significance: ***$P<0.001$; **$P<0.01$; *$P<0.05$; NS: not significant).

The range of finite rate of population increase compares closely with that reported for the annual *Phlox drummondii* (Schmidt & Levin 1985). Selection was consistently against the alien populations and of sufficient magnitude to account for genetic differentiation. Nevertheless the overall selection could be substantially less than that measured at certain constituent stages and different in direction from that at other stages. Such large variations in selection pressure at different stages of the life-history in an essentially predictable environment suggest that generalizing from any one stage could be very misleading. These measurements of selection based on large numbers of individual quadrats show also that, in a heterogeneous environment, it is possible to have large selection coefficients that are not statistically significant.

### Variation in selection from generation to generation

There is very little evidence concerning the long-term consistency of selection pressures on life-histories. How much do the weather or other stochastic variations in the environment modify the selection pressures on different life-history traits from generation to generation? Does a variation in selection pressure at one stage affect selection at later stages?

An earlier reciprocal transplant with *Salicornia* was carried out between November 1979 and October 1980, by essentially the same method, but on a smaller scale. The results were strikingly similar in some respects and different in others. A similarity was selection against the home population (0.27) at the establishment stage on the lower marsh: a difference was further selection against the home type in terms of fecundity (0.31). Very strong selection against the alien population on the lower marsh (0.80), acting on survival during the growth phase, more than offset both of these. On the upper marsh, selection was lower than in 1981–82 at the establishment phase, similar during the growth phase (0.38) and greater when acting on fecundity (0.63). Average whole life-cycle selection coefficients against the alien populations were 0.70 and 0.65 on lower and higher marshes respectively. It does seem that amelioration of the selection pressure at the establishment phase in a particular year can lead to intensification of the selection associated with subsequent survival or fecundity. There is clearly variation from year to year and even in an annual there is perhaps no *a priori* necessity for strong selection to occur every year.

### Variation in selection pressure in space

In a heterogeneous environment one of the important sources of small-scale environmental variation is likely to arise from the distribution of other, interfering species within the community. In the case of a salt marsh annual, the distribution of the matrix of long-lived perennial species might be expected to influence selection. Table 1.5 contrasts the selection coefficients against lower marsh plants transplanted to various microhabitats on the higher marsh: those with perennial vegetation, those cleared of it experimentally, and natural wet depressions and salt-pans. Again, reduced selection pressure on survival associated with experimental removal of the perennial vegetation led to selection operating more severely on fecundity; the effects on whole life-cycle selection were in fact not very different. Salt-pans and wet depressions are very adverse microhabitats of the generally adverse upper marsh, for lower marsh plants at all life-cycle stages.

## CONCLUDING REMARKS

We still have scant information as to the nature of the genetic differences responsible for the divergent life-history characteristics in closely related genotypes. Jefferies and Gottlieb (1982) examined the electrophoretic mobilities of thirty isoenzymes in fifteen populations of the *Salicornia europaea* aggregate. Twenty-four were identical in all individuals and only two types of plant could be distinguished on the basis of the two electrophoretic variants found in each of the other six isoenzymes. The upper marsh population at Stiffkey was predominantly (*c.* 90%) of one type and the lower marsh population predominantly (also *c.* 90%) of the other type. Further populations at Stiffkey that can be recognized by morphology or their response to transplantation, such as *S. pusilla* or *S. dolichostachya* (Smith 1985), were not distinguished in this analysis.

The monomorphic, apparently homozygous pattern found by Jefferies

TABLE 1.5. Stage-specific selection coefficients against the lower marsh population of *Salicornia europaea* agg. after transplantation to four different microhabitats on the upper marsh at Stiffkey, Norfolk.

| Stage | Cleared plots | Vegetated plots | Salt-pan | Depression |
|---|---|---|---|---|
| Survival of seeds to establishment | 0.24** | 0.43*** | 0.61 NS | 0.40*** |
| | 0.44** | 0.36*** | | |
| | 0.32* | 0.47*** | | |
| | 0.24** | 0.27*** | | |
| Survival during growth phase | 0.38 NS | 0.77*** | 0.57** | 0.65*** |
| | 0.37 NV | 0.77*** | | |
| | 0.10 NS | 0.66*** | | |
| Fecundity | 0.37* | 0.13 NS | 0.38 NS | 0.57*** |
| | 0.23 NV | 0.22 NV | | |
| Finite rate of population increase | 0.29* | 0.41* | 0.75 NS | 0.79*** |
| | 0.80*** | 0.54*** | | |

NV: not valid comparison for significance test as slopes of regressions significantly different; significance: ***$P<0.001$; **$P<0.01$; *$P<0.05$; NS: not significant.

and Gottlieb has certain parallels in the local population differentiation identified by Allard, Miller & Kahler (1978) in their studies of the allelic variability of enzyme loci in *Avena barbata* from mesic and xeric sites in California. It is possible that there are co-adapted gene complexes but, as Allard *et al.* point out, they could be markers fixed by inbreeding in a diploidised tetraploid; fixed heterozygosity can also result from gene duplication. The examination of restriction fragment length polymorphisms (RFLP) of DNA offers new opportunities for the analysis of genetic variability (e.g. Schaal, 1988). RFLP studies of *Salicornia* DNA are currently revealing much greater genetic diversity; different populations show characteristic distribution patterns of ribosomal DNA RFLPs (S. Noble, unpublished data).

Reciprocal transplantation is a powerful method of revealing the selection pressures that would operate on migrants entering particular environments and thus the forces that could maintain genetic differentiation of life-history traits. Clearly selection can vary greatly at different stages in the life-cycle and adequately replicated transplants can resolve the differences. The experiments with *Salicornia* draw attention to a number of assumptions and limitations of design in previous transplantation that may hinder progress. Some of these are purely technical. For instance, even selection coefficients that would normally be regarded as sufficiently large to account for genetic differentiation can fail to achieve statistical significance; we need more experiments that have been designed to allow selection coefficients to be tested. Similarly selection at any one stage in the life-cycle is not necessarily a good predictor of overall selection.

Another set of issues concerns the adequacy of the underlying paradigms for both selection by the environment and population regulation. There may be considerable variations in selection in time and space within an environment. In particular, where the selective forces are modulated by the weather or some other erratic influence, they may vary from year to year and hence from generation to generation. The consequences of such variations for the equilibrium size and genetic composition of populations are poorly understood. A change in selection pressure at one stage in the life-cycle may influence the strength of selection at subsequent stages, usually by negative feedback so as to be compensatory. Where density-dependent processes are believed to be involved in the regulation of populations it is essential that transplants should reflect the range of density normally encountered. Only by this approach can we determine the density at which the finite rate of increase is unity and therefore the equilibrium population sizes of transplants. Then we may ask whether genotypes with specific life-histories result in relatively larger or smaller populations in a particular environment (Antonovics & Via 1988).

The relationship between life-history variation and environment remains a difficult and challenging study. Even when we have established the existence of genetic differentiation and have made rigorous estimates of the strength of selection, the question may remain as to whether differentiation is occurring and being maintained currently, or whether the variability is part of an evolved response in populations that are now, for practical purposes, genetically or physically isolated.

## ACKNOWLEDGMENTS

We thank Diane Alden for much technical help in the field and laboratory. We are grateful to Dr Janis Antonovics and Dr Alan Gray for valuable discussion and comment. The National Trust and the Nature Conservancy Council kindly allowed work at Stiffkey, Norfolk. This work was supported by the Natural Environment Research Council (Grant Number GR3/4172).

## REFERENCES

**Abrahamson, W. G. & Gadgil, M. (1973).** Growth form and reproductive effort in goldenrods (*Solidago*, Compositae). *American Naturalist*, **107**, 651–61.

**Watkinson, A. R. & Gibson, C. C. (1985).** Life-history variation and the demography of plant populations. *Structure and Functioning of Plant Populations 2* (Ed. by J. Haeck & J. W. Woldendorp), pp. 105–13. North Holland Publishing Company, Amsterdam.
**Wilbur, H. M. (1976).** Life-history evolution of seven milkweeds of the genus *Asclepias*. *Journal of Ecology*, **64**, 223–40.

# 2. THE EVOLUTION OF REPRODUCTIVE SYSTEMS: A HIERARCHY OF CAUSES

P. H. GOUYON[1,2], C. J. GLIDDON[3] AND D. COUVET[1]

[1]CNRS CEPE BP 5051, F-34033 Montpellier cedex and
[2]CNRS GPDP, F-91190 Gif sur Yvette, France and
[3]School of Plant Biology, UCNW, Bangor LL57 2UW, UK

## SUMMARY

**1** The implicit evolutionary questions posed by previous workers concerning the evolution of sex are critically examined and the forces acting on the evolution of sex are discussed.

**2** Forces that give a long-term advantage cannot, *per se*, maintain sexual reproduction because the short-term disadvantage of sex, the cost of meiosis, will prevent the species from remaining sexually reproductive in the absence of short-term advantages.

**3** The existence of such short-term advantages has been suggested as sufficient to account for the existence of sexual reproduction and some have been demonstrated experimentally and/or theoretically.

**4** The diversity of these short-term causes, all leading to the same result (sexual reproduction), is unsatisfactory, given the ubiquity of sexual reproduction in the plant and animal kingdoms. Therefore, short-term forces, acting alone, are also rejected.

**5** A satisfactory explanation for the origin and maintenance of sexual reproduction is proposed that involves forces acting in both the long and short term. The forces controlling the extinction or survival of species determine which of the diverse outcomes of short-term forces will persist in the long term. That is, the explanation necessitates invoking a causality acting at a superior hierarchical level (more complex level of biological integration) that directs the way in which proximate causality can act (or has acted).

## INTRODUCTION

### Questions and answers

The plant kingdom has long been the object of interest to biologists searching for answers to the evolutionary questions posed by the incredible variety of breeding systems possessed by its constituent species. In attempting to answer these questions, a variety of approaches have been used. These range from encyclopaedic compilations of the variety of different breeding systems in plants (see e.g. Willson 1983; Richards 1986), through general questions applicable to a large subset of plant species (see e.g. Lloyd 1980a, b; Lande & Schemske 1985), to specific questions addressing particular problems associated with sexual reproduction in a given species. A large number of workers have adopted the latter approach and, with the intention of finding answers to the problem of the evolution of breeding systems, have posed such questions as:

1   Is an individual produced by one seed of *Anthoxanthum* twice as good as one tiller from the same plant?

2   Do ray florets permit more efficient reproduction than disk florets in *Senecio vulgaris*?

3   Do females of *Thymus vulgaris* produce many more seeds than hermaphrodites?

4   How does a female *Drosophila* choose her mate?

It is clear that there is a lack of general evolutionary interest in these questions and the answers that they provide. However, all the scientists whose questions are cited above are trying to answer more generally relevant questions. This illustrates an important point, namely that the challenge of science lies in the choice of a 'good' question (see Adams 1979). Once a good question is found, the answer is very likely to follow soon.

A good evolutionary question must (1) be generally relevant and (2) be tractable (i.e. have an answer). What determines the first point is very poorly understood; at a certain point in time, a scientific community accepts a topic as relevant and this chapter will not discuss how and why this is so. Concerning the second, although it seems trivial, it is very difficult to check *a priori*. Several examples of bad (although interesting) questions can be found in the history of science. For example, 'what are the two integers a and b such that a/b is the ratio between the area of a circle and the area of a related square?' is a question which was enormously popular although the correct question was 'do such numbers exist?'. Another kind of bad question is 'do birds have wings because they fly or do they fly because they have wings?'. This last question illustrates one of the main conceptual problems of biology. What is wrong with it is that it assumes a linear causality (which has proven its efficiency in physics) while, as Darwin has explained, causality is circular in biology (see Gouyon & Couvet 1985 for a more complete discussion on this point). This circularity is a fact for which we are not trained before becoming evolutionary biologists. It involves a change in the level of integration of the phenomenon. At a certain level, say the genome of individuals, a mechanism generates variability. This results in the existence of diversity between individuals within the group. At the group level, selection will then fix certain types of genetic information and eliminate others. Mutation plus development (acting at the molecular and individual levels) imply that birds fly because they have wings; selection (acting at the individual and population levels) implies that birds have wings because they fly. The circularity of causality in biology is not a property of one level; evolution by natural or artificial, or sexual selection is not an emergent property of one level; it is a property of the interaction between different levels. The most obvious case is the DNA-population interlevel interaction but there is no reason why similar processes would not occur at different scales. Diversity could be generated at the species level and selection could operate between species. For selection to be able to operate, *separate* sets of information must reproduce using common resources (Gouyon & Gliddon 1988). This is true for individual genomes within a population and for species within a habitat. However, it is certainly less true for populations within a species because populations do not contain, by the usual definition, *separate* sets of information (due to processes such as migration).

### The question

The questions cited in the last paragraph were tractable but not interesting in terms of providing answers to generally applicable evolutionary questions. Scientists are asking them because there is a belief that such questions can be related to a more fundamental (i.e. broadly evolutionarily relevant) question to which the answer either cannot be found directly or does not exist. The above examples were all related to a single question discovered, in its present form, by Williams & Mitton (1973) and further formalized by Williams (1975) and Maynard Smith (1978). The question is: 'Why reproduce sexually?'.

In order to try to solve this problem, scientists have cut it into parts: why do aphids, rotifers, *Anthoxanthum*, thyme, etc. reproduce sexually? Each of these sub-questions is supposed to be a way of answering the big question about the cause of sex. Our purpose in this chapter is to show that, because of the complex functioning of causality in biology, the change in the level at which the question is posed changes the answer in a drastic way. That is, such answers provide information about short-term responses in particular organisms to particular selection regimes and represent valuable contributions to our knowledge of the functioning of such species. However, these specific answers to sub-questions do not suggest a global answer to the general question.

*The answers*

This particular question has become so fashionable during the last decade that the quantity of publications on the subject is now immense. The synthesis which we will give in this chapter does not pretend to include all research on the subject. However, we believe that the data which are ignored in the present paper could be included easily without necessitating any major change in the main conclusion.

One of the striking peculiarities of the different texts found in the literature is that an important fraction provides an answer to a question (why does species X reproduce sexually?) but that this answer varies greatly from one paper to the other. In other words, the logical conclusion of the results available now is: each group of organisms which reproduces sexually has a good reason to do so but this reason varies from one group to another. Although logical, this conclusion is difficult to address in a paper and, while it has been formulated already concerning outcrossing (Abbott 1985, see detailed explanations in next paragraph), we have seen no paper globally examining this possibility. The reason for this is obvious, it should be shocking to an evolutionary biologist that a phenomenon as general as sex could have arisen independently from a number of different causes. We will try to present some of these independent causes and then to discuss the possibility of an explanation that all of them seem to converge towards the same 'goal', that is, the origin and maintenance of sex.

## WHY REPRODUCE SEXUALLY?

The problem can be divided into two parts: the origin and the maintenance of sex. Concerning the origin, two ideas seem to have invaded the scientific environment: (1) sex was inevitable because transformation by free nucleic acids occurs spontaneously and (2) this inevitable event would be favoured because it corresponds to a mechanism of repair against mutations. No experiments or formalized models have been published so that it is difficult to make any interesting comparisons about this point (for instance is meiosis a more recent transformation of mitosis or did the two processes appear simultaneously?).

On the contrary, concerning the maintenance of sex, the problem seems clear. In outbreeding anisogamous species, sexual reproduction involves a cost roughly equal to two. This cost was called 'cost of meiosis' by Williams (1975); Maynard Smith proposed 'cost of producing males' in a dioecious species but Charnov, Maynard Smith & Bull (1976) have shown that it would be approximately the same for hermaphrodites; the same value is also found when considering the 'cost of outcrossing' (Maynard Smith 1978, p. 126). This cost can be avoided through parthenogenesis on one hand and selfing on the other (selfing anyway leads to the ultimate loss of male function). Lots of intermediate situations between these two extremes can be found (see Maynard Smith 1978 for a review). Some parthenogenetic or selfing species exist; how is it possible that not all species are the same?

## Long-term forces

The old answer was that sex is maintained because it allows, through recombination, the possibility of the species evolving faster, of producing new genetic combinations in the future and thus being able to accommodate future changes in the environment. In addition, sexual reproduction provides a means of incorporating more quickly new beneficial mutations (Müller 1933). As shown by Williams (1975), this argument is teleological, Panglossian and group selectionist. Finalism can no longer be considered an insult, since evolutionists are becoming conscious that they are explaining finality rather than denying it: an attitude called teleonomy by some authors, e.g. Monod (1972)—the status of group selection is still not absolutely clear.

Another long-term advantage of sex has been proposed by Müller (1964) and formalized by Felsenstein (1974). As noted above for the origin, sex allows the repair of mutations through recombination with other genetic information. The above argument still holds concerning this pressure and we are confronted inevitably by the following question. How could a species be sufficiently self-aware to evolve such a trait which will perhaps be useful in the future but which, in the short term, has a handicap of two? It is necessary that in every sexually reproducing species, there must be a short-term advantage to sex and this short-term advantage has been found as will be shown below.

## Short-term forces

Different processes favouring, more or less immediately, individuals which reproduce sexually have been proposed on theoretical and/or experimental grounds. Presented here are some of them without developing the arguments for or against. We have tried to classify these hypotheses using methods ranging from the very convincing arguments of Williams (1975) to the postulation of esoteric mechanisms proposed for very particular cases.

*Genetic advantage to sex*

In 1973, Williams & Mitton suggested an explanation of the maintenance of sex using a model: the aphid–rotifer model. In this model, an animal lays an (average) equal number of sexually and asexually produced propagules in different sites. The progeny then reproduce asexually and the relative proportion of the offspring from sexually produced propagules is followed from generation to generation. When this proportion reaches two in three, the cost of meiosis is paid. The time needed to reach this point decreases when the number of propagules laid at first in the site increases. If this number is less than three (on average) for each of the two types of propagules, the cost of meiosis will never necessarily be paid. The time required varies from more than a hundred (asexual) generations, when there are three propagules of each type at the beginning in each site, to twelve when there are a hundred propagules of each type. Williams (1975) has further developed this model to the elm–oyster model where the number of propagules is so high that the cost of meiosis is paid in the first generation, leading to the selection in each generation, from among an enormous number of genotypes, those which are Sisyphean.

The processes responsible for the advantage of sex in these models are (1) sibling competition and (2) the fact that sexual and asexual individuals are drawn from the same distribution although the parents are derived from a population undergoing very severe selection. This last point is certainly a serious restriction to the applicability of the model to real organisms since it means either that the heritability of fitness in the considered species must be unity or that the environment must vary in such a drastic way that the value of a genotype in year $n$ gives no information about the value of this genotype in year $n+1$ (an eventual negative heritability of fitness is even proposed).

Nevertheless, in species producing a large number of progeny and not dispersing them very far, this process certainly acts in favour of sex; whether it is sufficient to explain the maintenance of sex in elms is a more difficult question which would be difficult to test. Note however that, from this point of view, it would be more advantageous to produce sexually non-dispersed propagules and asexually dispersed ones; a prediction which does not fit with the observations (see for instance Olivieri *et al.* 1983).

*Short-term evolutionary advantage*

As Sartre pointed out, 'L'enfer, c'est les autres'; the only hostile element in the environment to which an organism (or set of organisms) can never possibly be adapted is the other organisms. This leads to the thought that genetic diversity could be useful mainly in the sense of trying constantly to solve, at least for a short while, this eternally recurrent problem (Hofstadter, 1979). The discovery by phytopathologists and plant-breeders of gene for gene coevolutive *resistance* phenomena (Flor 1942) has led some plant geneticists (Levin 1975; Schmitt & Antonovics 1986) to believe that resistance to pathogens is the central problem encountered by plants and that this could explain why plants would need to produce offspring which are genetically different from themselves.

Van Valen (1973), using paleontological considerations, has proposed an explanation of the regular extinction of species within groups of ecologically related species by the Red Queen hypothesis, i.e. the effect of competition between these species leads constantly to new adaptations against the others. This large scale process could have a genetic equivalent on a smaller scale as proposed by Harper (1977, p. 768). This idea has spread less widely, perhaps because no gene-for-gene competition, *sensu stricto*, event has been discovered yet.

As far as we know, apart from the demographic importance of their effects and particularly the cycles they can induce, the evolutionary importance of predators has not been stressed in itself and predators have not yet been proposed as candidates for the process of maintenance of sex in their prey. However, the coevolutionary race between predator and prey has no reason to be less strenuous and to require less genetic diversity than the race with pathogens or competitors.

*Convenient developmental constraints*

On this point, we will be obliged to be very incomplete because the number of examples is incredibly large. Constraints are becoming fashionable in evolutionary biology (see e.g. Gould & Lewontin 1978) but apart from stating that 'species X cannot evolve trait Y', we are neither able to explain why nor able to prove that it will remain so eternally. The developmental constraints which conveniently maintain sex in species (as in humans), where none of the above models would predict its maintenance, are numerous and a small subset are presented below. Some species have absolutely no possibility of producing progeny without a meiotic event. This is the case in most vertebrates, although some fishes and some reptiles have lost this constraint and become parthenogenetic apparently without major damage.

Some species can produce asexual progeny but with different ecological potentialities. For example, many plants can produce tillers but need pollen and/or seeds to disperse their genes. Sexual reproduction is very often needed to produce resistant forms. This is true in plants and in animals: aphids, for instance, cannot produce parthenogenetic eggs (resistant to frost) and they produce eggs only in autumn. From this point of view, Williams' balance argument (there must be a short-term advantage to sex at least in species, like aphids, which practice both asexual and sexual reproduction and , thus, can choose between them) is dramatically modified for these species. In winter, they cannot avoid passing through the egg stage in temperate climates (in tropical climates it seems that wholly parthenogenetic species exist).

There are adaptations which cannot be realized in the absence of sexual reproduction. Another example is the cleansing of systemic viruses in plants. During the process of seed production, the viruses which are in the parental vascular system disappear. This cannot be done in a vegetative cell and has long been a real problem for crops vegetatively multiplied like potatoes (*in vitro* cultures are now able to solve it). Another kind of constraint can prevent hermaphroditic species from escaping sex by selfing. The resource allocation constraints can be such that they favour separate genders (Charnov, Maynard Smith & Bull 1976). Once these species are dioecious or gonochoric, they are definitely protected against any possibility of selfing.

*Esoteric physiological constraints in asexuals*

We are now coming to some quite peculiar examples. In a paper in 1982, Templeton violently criticizes the notion of cost of meiosis and all its relatives on the following basis. Some Hawaiian *Drosophilae* are able to reproduce parthenogenetically. If one compares the rate of reproduction of a parthenogenetic strain to that of a sexual strain, one finds that, despite the fact that parthenogens produce no males (which lay no eggs), their demography is inferior to that of sexual strains. Templeton concludes that there is no cost of meiosis. It could also be concluded that the cost of meiosis is compensated, in this particular case, by a physiological disadvantage due to the manner in which parthenogenesis is achieved in *Drosophila*.

A comparable result was found by Abbott (1985). Comparing outcrossing radiate (gynomonoecious: producing female and hermaphroditic florets) and selfing non-radiate (producing only hermaphroditic flowers) forms of *Senecio vulgaris*, he finds that the radiate plants possess the ability to produce more seeds. This ability is unrelated to either effects of heterosis or inbreeding depression. Here again, the 'cost of outcrossing' is compensated for by a physiological disadvantage to the inbreeder.

### *Bizarre genetic constraints*

In the above paper, Abbott cites another example of a case where outcrossing is maintained by a cause which 'is unlikely to be the result of an (inbreeding depression) effect': the case of gynodioecy (co-occurrence of female and hermaphroditic plants in a species). In thyme, for example, selfing is possible and some genotypes found in natural populations are able to produce good selfed progenies. However, this species remains primarily outcrossed because of the existence of a proportion ranging from 5 to 95% of females, with an average around 60%. We have shown (Gouyon & Couvet 1985; Couvet, Bonnemaison & Gouyon 1986) that the presence of these females is due to the existence of a nucleocytoplasmic conflict where cytoplasmic genes tend to produce females (which reproduce cytoplasms more efficiently) while nuclear genes tend to restore the male fertility (which is useful for nuclear gene reproduction) of these females. This confirms Abbott's view of a large independence between the cause of the existence of these females and their most important effect, namely outcrossing.

## DISCUSSION AND CONCLUSION

It is clear then that different authors have each found a possible short-term reason for their species to reproduce sexually. As stated above, some examples are certainly missing but adding them would probably not change the following point: it is impossible to assume that only one of these causes is that which accounts for the existence of sexual reproduction. We are thus obliged to admit that a number of (sometimes very particular) causes are acting in the short term in each group and that all of these independent causes lead to the same result: sexual reproduction. The fact that some authors have expected that they could extrapolate the particular cause they had discovered to the whole living world simply comes from the fact that they have ignored the hierarchy of causes which is an essential element in evolution (concerning this hierarchy, see Gouyon & Gliddon 1987).

Taking this into account, one can formulate the following hypothesis. Throughout evolution, the make-up of the different species which existed either could or could not provide a short-term advantage to sex. The species which were such that sex was favoured by whatever short-term process got, incidentally, the long-term advantage of sex. Most of the others went extinct. The extant species would thus be those which possess a constitution which immediately favours sex. Group selection, in this case, is not acting against individual selection; group selection sorts the groups in which individual selection had beneficial effects. Group selection, as used here, is different from the definition given in Maynard Smith (1978, p. 2). It is used in the sense of the 'theory of biased extinction' (Williams 1975, p. 156). Williams (loc. cit., pp. 156–60) has criticized the validity of the paleontological evidence on which this theory is based and, indeed, has refused to accord the status of 'selection' to forces acting at this level. However, Maynard Smith (1978, pp. 51–4) argues against Williams' assertions and states 'The facts fully support the conventional view that parthenogenetic varieties are doomed to early extinction. The evidence seems to me so strong that extinction is an almost inevitable consequence of a lack of evolutionary potential . . .' One could even imagine that group selection, by this process, has progressively selected the types of organization which did not permit the existence of asexual reproduction, thus leading to the situation found in higher animals. That is, group selection directs the way in which individual selection is allowed to act.

From that point of view, the question 'Why reproduce sexually?' is not a good one. A good question would be 'Why do aphids reproduce sexually?' with the possible answer 'Because they cannot produce resistant forms (i.e. eggs) in another way'. Another good question would be 'Why do most species reproduce sexually?'; the corresponding answer being: 'Because most of the others went extinct'. These two questions do not address the same level (Gouyon & Gliddon 1987) so that, although they seem to be related, they receive very different answers. Note that, ideally, in order to completely understand sex in living organisms, one would need to know all short-term forces (proximate causes) in each species plus the long-term forces (final causes). In fact, it is clear that knowing each proximate cause is not essential while ignoring one final cause would really be a handicap (two only have been presented here, they both were presented by Müller who clearly was very interested in final causes).

Therein lies the difficulty with the hypothesis presented in this paper: it is not yet amenable to experimental validation. Further research on the means of investigation of group selection is needed in order to allow research into the evolution of reproductive systems to escape the fashionable ghetto of particular causes in which it is presently confined. Only when we are able to explore the complex field of causality involved, will we be able to define the suite of all the determinants of this central phenomenon of life.

# REFERENCES

**Adams, D. (1979).** *The Hitch Hikers' Guide to the Galaxy.* Pan Books, London.

**Abbott, R. J. (1985).** Maintenance of a polymorphism for outcrossing frequency in a predominantly selfing plant. *Structure and Functioning of Plant Populations* 2 (Ed. by J. Haeck & J. W. Woldendorp), pp. 277–86. North-Holland Publishing Company, Amsterdam.

**Charnov, E. L., Maynard Smith, J. & Bull, J. J. (1976).** Why be an hermaphrodite? *Nature,* **263,** 125–6.

**Couvet, D., Bonnemaison, F. & Gouyon, P. H. (1986).** The maintenance of females among hermaphrodites: the importance of nuclear–cytoplasmic interactions. *Heredity,* **57,** 325–30.

**Felsenstein, J. (1974).** The evolutionary advantage of recombination. *Genetics,* **78,** 737–56.

**Flor, H. H. (1942).** Inheritance of pathogenicity in *Melampsora lini. Phytopathology,* **32,** 653–69.

**Gould, S. J. & Lewontin, R. C. (1979).** The spandrels of San Marco and the Panglossian paradigm. *Proceedings Royal Society of London Series B,* **205,** 581–98.

**Gouyon, P. H. & Couvet, D. (1985).** Selfish cytoplasm and adaptation: variations in the reproductive system of Thyme. *Structure and Functioning of Plant Populations* 2 (Ed. by J. Haeck & J. W. Woldendorp), pp. 219–319. North-Holland Publishing Company, Amsterdam.

**Gouyon, P. H. & Gliddon, C. J. (1988).** Genetics of information and the evolution of avatars. *Population Genetics and Evolution* (Ed. by G. de Jong) pp. 119–23. Springer Verlag, Berlin.

**Harper, J. L. (1977).** *Population Biology of Plants.* Academic Press, London.

**Hofstadter, D. R. (1979).** *Gödel, Escher, Bach: An Eternal Golden Braid.* Basic Books inc., New York.

**Levin, D. A. (1975).** Pest pressure and recombination systems in plants. *American Naturalist,* **109,** 437–51.

**Lande, R. & Schemske, D. (1985).** The evolution of self-fertilisation and inbreeding depression in plants. I and II. *Evolution,* **39,** 24–52.

**Lloyd, D. G. (1980a).** Demographic factors and mating patterns in angiosperms. *Demography and Evolution in Plant Populations* (Ed. by O. T. Solbrig), pp. 67–88. Blackwell Scientific Publications, Oxford.

**Lloyd, D. G. (1980b).** Benefits and handicaps of sex. *Evolutionary Biology,* **13,** 69–111.

**Maynard Smith, J. (1978).** *The Evolution of Sex.* Cambridge University Press, Cambridge.

**Monod, J. (1972).** *Chance and Necessity: an Essay on the Natural Philosophy of Modern Biology.* Collins, London.

**Müller, H. J. (1932).** Some genetic aspects of sex. *American Naturalist,* **66,** 118–38.

**Olivieri, I., Swann, M. & Gouyon, P. H. (1983).** Reproductive system and colonizing strategy of two species of *Carduus* (Compositae). *Oecologia (Berlin),* **60,** 114–17.

**Richards, A. J. (1986).** *Plant Breeding Systems.* Allen & Unwin, London.

**Schmitt, J. & Antonovics, J. (1986).** Experimental studies of the evolutionary significance of sexual reproduction IV. Effect of neighbor relatedness and aphid infestation of seedling performance. *Evolution,* **40,** 830–36.

**Templeton, A. R. (1982).** The prophecies of parthenogenesis. *The Evolution and Genetics of Life-history Strategies* (Ed. by H. Dingle & J. P. Hegman), pp. 75–101. Springer Verlag, Berlin.

**Van Valen, L. (1973).** A new evolutionary law. *Evolutionary Theory,* **1,** 1–30.

**Williams, G. C. (1975).** *Sex and Evolution.* Princeton University Press, Princeton.

**Williams, G. C. & Mitton, J. B. (1973).** Why reproduce sexually? *Journal of Theoretical Biology,* **39,** 545–54.

**Willson, M. F. (1983).** *Plant Reproductive Ecology.* Wiley Interscience, New York.

# 3. PLASTICITY, CANALIZATION AND EVOLUTIONARY STASIS IN PLANTS

## DONALD A. LEVIN

*Department of Botany, University of Texas, Austin, Texas 78713, USA*

## SUMMARY

**1** A plant interacts with its environment through its phenotype. The phenotype may be altered by developmental responses to environmental change; it shows plasticity. The profile of phenotypes produced by a genotype across environments (the 'norm of reaction') is subject to evolutionary change.

**2** Canalization involves resistance of the phenotype to change resulting from fluctuations in the environment. Developmental canalization also may buffer the phenotype against genetic change.

**3** Although developmental canalization and phenotypic plasticity have opposing effects, they may both reduce the coupling between genotype and phenotype. This may reduce the impact of selection, which acts on phenotypes, and promote evolutionary stasis.

**4** I discuss the constraints imposed by plasticity and canalization on plant microevolution and consider their effect on the potential for 'somatic evolution'.

## INTRODUCTION

A plant interacts with the environment through its phenotype. The capacity of an individual to survive and reproduce in different environments may be enhanced if the phenotype can be modified or conserved as needed. The phenotype may be altered by developmental responses to environmental change; i.e. the individual may exhibit phenotypic plasticity (Bradshaw 1965). The profile of phenotypes produced by a genotype across environments is its 'norm of reaction' (Schmalhausen 1949). The 'norm of reaction' of a genotype is like a cloud in multidimensional space. It has a centre and an area. The phenotypic expression of a given genotype varies around the centre and has the potential to occur anywhere in the area. This chapter will deal exclusively with changes in the position of the norm centre, i.e. where the norm centre lies in phenotypic space. Phenotypic evolution is change in this centre. The evolutionary flexibility of a character lies in the ability of the norm centre to change its position in phenotypic space in response to selection.

The phenotype may be buffered from or resistant to fluctuations in the environment, in which case the phenotype is canalized (Waddington 1957). Developmental canalization also may buffer the phenotype against genetic change, so that genetic differences among individuals are not manifested. Both the level of phenotypic plasticity and developmental canalization are genetically based (Bradshaw 1965, 1974; Levin 1970; Bachmann 1983; Schlichting 1986).

Although phenotypic plasticity and developmental canalization have opposing effects, they have one very important feature in common. They both may reduce the association between the phenotype and the genotype. This is depicted in Fig. 3.1. In the absence of both regulatory mechanisms, a given genotype has a constant phenotype over a range of environments, and different genotypes exhibit different phenotypes in the same environment. With plasticity in the extreme, a given genotype may exhibit a range of phenotypes across an environmental gradient, whereas all genotypes in one environment will display the same phenotype. With canalization in the

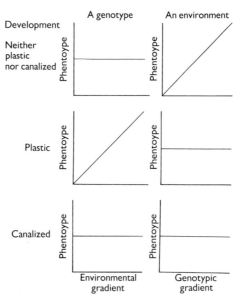

FIG. 3.1. The relationship between the phenotype and environmental and genotypic gradients when development is unmodulated, plastic, and canalized. When the environmental gradient is considered the genotype is constant, when the genotypic gradient is considered the environment is constant.

extreme, the phenotype will be constant in the presence of environment and genetic variation.

By reducing the coupling between the phenotype and the genotype, plasticity and canalization may reduce the impact of selection, which acts on phenotypes (Wright 1931; Bradshaw 1965; Stearns 1982). The individual adjustment to environmental forces (in the case of plasticity) and the retention of a given phenotype in the face of genetic variation (in the case of canalization) reduce the exposure of alternative genotypes to natural selection, and thus reduce the selective differentials between genotypes.

The purpose of this chapter is to discuss some constraints imposed by plasticity and canalization on plant microevolution. This chapter will also deal with the effects of these mechanisms on the potential for 'somatic evolution'. The phenotype may be altered over time by a combination of somatic mutation and somatic selection, followed by differential ramet or branch growth (Whitham & Slobodchikoff 1981; Gill 1986). I assume that plasticity reduces the phenotypic manifestation of genetic differences, so that phenotypes converge toward a common form dictated by the environment. This assumption is often met (Johannsen 1911; Clausen 1951; Bradshaw 1965).

## PHENOTYPIC PLASTICITY

Plant populations grow on environmental mosaics. If selection is strong, there will be some correspondence between the spatial pattern of gene frequencies and the spatial pattern of the environment. Because plasticity reduces the impact of selection, the genotype-environment accord will vary inversely with the plasticity of a character. Therefore, the greater the level of plasticity, the more the organization of genetic variation within populations depends on the pattern of gene dispersion. Restricted gene flow may incidentally allow the development of a spatial genetic pattern, but it will be independent of the environmental pattern (Endler 1977). If there is a high covariance between genotypic and environmental mosaics (e.g. Jain & Bradshaw 1966; Hamrick & Holden 1977; Nevo *et al.* 1986), the expression of the character under selection probably is dictated largely by the genotype.

Plasticity reduces the correlation between a character's genotypic and phenotypic variation within populations. Accordingly, the greater the level of plasticity the greater will be the difference between the amounts of phenotypic and genetic variation observed in the field. Correlatively, the greater the level of plasticity the more the spatial pattern of phenotypes within and among populations may be dependent on the environmental pattern, and the less it may be dependent on the organization of genotypes in space.

The level of character plasticity not only affects the response of subpopulations and populations to diversifying selection, but also affects the adaptability of different breeding systems. In a species with relatively high mean plasticity, the products of self-fertilization may be as variable in phenotype as the products of outcrossing, when both are grown in the same heterogeneous environment. Accordingly, plasticity diminishes the value of outcrossing, and sexual reproduction in general. The more plastic the species the less genetic diversity among progeny is a prerequisite for their phenotypic diversity. Outcrossing species with high levels of plasticity might adopt self-fertilization, as an alternative reproductive mode, because selfing would increase reproductive assurance without a concomitant loss in progeny phenotypic variation. Phenotypic plasticity would facilitate a shift from outcrossing to partial self-fertilization because inbreeding depression would be reduced. Outcrossing species with high levels of plasticity also might adopt apomixis as an alternative reproductive mode, because it would increase reproductive assurance with little loss in phenotypic variation among progeny.

If we turned the argument around, we would also expect species with restricted recombination systems to be more developmentally plastic than species with open recombination systems. Progeny of inbreeders and facultative apomicts have narrower distributions of fitness potential than progeny of outbreeders. Therefore, the former are less likely to achieve genetic adaptation to environmental heterogeneity, and are likely to employ developmental solutions more extensively than outbreeders. The premium on plasticity ostensibly increases as the recombination system becomes more restricted. Data on plasticity in relation to recombination systems are meagre. Species with restricted recombination systems seem to be more plastic than their outcrossing relatives (Jain 1979; Silander 1985; Richards 1986).

## DEVELOPMENTAL CANALIZATION

Developmental canalization suppresses the phenotypic effects of genetic variation, and thus renders a broad range of genotypes indistinguishable. This in turn reduces the potential for selective differentiation in time and space, and reduces the advantage of outcrossing.

Floral architecture and merism are the most highly canalized characteristics of flowering plants (Berg 1959; Bradshaw 1965), and thus have considerable utility in taxonomy. In species with highly canalized flowers, the phenotypic variability for selective responses to changes in pollinator type and availability may not be available, and without it no change is possible. Thus the shift from one zoophilous floral syndrome to another, or from a zoophilous syndrome to one associated with wind- or self-pollination may lag far behind the change in pollination environment. The evolutionary flexibility of floral characteristics is sacrificed for the sake of immediate adaptation to a pollen vectors.

Canalized characters may evolve when canalization is incomplete, and the underlying genetic variation is expressed. Heritable variation for 'invariant' characteristics has been demonstrated in *Linanthus androsaceus* (Huether 1968), *Antirrhinum majus* (Stubbe 1963), *Nicotiana rustica* (Paxman 1956), *Triticum aestivum* (Frankel, Shineberg & Munday 1974) and *Gossypium hirsutum* (Wilson & Stapp 1979). This variation may be released by environmental stress (Grant 1956; Huether 1968; Heslop-Harrison 1959), and inbreeding or interspecific hybridization (Levin 1970; Bachmann 1983). The penetrance and expressivity of favourable expressions may be enhanced and subsequently stabilized by concomitant selection for one or more major genes and a constellation of modifiers. Increased penetrance of unstable expressions through selection has been achieved in *Antirrhinum* (Stubbe 1959, 1963), *Linanthus* (Huether 1968), mice (Dun & Fraser 1959) and *Drosophila* (Rendel 1959). Once an expression is exposed, it may be canalized and replace its predecessor as the invariant character (e.g. *Drosophila*, Waddington 1961; mice, Kindred 1967).

The breakdown of developmental canalization may set the stage for the emergence of a new species or genus (Levin 1970; Stebbins 1974). The expressions may be so deviant that their fixation imposes ethological or mechanical barriers between populations with the original and novel phenotypes. The importance of phenodeviants in specific and transpecific evolution depends on the extent to which the floral apparatus is altered. In *Gilia millifoliata-achilleifolia* (Grant 1956) and *Ipomopsis aggregata* (Ellstrand 1983), flowers with abnormal merism tend to have coordinated changes for the normally pentamerous calyx, corolla and androecium. Other members of the Polemoniaceae show deviant merisms in all floral parts (Grant 1956). On the other hand, in *Phlox drummondii* meristic abnormalities in one flower part number usually is not associated with meristic abnormalities in other parts (Levin, unpublished data). Covariance in stamen and corolla size has been described in *Nicotiana rustica* (Paxman 1956). These data suggest that simultaneous multiple character evolution is possible.

The level of developmental instability may vary widely among populations or cultivars of the same species. In some *Phlox drummondii* cultivars, 8% of the flowers have abnormal numbers of parts, whereas in others 47% of the flowers have at least one abnormal number (Levin, unpublished data). In

*Linanthus androsaceus*, populations averaged from 0.5 to 4% of flowers with abnormal petal numbers (Huether 1969); the range in *Ipomopsis aggregata* was from 2 to 33% (Ellstrand 1983). Differences in instability among populations or cultivars may reflect differences in the level of unexpressed genetic variation for meristic characters or differences in the regulation of this variation.

## DEVELOPMENTAL MODULATION AND EVOLUTIONARY STASIS

Phenotypic plasticity and developmental canalization are properties of the morphogenetic system which reduce the phenotypic manifestation of genotypic differences. As a consequence, they reduce a population's ability to track environmental change genetically from one generation to another, and to exploit ecological opportunity by genetic change. Thus, plasticity and canalization may foster the genetic stasis of populations and species (Stearns 1982). Stasis also may result from a difficulty in recruiting and fixing novel advantageous genes. The probability of fixing a new favourable gene in a local population is dependent on the effective size of the population and on the selective advantage of the gene (Wright 1969). The greater the level of plasticity or canalization, the less will be the selective advantage of the new gene, and the higher will be the probability that it will be lost from the population genetic drift.

An advantageous gene will increase in one part of a population and then gradually spread. The velocity of the wave of increase of gene frequency within a population is a positive function of the level of selection and of gene dispersion (Fisher 1937). The greater the plasticity or canalization of a character, the more the advancing wave would be retarded, and the slower would be the rate of gene substitution, because the selective differential between genotypes would be reduced. The spread of an advantageous gene from one population to another also is a function of its selective advantage and the level of gene flow between populations (Slatkin 1976). The spread of such a gene across a species would be retarded if the character it controlled were plastic or canalized, because its selective advantage would be reduced.

Phenotypic plasticity may interfere with the ecological diversification of species. Although plasticity may allow a species to establish a foothold in a novel habitat, it is likely to suffer extinction there when the environment changes. Exploitation of novel habitats thus will be transient. Ecological diversification of long-standing is most likely to occur when the centre of the norm of reaction is dependent upon genetic change such as that attending the invasion of heavy metal soils by some grasses (Bradshaw & McNeilly

1981). The greater the level of plasticity the less responsive will be the norm centre to natural selection. The ecological diversity or amplitude of species is dependent on its ability to evolve populations adapted to different environments (Bradshaw 1984). It is important to determine whether the wide ecological amplitude of some species is due to rich gene pools for characteristics related to habitat tolerance.

Developmental adaptive modes may doom species to phyletic conservatism. Phenetic diversification and speciation by adaptive mechanisms may proceed very slowly. Left to developmental devices alone, species in changing environments face extinction. Only when the genetic variation masked by plasticity and canalization is expressed in the phenotype, and thus is accessible to selection, will adaptation to major environmental change be possible.

The mobilization of hidden variation may lead to a major and rapid change in the genetic composition of the species. The change may be so pronounced that it leads to the formation of a new species or genus. The newly derived entity may subsequently be buffered by the re-establishment of plasticity and canalization, thereby entering another period of stasis. Species whose phenotypes poorly reflect their underlying genotypes are prime candidates for episodes of punctuated evolution followed by long periods of stasis (Stearns 1982; West-Eberhard 1986).

The genetic stasis of species is not the result of a depauperate gene pool. By buffering a gene pool from directional selection, phenotypic plasticity and developmental canalization protect a gene pool from the erosion of genetic variation (Wright 1931; Schmalhausen 1949; Sultan 1987). They also may reduce the erosion of genetic variation by genetic drift, because they decrease the variance of progeny number, which in turn increases the effective population size (Nei & Murata 1966). Indeed, plasticity and canalization may promote an accumulation of genetic variation, because the opposition to the effects of mutation and migration by selection is reduced.

To the extent that developmental mechanisms foster phyletic conservatism, they also foster the integration of species, which in a sense is phyletic conservatism in space. The more developmental mechanisms reduce the phenotypic manifestations of genotypic differences, the less populations of a species will undergo selective differentiation when they occupy different habitats, even if they display different phenotypes in them. The reduction of interpopulation differences when plants are grown in a common environment has been demonstrated in several species (Abbott 1976; Akeroyd & Briggs 1983; Fowler, Zasada & Harper 1983; McGraw & Antonovics 1983; Seliskar 1985).

If plasticity retards adaptive radiation, then divergence during speciation

# 4. SOMATIC VARIATION AND GENETIC STRUCTURE IN PLANT POPULATIONS

BARBARA A. SCHAAL

*Department of Biology, Washington University, St Louis, Missouri 63130, USA*

## SUMMARY

**1** Restriction site analysis of DNA in native plant populations has revealed high levels of variation in ribosomal DNA of many plant species.

**2** This variation occurs at different levels: within and among individuals, populations and species.

**3** These studies indicate the importance of population processes such as gene flow, drift, or selection in influencing variation at the DNA level.

**4** DNA analysis can be used to study other processes such as somatic mutation or rapid genomic change in response to environmental stress.

**5** Somatic variation in rDNA among the ramets of a clone has been documented in *Solidago altissima*.

**6** Changes in rDNA cistron number occur in *Brassica campestris* in response to a change in growth conditions.

**7** Such processes may potentially be important in influencing the ecological genetics of plant species.

## INTRODUCTION

The evolutionary potential of populations and species is directly related to their genetic structure. Plant genetic population structure encompasses both the levels of genetic variability and the distribution or apportionment of this variation. Emphasis in the past was on determining the levels of genetic variation within species and elucidating the population factors such as mating system, population size or subdivision, which influence genetic structure. Much of this information has been obtained from studies of morphology or biochemical markers such as allozymes. The development of molecular techniques which incorporate DNA sequences as cloned probes affords the possibility of looking at potentially any DNA sequence. The study of plant genetic structure using DNA sequences is rapidly expanding and has provided information on phylogeny (Sytsma & Schaal 1985), species origin (Sytsma & Gottlieb 1986), and genetic variation and subdivision within populations (Saghai-Maroof *et al.* 1984).

47

The direct analysis of DNA sequences also allows for the study of processes which have previously been difficult to examine due to technical limitations. One such area is that of somatic variation of DNA sequences within plants. Somatic variation can influence the genetic structure of plant populations by adding new genetic variants to the gene pool (Klekowski 1976) and there is speculation that some of these mutations can increase adaptation, specifically in the avoidance of insect predation (Whitham & Slobodchikof 1981; Gill & Halverson 1984). Moreover, there is a good evidence that somatic genome change may be a common and adaptive response of plants to environmental stress (Walbot & Cullis 1983, 1985; McClintock 1984) and that some of these genomic changes can be inherited, thus entering into a population's gene pool (Cullis 1986). Somatic variation and mutation may be of particular importance in the adaptation and evolution of clonal plants. We will survey the kinds of somatic variation encountered in both plants and animals, examine the evidence for somatic variation in native plant populations, and finally report on some preliminary work on somatic variation currently being conducted at Washington University.

## KINDS OF SOMATIC VARIATION

Among the fundamental tenets of genetics has been the concept that genes are unalterable during the life-cycle of an individual. Genes were considered to have fixed order on the chromosomes, like 'beads on a string', and were passed unchanged from parent to offspring with the rare exception of germplasm mutation. Recent advances in molecular genetics clearly demonstrate that both these assumptions of linearity and unalterability are incorrect. Somatic changes in DNA during the life of an organism appear to be common and can be of several types, including amplification, rearrangements and mutations. Ribosomal DNA, the sequences that code for the 18S and 28S RNA subunits of the ribosome, are amplified during development in many species (Long & Dawid 1980). Preferential amplification of rDNA variants is observed in many instances, and can lead to the *bobbed* (Ritossa 1976) or *scabrous* (Frankham 1980) phenotypes in *Drosophila melanogaster*, and to *abnormal abdomen* in *D. mercatorum* (DeSalle & Templeton 1986). DNA amplification can occur in response to external environmental agents such as toxins or drugs (Schimke 1982).

Rearrangements of DNA sequences also occur during development such as in the generation of functional immunoglobulin genes. Here, two types of somatic recombination are necessary for the generation of complete, functional immunoglobulin heavy chain genes. Random DNA rearrangements occur for the constant, variable and joining regions of the protein (Alberts *et al.* 1983). Similarly regions of the B-chain polypeptide of the human T-cell antigen receptor rearrange during somatic development to produce an active gene (Duby *et al.* 1985). Other types of regular genomic changes are observed in the production of H1 versus H2 flagellar proteins in *Salmonella* (Silverman & Simon 1983) and the switching of mating type in yeast (Haber 1983). Transposition is a well-known example of somatic genome change and has been extensively documented (e.g. Shapiro 1983). Transposition is responsible for much of the sequence variation in the flanking regions of the Alcohol dehydrogenase 1 gene of maize (Sachs *et al.* 1986). In the *Antirrhinum* plant, transposition events render the *pallida* gene hypervariable (Coen, Carpenter & Martin 1986).

## SOMATIC VARIATION IN PLANTS

Somatic genome changes are clearly important in any consideration of gene structure and function. This chapter is concerned with an additional aspect of somatic variation: how somatic changes in genomes influence the levels and apportionment of genetic variability in plant populations.

In several cases, such as chromosomal loss associated with heat shock or gene amplification associated with toxins, the somatic genome changes are a response to an environmental challenge. A well documented case of environmentally induced changes in DNA is that for cultivated flax, *Linum usitatissimum*. Durrant (1958, 1962, 1971) discovered that heritable changes in plant stature could be induced by growing plants under different nutrient conditions. The progeny of these plants maintained the stature of the parent plants. A number of other traits vary with induction of *Linum* genotrophs including height, seed pubescence, peroxidase isozymes, and of particular interest, DNA content (Cullis 1977). Nuclear DNA amounts of large and small genotrophs differ on average by 16%. DNA renaturation studies show a large increase in the mid-repetitive DNA, and smaller changes in both single copy and highly repetitive DNA. Cistron numbers for both the 18 and 28S rDNA and 5S rDNA vary among the genotrophs (Cullis 1977; Goldsbrough & Cullis 1981). Change in rDNA cistron number occurs within a single individual and has been followed in apical buds (Cullis & Charlton 1981). The specific ribosomal repeat type does not appear to change; both the repeat length and internal restriction sites are the same for large and small genotrophs (Goldsbrough & Cullis 1981). On the other hand, there is preferential amplification of a 5S rDNA variant marked by a *Taq*I site (Cullis 1986). A similar situation occurs in maize, where the number of rDNA repeats varies during development (Phillips, Wang & Knowles 1983). This variation in rDNA cistron number can, in some cases, be transmitted through the gametes; some viable gametes can contain only 1–2% of the total number of rDNA repeats (Phillips, Wang & Knowles 1983). Likewise in *Drosophila*, the number of rDNA repeats varies in response to the environment and can be passed on through the gametes (Frankham 1980). Ribosomal genes are clearly capable of varying somatically in response to environmental heterogeneity. There is no information on how common such environmentally induced variation is in nature.

Studies of several plant species have shown that total nuclear DNA content varies among individuals, and that this variation often is associated with plant habitat. In *Picea glauca* and *P. sitchensis* nuclear DNA content varies up to 50% and is associated with latitudinal variation (Miksche 1968, 1971). In *Microseris douglasii* significant differences in DNA content are found within and among populations (Price & Bachmann 1975). Populations with the highest DNA content occupy the most mesic environments with rich soils, while populations with the lowest DNA contents have poor soils and low precipitation. Within populations, in years of low rainfall, mean DNA content per individual declines (Price *et al.* 1983). Recently Cavellini *et al.* (1986) found large differences in DNA amounts among cultivars of *Helianthus*. Furthermore, they found up to 25% variation in total DNA amounts per nucleus for seedlings derived from the seeds of a single flower head. These studies suggest genomic changes may in fact be a response to environmental variation in these species.

In addition to the generation of variation by genomic changes, DNA sequences can also vary somatically due to mutation. Somatic mutations can occur throughout the genome during the life of an organism. Somatic mutations have been implicated in loss of function associated with ageing (Strekler 1977). Somatic mutations can occur during transposition (Shapiro 1983; Coen, Carpenter & Martin 1986) and during the insertion of proviruses (King *et al.* 1985). Somatic mutation can have an important functional role as well. Random somatic mutation occurs in the hypervariable region of the immunoglobulin gene and is correlated to antibody diversity (Alberts *et al.* 1983). Somatic mutations that affect morphological traits are well known in plants (e.g. Darwin 1889). Many cultivated varieties of plants are the result of somatic mutations (Hartman & Kester 1975). Since the development of tissue culture techniques, the degree of genetic variation among somatic cells has become apparent. Somaclonal variation of nuclear and cytoplasmic DNA can be heritable (Larkin *et al.* 1985), and single gene mutations can be recovered from tissue culture (Evans & Sharp 1983). The importance of somatic mutation in tissue culture is widely recognized and somaclonal variation has become the basis for potential crop improvement (Larkin *et al.* 1984).

The role of somatic mutations in maintaining and augmenting genetic diversity in native, non-cultivated plant species is not known. Several studies have pointed out the potential importance of somatic mutation for the population genetics and ecology of plant species (Whitham & Slobodchikof 1981; Walbot & Cullis 1983, 1985; Gill & Halverson 1984). These studies suggest that somatic mutations may be common and result in variable phenotypes. Gill & Halverson (1984) suggest that this phenotypic variability could possibly result in several toxic secondary compounds and thus reduce herbivore damage. Such ideas are speculative; there are very few data to test these hypotheses. Theoretical studies suggest that somatic mutation may be common (Antolin & Strobeck 1985; Klekowski & Kazarinova-Fukshansky 1984a, b; Klekowski, Mohr & Kazarinova-Fukshansky 1986). However, studies of somatic variation in native plant species are rare. A notable exception is the empirical work of Klekowski (1976) who examined the rate and accumulation of somatic mutations in the asexually reproducing fern *Osmunda* by an analysis of gametophytic mutations. Klekowski was able to trace the spread of somatic mutations through a clone and compare mutation rates among populations of different environments. Such studies are rare, partly for lack of methods for the determination of small genetic changes.

## CURRENT STUDIES

The examination of native plant species for somatic variation, either via induced genomic change or by somatic mutation is a subject of great interest. Several workers have analysed ribosomal DNA which is highly variable in plant species. The sequence is tandemly repeated and a plant's genome can contain from 3000 to over 10000 copies (Long & Dawid 1980). The rDNA sequence consists of coding sequences for the 18S, 5.8S, 28S ribosomal RNA. Separating these sequences is a non-transcribed spacer region (NTS) which is of variable length. The coding sequences tend to be evolutionarily conservative and phylogenetically informative (Sytsma & Schaal 1985). The NTS is highly variable within species, populations and individuals. Variation often is in the form of length heterogeneity. The NTS consists of a series of small, repeated sequences that vary in size, depending on species from approximately 100 to 350 base pairs (Rogers, Honda & Bendich 1986; Jorgenson *et al.* 1987). The presence of variable numbers of these repeat sequences generates length heterogeneity in ribosomal DNA. Ribosmal DNA can also vary in copy number among plants within populations and among lineages of corn (Rivin, Cullis & Walbot 1986). Finally, ribosomal

Ribosomal DNA in *Solidago* varies significantly among plants ($P<0.05$), and ranges from 0.62–3.61% of the total genome.

The frequency of somatic variation in *Solidago altissima* clones has been studied (Schaal & Helenurm, unpublished data). Eight *Solidago* clones were analysed from a recently abandoned (<20 years) old field. The clones were small, less than 0.5 m across and individual stems were still connected by rizomes. Each clone was sampled at four equidistant compass points. The different parts of single clones were compared to determine if there were any differences in either restriction sites or in the relative proportions of a repeat type within the clone. Again, fifteen restriction endonucleases were used to survey for intraclonal variation (see Table 4.1). Two of them, *Eco*RI and *Eco*RV, showed differences among the parts of single clones. DNA isolation, digestion with *Eco*RI or *Eco*RV, blotting, and hybridization were repeated; the differences in rDNA among parts of the clones appeared consistently. Either new restriction sites occurred in these portions of the clones, or the repeats which contain these restriction sites are in very low frequency, below the threshold of detection in other parts of the clone, and have been preferentially replicated in these portions of the clones. Fig. 4.2 shows the clone, a diagram of the rDNA restriction fragment pattern, and rDNA map of the restriction site variation. One of the clones appears to have undergone differentiation in rDNA copy number. Such genetic variation within a clone is suggestive either of somatic mutation, or of induction of rapid genomic change. Such differences can potentially influence the adaptation of a plant; whether this somatic variation is inherited and enters into the gene pool remains to be determined.

Preliminary studies on the induction of changes in rDNA copy number in *Brassica campestris* have been conducted by H. Hollacher and A. Colwell of Washington University. *B. campestris* is an annual, with cultivars that have extremely short life-cycles; flowering can occur 7 days after germination.

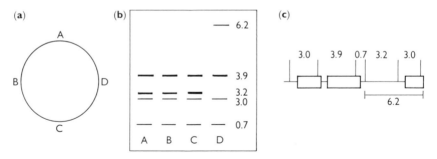

FIG. 4.2.   Ribosomal DNA variation within a clone of *Solidago altissima*. (a) Sampling points within clones; (b) *Eco*RV digest of rDNA; (c) ribosomal DNA map of intraclonal difference.

Because of these rapid cycling cultivars of diverse genetic backgrounds, *B. campestris* is a good experimental organism in which to examine rapid genomic changes. Seeds for the study were 'Crucifer Genetic Cooperative' (stock number 66), a rapid cycling, self-compatible lineage, with both *B. campestris* cytoplasm and genome. This stock has been selected for very short life-span, and the growth conditions under which it has been selected include high light flux and high nutrient availabilities (Williams & Hill 1986).

Since the stock had been growing for many generations of strong selection, it seemed most likely to be genetically adapted to high light and nutrients. For induction of genomic changes we placed the plants after 1 week of high light into a growth chamber with low light levels for 14 hours per day, at 21°C and no supplementary nutrients. We assumed that such a large shift in environment would be a stress to the plants and increase the likelihood of rapid genomic changes. The plants responded morphologically to the lower light and nutrient levels as would be expected; plants were tall and spindly and the life-cycle was much longer (4 weeks to flower rather than 7 days) than in the control which remained in the greenhouse.

Leaf tissue was sampled after 2 weeks and at 8 weeks during peak flowering. Genomic DNA was isolated and the percentage of rDNA in the genome determined by a slot blotting method, using a heterologous soybean probe. A total of sixteen plants were studied for rapid genomic changes. Both

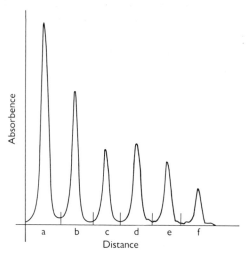

FIG. 4.3.  Change in ribosomal DNA amount before and after induction experiment. Standardized ribsomal DNA amounts of a single plant before the induction experiment are peaks a, b and c. The three peaks represent a dilution series. Peaks d, e and f are ribosomal DNA amounts in the same plant after induction. The same dilution series applies to d–f as a–c.

early and late DNA and a dilution series of these DNAs were run twice on a single filter, in addition to a DNA concentration series. Thirteen of the sixteen plants showed a reduction in rDNA amounts from early to late in the life-cycle ($P<0.01$, Wilcoxon sign rank test). Some of the plants showed dramatic reductions in rDNA, with less than half of the original amount remaining late in the life-cycle. Fig. 4.3 shows a densitometer tracing of a plant which has lost rDNA cistrons during the experimental treatment. We feel that these data strongly suggest rapid genomic change. This study is currently being replicated and the inheritance of these genomic changes determined.

## CONCLUSIONS

There is a great deal of evidence from the literature of molecular biology for the occurrence of somatic changes in the genome. Some of these changes are regular features of development, others are random somatic mutations and still others are programmed responses to environmental stress. The significance of these processes in population biology is yet to be determined. Indirect studies of total DNA content in native, non-cultivated species suggest genomic responses to environmental variation. Studies of cultivated species such as flax suggest such changes can enter into the gene pool. Our preliminary work on *Solidago*, *Brassica* and also *Taraxacum* indicates that somatic variation in DNA sequences may in some species be important factors in influencing the pattern and levels of genetic variation.

## ACKNOWLEDGMENTS

I thank Kaius Helenurm and Hope Hollocher for allowing me to use their unpublished data, E. Zimmer for the rDNA probe, and an anonymous reviewer for comments. This work was supported by NSF grant BSR 8207020.

## REFERENCES

Alberts, B., Bray, D., Lewis, J., Raff, M., Roberts, K. & Watson, J. D. (1983). *Molecular Biology of the Cell*. Garland, New York.

Antolin, M. V. & Strobeck, C. (1985). The population genetics of somatic mutation in plants. *American Naturalist*, 126, 52–62.

Cavallini, A., Zolfino, C., Cionini, G., Cremonini, R., Natali, L., Sassoli, O. & Cionini, P. G. (1986). Nuclear DNA changes within *Helianthus annuus* L. cytophotometric, karyological and biochemical analyses. *Theoretical and Applied Genetics*, 73, 20–6.

Coen, E. S., Carpenter, R. & Martin, C. (1986). Transposable elements generate novel spatial patterns of gene expression in *Antirrhinum majus*. *Cell*, 47, 285–96.

**Cullis, C. A. (1977).** Molecular aspects of the environmental induction of heritable changes in flax. *Heredity*, **38**, 129–54.

**Cullis, C. A. (1986).** Plant DNA variation and stress. *Genetics, Development and Evolution* (Ed. by J. Gustafson, G. Stebbins & F. J. Ayala), pp. 143–56. 17th Stadler Symposium, Plenum Press, New York.

**Cullis, C. A. & Charlton, L. M. (1981).** The induction of ribosomal DNA change in flax. *Plant Science Letters*, **20**, 213–17.

**Darwin, C. (1889).** *The Different Forms of Flowers on Plants of the Same Species.* Appleton, New York.

**DeSalle, R. & Templeton, A. (1986).** The molecular through ecological genetics of abnormal abdomen. III. Tissue-specific differential replication of ribosomal genes modulates the abnormal abdomen phenotype in *Drosophila mercatorum*. *Genetics*, **112**, 877–86.

**Duby, A., Klein, K., Murre, C. & Seidman, J. (1985).** A novel mechanism of somatic rearrangement predicted by a human T-Cell antigen receptor B-Chain complementary DNA. *Science*, **228**, 1204–6.

**Durrant, A. (1958).** Environmental conditioning of flax. *Nature*, **181**, 928–9.

**Durrant, A. (1962).** The induction of heritable changes in *Linum*. *Heredity*, **17**, 27–61.

**Durrant, A. (1971).** The induction and growth of flax genotrophs. *Heredity*, **27**, 277–98.

**Evans, D. A. & Sharp, W. (1983).** Single gene mutations in tomato plant regenerated from tissue culture. *Science*, **221**, 949–51.

**Frankham, R. (1980).** Origin of genetic variation in selection lines. *Selection Experiments in Laboratory and Domestic Animals* (Ed. by A. Robertson), pp. 56–68. Commonwealth Agricultural Bureau, Farnham Royal.

**Gill, D. E. & Halverson, T. G. (1984).** Fitness variation among branches within trees. *Evolutionary Ecology* (Ed. by B. Sharrocks), pp. 105–16, Blackwell Scientific Publications, Oxford.

**Goldsbrough, P. B., Ellis, T. & Lomonossof, G. (1982).** Sequence variation and methylation of the flax 5S RNA genes. *Nucleic Acids Research*, **10**, 4501–14.

**Haber, J. E. (1983).** Mating-type genes of *Saccharomyces cerevisiae. Mobile Genetic Elements* (Ed. by J. A. Shapiro), pp. 559–619. Academic Press, New York.

**Hartman, H. T. & Kester, D. E. (1975).** *Plant Propagation: Principles and Practices.* Prentice-Hall, Englewood Cliffs, New York.

**Hilton, C. D., Markie, P., Corner, B., Rikkerink, E. & Poulter, R. (1985).** *Genetics*, **112**, 162–8.

**Jorgensen, R. A., Cuellar, R. E., Thompsen, W. F. & Kavanagh, T. A. (1987).** Structure and variation in ribosomal RNA genes of pea: characterisation of a cloned rDNA repeat and chromosomal rDNA variants. *Plant Molecular Biology*, **8**, 3–12.

**King, W., Patel, M., Lobel, L., Coff, S. & Nguyen-Huu, M. (1985).** Insertion mutagenesis of embryonal carcinoma cells by retroviruses. *Science*, **228**, 554–8.

**Klekowski, E. J. (1976).** Mutational load in a fern population growing in polluted environment. *American Journal of Botany*, **63**, 1024–30.

**Klekowski, E. J. & Kazarinova-Fukshansky, N. (1984a).** Shoot apical meristem and mutation: fixation of selectively neutral cell genotypes. *American Journal of Botany*, **71**, 28–34.

**Klekowski, E. J. & Kazarinova-Fukshansky, N. (1984b).** Shoot apical meristem and mutation: selective loss of disadvantageous cell genotypes. *American Journal of Botany*, **71**, 28–34.

**Klekowski, E. J., Mohr, H. & Kazarinova-Fukshansky, N. (1986).** Mutation, apical meristems and developmental selection in plants. *Genetics Development and Evolution* (Ed. by J. Gustafson, G. Stebbins & F. J. Ayala), pp. 157–82. 17th Stadler Symposium, Plenum Press, New York.

**Larkin, P., Ryan, S., Brettell, R. & Scowcroft, W. (1984).** Heritable somaclonal variation in wheat. *Theoretical and Applied Genetics*, **67**, 443–55.

**Long, E. O. & Dawid, I. B. (1980).** Repeated genes in eukaryotes. *Annual Review of Biochemistry*, **49**, 727–64.

McClintock, B. (1984). The significance of response of the genome to challenge. *Science*, **226**, 792–801.

Miksche, J. P. (1968). Quantitative study of intraspecific variation of DNA per cell in *Picea glauca* and *Pinus banksiana*. *Canadian Journal of Genetics and Cytology*, **10**, 590–600.

Miksche, J. P. (1971). Intraspecific variation of DNA per cell in *Picea sitchensis* (Bong.) Carr. Provenances. *Chromosoma*, **41**, 29–36.

Phillips, R. L., Wang, A. S. & Knowles, R. V. (1983). Molecular and developmental cytogenetics of gene multiplicity in maize. *Stadler Genetics Symposium*, **15**, 105–18.

Price, H. J. & Bachmann, K. (1975). DNA content and evolution in the microseridinae. *American Journal of Botany*, **62**, 262–7.

Price, H. J., Chambers, K. & Bachmann, K. (1981). Geographic and ecological distribution of genomic DNA content variation in *Microseris douglasii* (Asteraceae). *Botanical Gazette*, **142**, 415–26.

Price, H. J., Chambers, K., Bachmann, K. & Riggs, J. (1983). Inheritance of nuclear 2C DNA content variation in intraspecific and interspecific hybrids of *Microseris* (Asteraceae). *American Journal of Botany*, **70**, 1133–8.

Ritossa, F. (1976). The *bobbed* locus. *The Genetics of Drosophila. Vol 1b* (Ed. by M. Ashburner & E. Novitski), pp. 801–46. Academic Press, London.

Rivin, C. J., Cullis, C. A. & Walbot, V. (1986). Evaluating quantitative variation in the genome of *Zea mays*. *Genetics*, **113**, 1009–19.

Rogers, S. O., Honda, S. & Bendich, A. J. (1986). Variation in the ribosomal RNA genes among individuals of *Vicia faba*. *Plant Molecular Biology*, **6**, 339–45.

Sachs, M., Dennis, R., Gerlach, W. & Peacock, W. (1986). Two alleles of maize *Alcohol dehydrogenase 1* have 3' structural and poly(A) addition polymorphisms. *Genetics*, **113**, 449–67.

Saghai-Maroof, M. A., Soliman, K. M., Jorgensen, R. A. & Allard, R. W. (1984). Ribosomal DNA spacer length polymorphisms in barley: Mendelian inheritance, chromosomal location, and population dynamics. *Proceedings of the National Academy of Sciences (USA)*, **81**, 8014–18.

Schimke, R. R. (1982). *Gene Amplification.* Cold Spring Harbor Laboratory, Cold Spring Harbor, New York.

Shapiro, J. A. (1983). *Mobile Genetic Elements.* Academic Press, New York.

Silverman, M. & Simon, M. (1983). Phase variation and related systems. *Mobile Genetic Elements* (Ed. by J. Shapiro). pp. 537–57. Academic Press, New York.

Strekler, B. L. (1977). *Times, Cells and Aging.* Academic Press, New York.

Sytsma, K. J. & Gottlieb, L. D. (1986). Chloroplast DNA evidence for the derivation of the genus *Heterogaura* from a species of *Clarkia* (Onagraceae). *Proceedings of the National Academy of Sciences (USA)*, **83**, 5554–7.

Sytsma, K. J. & Schaal, B. (1985). Phylogenetics of the *Lisianthius skinneri* (Gentianaceae) species complex in Panama utilizing DNA restriction fragment analysis. *Evolution*, **39**, 594–608.

Walbot, V. & Cullis, C. (1983). The plasticity of the plant genome—Is it requirement for success? *Plant Molecular Biology Reporter*, **1**, 3–11.

Walbot, V. & Cullis, C. (1985). Rapid genomic change in higher plants. *Annual Review of Plant Physiology*, **36**, 367–440.

Williams, P. H. & Hill, C. B. (1986). Rapid-cycling populations of *Brassica*. *Science*, **232**, 1385–9.

Witham, T. G. & Slobodchikof, C. N. (1981). Evolution by individuals, plant herbivore interactions, and mosaics of genetic variability: the adaptive significance of somatic mutations in plants. *Oecologia*, **49**, 287–92.

# 5. VARIATION IN THE PERFORMANCE OF INDIVIDUALS IN PLANT POPULATIONS

## JACOB WEINER

*Department of Biology, Swarthmore College, Swarthmore, PA 19081, USA*

## SUMMARY

1 Within plant populations there is much variation in size. This variation is extremely important, because size is correlated with both survivorship and fecundity, and therefore with Darwinian fitness.

2 Size variability can be evaluated with measures of inequality, such as the coefficient of variation or the Gini coefficient. Variation in size results from the interactions between several factors, about which very little is known.

3 Monocultures grown at high densities show greater variability in size than those grown at lower densities, supporting the hypothesis that competition is asymmetric. There is experimental evidence that this asymmetry is primarily due to competition for light. During density-dependent mortality (self-thinning) size variation among the survivors decreases, and this is also consistent with the hypothesis that asymmetric competition (and density-dependent mortality itself) are driven by shading.

4 Age differences are a primary determinant of size variation in populations with large variation in age. Even small age differences can act indirectly by establishing size differences, which then become exaggerated by competition or other factors. The fact that plant growth is sigmoidal has important implications for size differences in populations comprised of mixed age groups.

5 Herbivores may increase size variation by generating size differences which can persist or become exacerbated, but they may also reduce variation by removing large individuals or by reducing the intensity of competition. The most important effects of herbivory and disease on size variation may occur through interactions with competition.

6 There are two contrasting views of the role of size variation in the evolution of plant populations. Size differences may be a reflection of genotypic differences, in which case size (and resultant fecundity) variation will be a vehicle by which natural selection acts. Alternatively, size differences may be due to non-genetic factors, and result in genetic drift.

Although the history of population modelling requires that we start with exponential models, I believe that exponential models of plant growth are more misleading than helpful. Plant growth is exponential for a very short period, as can be seen by looking at whole-plant data (Hunt 1982), and exponential models are incapable of simulating even the most general behaviours of plant populations such as density-yield relationships. For example, in exponential models of plant growth, total yield decreases with increasing density after a certain period of growth, and the higher the density, the sooner the yield of one plant surpasses the combined yield of several competing plants! Some researchers (e.g. Turner & Rabinowitz 1983; Benjamin & Hardwick 1986; Huston & DeAngelis 1987) continue to use exponential models of plant growth, perhaps because exponential growth is mathematically tractable. Sigmoidal models of plant growth, despite their limitations, represent a vast improvement over exponential models in their ability to simulate plant behaviour.

What are the implications of sigmoidal growth for modelling the effects of age on size distributions? If we assume (1) plant growth is sigmoidal (e.g. 'logistic'), (2) recruitment of seedlings is continuous and uniform, and (3) relative growth rates and maximum sizes are normally distributed, the outcome is a bimodal distribution of plant sizes. There is a broad peak reflecting those individuals which have achieved their maximum size, and a narrow peak representing very young individuals (Fig. 5.5). The valley between the peaks is a reflection of the fact that in sigmoidal growth, the growth rate is fastest at intermediate sizes. Put another way, plants spend a relatively short period of time at intermediate sizes. If age is normally

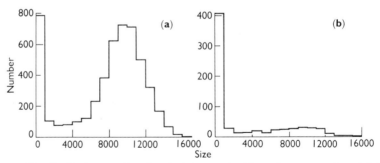

FIG. 5.5 Size distributions resulting from logistic growth with continuous recruitment. Initial sizes, instantaneous growth rates and carrying capacities for 5000 individuals are normally distributed with means of 1, 1, and 10 000 respectively, and coefficient of variation of 20%. (a) All plants live to age 40: continuous recruitment, sigmoid growth, no mortality; (b) constant per capita death-rate (0.15 individuals/individual/unit time): continuous recruitment, sigmoid growth and constant mortality.

distributed and growth is sigmoidal, most plants will diverge in size during the 'exponential' part of their growth, but they will tend to converge as they reach their maximum sizes. The greatest variability in size will occur at intermediate sizes. Remember, however, that these models assume there is no competition between plants. If we add competition to such models the results can be fundamentally altered. Intuitively, if there is large variation in age, asymmetric competition can be expected to produce bimodality, or high mortality of late recruits. Two-sided competition may reduce the effects of age variation on size variation by slowing the growth of older plants, although this has not yet been shown. Our understanding of the interactions between age, competition and size is underdeveloped, and much in need of models and experimental data.

Two sorts of data which bear on the role of age differences in determining variation in performance have been collected. There have been several studies which have looked at the influence of time of emergence on size in fields and glasshouse populations. In most cases, time of germination appears to have a major effect in determining final size or fecundity (e.g. Black & Wilkerson 1963, Ross & Harper 1972; Howell 1981; Waller 1985; Firbank & Watkinson 1987). Often, time of germination is the most important among a group of independent variables in accounting for variation in performance. It is often rank order of emergence, rather than age itself, which is most highly correlated with final size. This suggests that the effects of germination time are mediated by competition, since rank order of emergence may reflect position in the size hierarchy.

The other sort of data which has been collected relating age to plant size are distributions of age and size of trees in temperate forests. Age can be determined from ring counts and size is usually measured as stem diameter. Unfortunately, most of these data are static (i.e. collected at one point in time), and this makes any inferences about the dynamic processes which result in the observed distributions very weak. Data on size–age distributions over periods of even a few years would greatly increase our ability to study the concomitant changes in age and size of individuals which determine population size structure. To date, researchers have either focused on age-dependent survival and fecundity (classical demography) or size, but not both (but see Law 1983). In a given situation, information on both a plant's size and its age may enable us to predict its future performance with a high degree of accuracy.

Some conifer stands in the Rocky Mountains which have different age distributions appear to have similar size distributions (Knowles & Grant 1983). This suggests that there are ecological factors which are independent of age that act to determine population size structure. Diameter distributions of lodgepole pine (*Pinus contorta*) in the Sierra Nevada seem to show a good fit to an exponential decay (Fig. 5.6, Parker 1986), although in such a fit most of the variance is due to the first few columns of the size distribution. These are climax forests in which regeneration occurs in canopy gaps. Since a size distribution is a function of recruitment, growth and mortality, static data tend to be consistant with many alternative hypotheses. An attractively

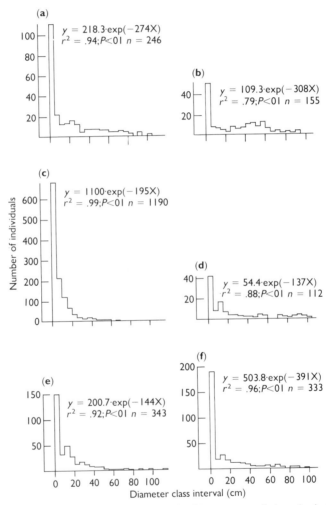

FIG. 5.6 Diameter class distributions for six stands of *Pinus contorta* (lodgepole pine) in the Sierra Nevada. Non-linear regression equations for negative exponential curves are given (y = number of stems per diameter class and x = diameter class midpoint) with their coefficients of determination and levels of significance (data from Parker 1986).

simple hypothesis to explain an exponentially decaying distribution of diameter in forests would be continuous, uniform recruitment, linear (arithmetic) growth in diameter, and a constant per capita death-rate. An equally extreme but less plausible alternative hypothesis would be continuous recruitment (at a lower rate), exponential growth in diameter and no mortality. In between these two extremes lies a continuum of compromise hypotheses. Dynamic demographic and growth data will be required to test these and other alternatives.

Differences in size due to age may, in many cases, have different implications for ultimate performance (i.e. final size or total reproductive output) from differences in size due to other variables. A plant which is small simply because it is young has greater prospects than a plant of the same size which is old but has suffered extensive suppression, herbivory or disease. On the other hand, when size differences are due to competitive suppression of slightly younger plants, relative size may be a good reflection of relative performance. In this case the difference in size is not a direct effect of age but an interaction between age and other factors. From the evolutionary perspective, we want to look at the cumulative performance of an individual over its life, not simply its size at one point in time. Again, the dynamic, demographic approach is necessary if we are to obtain the type of data needed to study this. Non-destructive measures of size are greatly needed so we will be able to follow the performance of individuals over time.

## HERBIVORY, PARASITES AND PATHOGENS

### Herbivory

The study of herbivory is one of the most exciting areas of active research in plant ecology (Crawley 1983), and researchers are now beginning to study the effects of herbivory on plant size variability (e.g. Windle & Franz 1979; Dirzo & Harper 1980; Cottam 1986). The effects of herbivory can be understood in terms of (1) the initial size distribution of plants, (2) the effect of herbivores on this size distribution (how herbivory modifies the distribution) and (3) the interaction of herbivory and other processes, e.g. herbivore attack may change subsequent patterns of growth and interference.

Many researchers (e.g. Harper 1977; Dirzo 1984) seem to feel that herbivory will increase variation in individual plant performance, but this is not necessarily the case (Table 5.2). If herbivory is random with respect to biomass, i.e. if each unit of biomass is equally likely to be consumed, and if the loss of parts does not result in mortality, then size variability will not be affected: the distribution of sizes will be shifted proportionally. If herbivory

Ford, E. D. & Diggle, P. J. (1981). Competition for light in a plant monoculture modelled as a spatial stochastic process. *Annals of Botany*, **48**, 481–500.

Gottlieb, L. D. (1977). Genotypic similarity of large and small individuals in a natural population of the annual plant *Stephanomeria exigua* ssp. *coronaria* (Compositae). *Journal of Ecology*, **65**, 127–34.

Hara, T. (1984a). A stochastic model and the moment dynamics of the growth and size distribution in plant populations. *Journal of Theoretical Biology*, **109**, 173–90.

Hara, T. (1984b). Dynamics of stand structure in plant monocultures. *Journal of Theoretical Biology*, **110**, 223–39.

Hara, T. (1986). Growth of individuals in plant populations. *Annals of Botany*, **57**, 55–68.

Harper, J. L. (1977). *Population Biology of Plants*. Academic Press, London.

Heywood, J. S. (1986). The effect of plant size variation on genetic drift in populations of annuals. *American Naturalist*, **137**, 851–61.

Heywood, J. S. & Levin, D. A. (1984). Genotype-environment interactions in determining fitness in dense, artificial populations of *Phlox drummondii*. *Oecologia*, **61**, 363–71.

Holliday, R. J. (1960). Plant population and crop yield. *Field Crop Abstracts*, **13**, 159–67.

Howell, N. (1981). The effect of seed size and relative emergence time on fitness in a natural population of *Impatiens capensis* Meerb. (Balsaminaceae). *American Midland Naturalist*, **105**, 312–20.

Hunt, R. (1982). *Plant growth curves*. University Park Press, Baltimore.

Huston, M. A. & DeAngelis, D. L. (1987). Size bimodality in monospecific populations: a critical review of potential mechanisms. *American Naturalist*, **129**, 678–707.

Knowles, P. & Grant, M. C. (1983). Age and size structure analysis of engelmann spruce, ponderosa pine, lodgepole pine and limber pine in Colorado. *Ecology*, **64**, 1–9.

Knox, R. (1987). *Hypothesis testing in plant community ecology: analysis of long-term experiments and regional vegetation data*. D. Phil. thesis, University of North Carolina.

Koyama, H. & Kira, T. (1956). Intraspecific competition among higher plants. VIII. Frequency distribution of individual plant weight as affected by the interaction between plants. *Journal of the Institute of Polytechnics, Osaka City University, Series D.*, **7**, 73–94.

Law, R. (1983). A model for the dynamics of a plant population containing individuals classified by age and size. *Ecology*, **64**, 224–230.

Levins, R. & Lewontin, R. (1985). *The Dialectical Biologist*. Harvard University Press, Cambridge, Massachusetts.

Loehle, C. (1983). Evaluation of theories and calculation tools in ecology. *Ecological Modelling*, **19**, 239–47.

Miller, T. E. (1987). Effects of emergence time on survival and growth in an early old-field plant community. *Oecologia*, **72**, 272–78.

Mithen, R., Harper, J. L., and Weiner, J. (1984). Growth and mortality of individual plants as a function of 'available area'. *Oecologia*, **62**, 57–60.

Morris, E. C. & Myerscough, P. J. (1984). The interaction of density and resource levels in monospecific stands of plants: a review of hypotheses and evidence. *Australian Journal of Ecology*, **9**, 51–62.

Parker, A. J. (1986). Persistence of lodgepole pine forests in the Central Sierra Nevada. *Ecology*, **67**, 1560–67.

Paul, N. D. & Ayers, P. G. (1986). The impact of a pathogen (*Puccinia lagenophore*) on populations of groundsel (*Senecio vulgaris*) overwintering in the field. *Journal of Ecology*, **74**, 1069–84.

Paul, N. D. & Ayers, P. G. (1987) Water stress modifies intraspecific interference between rust (*Puccinia lagenophore* Cooke)-infected and healthy groundsel (*Senecio vulgaris* L.). *New Phytologist*, **106**, 555–66.

Ross, M. A. & Harper, J. L. (1972). Occupation of biological space during seedling establishment. *Journal of Ecology*, **60**, 77–88.

**Samson, D. A. & Werk, K. S. (1986).** Size-dependent effects in the analysis of reproductive effort in plants. *American Naturalist*, **127**, 667–80.

**Sen, A. (1973).** *On Economic Inequality*. Clarendon Press, Oxford.

**Solbrig, O. T. (1981).** Studies on the population biology of the genus *Viola*. II. The effect of size on fitness in *Viola sororia*. *Evolution*, **35**, 1080–93.

**Sprugel, D. G. (1984).** Density, biomass, productivity, and nutrient-cycling changes during stand development in wave regenerated balsam fir forests. *Ecological Monographs*, **54**, 165–86.

**Tilman, D. (1987).** The importance of the mechanisms of interspecific competition. *American Naturalist*, **129**, 769–74.

**Turner, M. D. & Rabinowitz, D. (1983).** Factors affecting frequency distributions of plant mass: the absence of dominance and suppression in *Festuca paradoxa*. *Ecology*, **64**, 469–75.

**Vandermeer, J. (1984).** Plant competition and the yield-density relationship. *Journal of Theoretical Biology*, **109**, 393–99.

**Waller, D. M. (1985).** The genesis of size hierarchies in seedling populations of *Impatiens capensis* Meerb. *New Phytologist*, **100**, 243–60.

**Watkinson, A. R. (1980).** Density-dependence in single-species populations of plants. *Journal of Theoretical Biology*, **83**, 345–57.

**Weiner, J. (1986).** How competition for light and nutrients affects size variability in *Ipomoea tricolor* populations. *Ecology*, **76**, 1425–27.

**Weiner, J. & Solbrig, O. T. (1984).** The meaning and measurement of size hierarchies in plant populations. *Oecologia*, **61**, 334–36.

**Weiner, J. & Thomas, S. C. (1986).** Size variability and competition in plant monocultures. *Oikos*, **47**, 211–22.

**Weiner, J. & Whigham, D. F. (1988).** Size inequality and self-thinning in wild-rice (*Zizania aquatica*) *American Journal of Botany*, **73**, 445–48.

**Weller, D. E. (1987).** A reevaluation of the −3/2 power rule of plant self-thinning. *Ecology*, **57**, 23–43.

**Westoby, M. (1982).** Frequency distributions of plant size during competitive functions. *Annals of Botany*, **50**, 733–35.

**Westoby, M. (1984).** The self-thinning rule. *Advances in Ecological Research*, **14**, 167–225.

**White, J. (1980).** Demographic factors in populations of plants. *Demography and Evolution in Plant Populations* (Ed. by O. T. Solbrig), pp. 21–48. Blackwell Scientific Publications, Oxford.

**Williams, G. C. (1975).** *Sex and Evolution*. Princeton University Press, Princeton.

**Windle, P. N. & Franz, E. H. (1979)** Plant population structure and aphid parasitism in barley monocultures and mixtures. *Journal of Ecology*, **16**, 259–68.

**Yoda, K., Kira, T., Ogawa, H. & Hozumi, H. (1963).** Self-thinning in overcrowded pure stands under cultivated and natural conditions. *Journal of Biology, Osaka City University*, **14**, 107–29.

# 6. MORPHOLOGICAL PLASTICITY, FORAGING AND INTEGRATION IN CLONAL PERENNIAL HERBS

MICHAEL J. HUTCHINGS AND ANDREW J. SLADE

*School of Biological Sciences, University of Sussex, Falmer, Brighton, Sussex BN1 9QG, UK*

## SUMMARY

**1** The term foraging is defined and investigations are described which illustrate differences in the foraging behaviour of clonal herbs, particularly *Glechoma hederacea*, under different levels of nutrient and light supply.

**2** Differences are described between the structure of parts of clones formed in the proximity of neighbouring clones and parts of clones developing in isolation from neighbours.

**3** Some preliminary hypotheses are presented concerning the physiological causes of the alterations seen in clone morphology under different conditions.

**4** The effects of physiological integration between connected clonal ramets upon clone morphology, and upon growth and proliferation of different parts of the same clone, are illustrated. Explanations are sought for these effects based on differences in the extent of acropetal and basipetal patterns of translocation within the clone.

## INTRODUCTION

The concept of foraging, which has been used for some time by animal behaviourists, has recently been adopted by plant ecologists (Grime 1979; Bell 1984). A rigorous definition of the term foraging is, however, not easy to find in the literature. Slade & Hutchings (1987a) have recently defined it as the process whereby an organism searches or ramifies within its habitat in the activity of acquiring essential resources. Foraging behaviour in both plants and animals is genetically controlled. However, organisms may display environmentally induced plasticity in their manner of foraging. It should be recognized that this plasticity is externally imposed, in order to guard against teleological usage of the term.

Bell (1984) has drawn attention to a fundamental distinction between the manner in which foraging is accomplished in animals and plants. Whereas most foraging animals move about from place to place in their habitats,

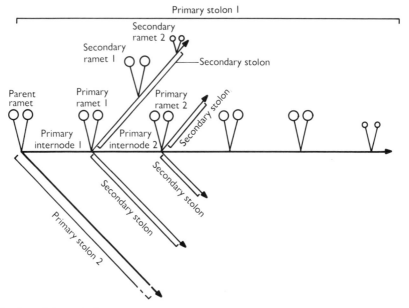

Fig. 6.1.   Schematic diagram of the basic vegetative morphology of *Glechoma hederacea*. See text for further details.

Under greenhouse conditions clones of *G. hederacea* rarely flower. Each ramet produced by an experimental clone of *G. hederacea* can be rooted in a separate pot, without damaging its connections with other ramets. Ramets can then be subjected independently to any combination of growing conditions, allowing great flexibility in the planning of experimental treatments.

Experiments have been conducted to analyse the foraging behaviour of *G. hederacea* clones grown under different levels of either nutrients or light. Throughout these experiments, excised parent ramets received an ample supply of both nutrients and light, and within a short period they began to produce two primary stolons. In the experiment to test the effects of nutrients on foraging, the daughter ramets on these stolons were each allowed to establish in sand in a separate pot, and either watered with full nutrient solution (soluble 'Phostrogen' at the recommended strength of $2.5 \, g \, l^{-1}$ of water) or with distilled water. In a third treatment the ramets on one primary stolon of the experimental clones were given full strength 'Phostrogen' while the ramets on the other primary stolon were watered with distilled water (this is referred to below as the split treatment). Secondary ramets were treated with the same nutrient supply as their parent ramets. The treatments were replicated to allow six harvests of six clones per treatment at 2 week intervals.

In the experiment to analyse the effects of shade on foraging behaviour, the daughter ramets were potted individually in sand with full strength nutrient solution and either grown in full light or under shading to reduce incident light by 75%. After 12 weeks of growth five clones were harvested from each treatment.

In each experiment, marked contrasts were seen in the morphologies of clones grown under different conditions. Some of the results are given in Table 6.1 (see also Slade & Hutchings 1987a, b). Many of the qualitative morphological differences between clones grown under high and low resource supply were the same, regardless of whether nutrient level or light intensity was the experimental variable (Table 6.2). In the treatments applied in these experiments, shading generally caused more dramatic alterations in clone structure than reduction in nutrient availability (Slade & Hutchings 1987c). Under low resource supply, stolons had long internodes and fewer branches (Fig. 6.2a, b). The ramets produced were small. Investment of biomass per unit length of stolon was significantly lower under shade than under high light conditions. A similar, but far less marked alteration was seen under low nutrient conditions (Table 6.1). Proportional allocation of biomass to stolons was higher under both low light and low nutrients than under resource-rich

TABLE 6.1. Mean values ($\pm$ S.E.) of various characteristics of *Glechoma hederacea* grown for 12 weeks under different conditions.

| | High light, high nutrients | High light, low nutrients | Low light, high nutrients | Significance |
|---|---|---|---|---|
| Mean internode length (cm) | $6.2 \pm 0.2$ (a) | $6.9 \pm 0.1$ (b) | $9.8 \pm 0.4$ (c) | $P < 0.05$ |
| Number of stolon branches per clone | $38 \pm 3$ (a) | $21 \pm 2$ (b) | $4 \pm 1$ (c) | $P < 0.01$ |
| Mean dry weight per unit length of stolon ($g \, cm^{-1}$) | $(30.6 \pm 2.05) \times 10^{-4}$ (a) | $(24.7 \pm 1.0) \times 10^{-4}$ (b) | $(11.5 \pm 0.9) \times 10^{-4}$ (c) | $P < 0.05$ |
| Mean dry weight per ramet (g) | 0.17 (a) | 0.15 (b) | 0.07 (c) | $P < 0.05$ |
| Percentage allocation of dry weight to | | | | |
| Leaves | $39.1 \pm 1.6$ (a) | $31.2 \pm 0.3$ (b) | $40.1 \pm 0.1$ (a) | $P < 0.01$ |
| Petioles | $14.8 \pm 0.5$ (a) | $11.3 \pm 0.2$ (b) | $26.6 \pm 0.4$ (c) | $P < 0.001$ |
| Stolons | $16.5 \pm 0.6$ (a) | $20.6 \pm 0.6$ (b) | $18.4 \pm 1.0$ (a, b) | $P < 0.005$ |
| Roots | $29.6 \pm 0.7$ (a) | $36.9 \pm 0.9$ (b) | $14.9 \pm 1.1$ (c) | $P < 0.001$ |

Different letters denote statistical differences at the level of significance indicated. Results based on Slade & Hutchings (1987c).

TABLE 6.2. Summary table of responses by *Glechoma hederacea* to availability of light and soil nutrients. (For detailed description of the treatments applied see text and Slade & Hutchings 1987c.)

| Characteristics measured | Treatment applied | | |
|---|---|---|---|
| | + Light + Nutrients | + Light − Nutrients | − Light + Nutrients |
| Number of ramets per clone | ⎯⎯⎯⎯⎯⎯⎯Decreases⎯⎯⎯⎯⎯⎯⎯⎯→ | | |
| Number of stolon branches per clone | ⎯⎯⎯⎯⎯⎯⎯Decreases⎯⎯⎯⎯⎯⎯⎯⎯→ | | |
| Internode lengths | ⎯⎯⎯⎯⎯⎯⎯Increases⎯⎯⎯⎯⎯⎯⎯⎯→ | | |
| Total clone dry weight | ⎯⎯⎯⎯⎯⎯⎯Decreases⎯⎯⎯⎯⎯⎯⎯⎯→ | | |
| Dry weight per unit length of stolons | High | Intermediate | Low |
| Dry weight (per clone) of | | | |
|   Leaves | High | Low | Very low |
|   Roots | ←⎯⎯⎯No difference⎯⎯⎯→ | | Low |
|   Petioles | High | Low | Very low |
|   Stolons | ←⎯⎯⎯No difference⎯⎯⎯→ | | Low |
| Proportional allocation of weight to | | | |
|   Leaves | High | Low | High |
|   Petioles | Intermediate | Low | High |
|   Stolons | Low | High | High |
|   Roots | Intermediate | High | Low |
| Leaf area per clone | ⎯⎯⎯⎯⎯⎯⎯Decreases⎯⎯⎯⎯⎯⎯⎯⎯→ | | |
| Specific leaf area | High | Low | High |
| Chlorophyll content of leaves | High | Low | Not measured |
| Growth form | Phalanx⎯⎯⎯⎯⎯⎯⎯⎯⎯⎯⎯⎯⎯→Guerilla | | |

conditions. These plastic alterations in structure result in less energy being invested in local foraging for resources in sites from which the rewards will be low, and more energy being devoted to activities which may lead, in a heterogeneous environment, to placement of feeding sites in more rewarding positions. Clones receiving a low nutrient supply, also devoted more biomass to roots, and less to leaves and petioles (Fig. 6.3), than clones in resource-rich conditions, whereas clones grown under shade allocated more biomass to petioles and less to roots. Although there was no significant difference in proportional allocation of biomass to leaves in the two light treatments, specific leaf area was nearly three times greater under shade than in full light (Slade & Hutchings 1987c).

Despite the marked changes in stolon internode lengths, clone biomass and proportional allocation of biomass, there was no significant difference in the total length of primary stolons produced when nutrient availability was reduced by a factor of eight and when light intensity was reduced by a factor of four. However, there were marked differences in the total length of

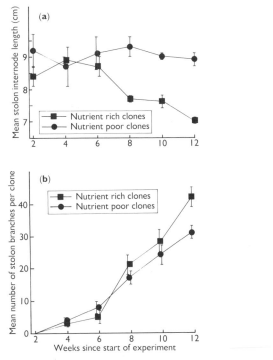

FIG. 6.2.   (a) Mean stolon internode lengths and (b) mean number of stolon branches ($\pm$S.E., $n=6$) for clones of *Glechoma hederacea* after different periods of growth under greenhouse conditions in an experiment in which nutrient supply was the variable.

secondary stolons; the length of secondary stolons produced by clones grown in high light and under high nutrient level was over thirteen times greater than that produced by clones grown under low light, and nine times greater than that produced by clones grown in low nutrient conditions respectively (Fig. 6.4a). Total stolon length in the three treatments therefore differed entirely as a result of the differences in branching of clones. Production of ramets on primary and secondary stolons displayed similar trends (Fig. 6.4b). Despite very marked reductions in biomass of clones under resource-poor conditions, their capacity for linear spread was not at all impaired, whereas their tendency to spread and establish feeding sites in two dimensions was substantially reduced.

Thus, under resource-poor conditions, there is a reduction in the frequency of feeding sites per unit length of stolon and a lower frequency of stolon branching. Feeding sites are therefore established at lower density by clones growing under adverse conditions. The plant forages more widely in a

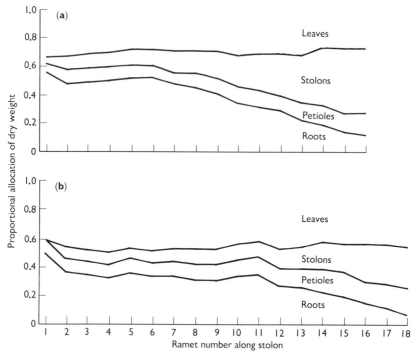

FIG. 6.3. Proportional allocation of biomass by ramets along primary stolons of *Glechoma hederacea* in a greenhouse experiment for (a) clones not supplied with nutrients, (b) clones supplied with nutrients.

resource-poor habitat when it has achieved a given biomass, and it forages in a less intensive fashion. The possession of spacers (stolons or rhizomes) by herbaceous clonal species thus provides a mechanism for either enabling exploitation of favourable patches of habitat, or escape from adverse conditions (Turkington & Cavers 1978; Aarssen, Turkington & Cavers 1979).

Clegg (1978) introduced the terms 'phalanx' and 'guerilla' to distinguish opposite ends of a continuum of growth forms in clonal plant species. Phalanx clones are highly branched, they expand slowly and have densely packed ramets which come into contact mainly with other ramets belonging to the same clone. Guerilla clones are less branched, more invasive, and their ramets frequently encounter parts of other plants. Because the average diameter of ramets in phalanx clones is in many cases similar in magnitude to their internode lengths, a high proportion of the ground occupied by a phalanx clone is covered by its own leaves. By comparison, many guerilla clones have internode lengths much greater than the diameter of their ramets, resulting in the percentage cover of the ground occupied by a guerilla clone

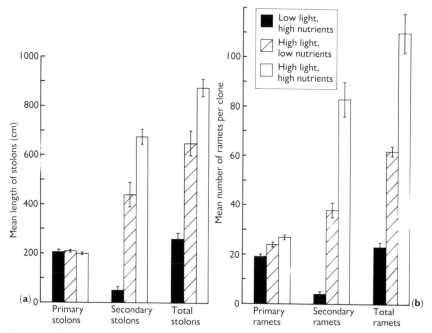

FIG. 6.4. (a) Mean length of primary, secondary and all stolons, and (b) mean number of primary, secondary and all ramets, produced by clones of *Glechoma hederacea* in a greenhouse experiment. Vertical bars = ±S.E., $n = 6$.

being low. In these terms, the alterations in morphology which have been shown to accompany a reduction in resource availability, result in a change from more phalanx to a more guerilla growth form. Whether the same directional changes will be seen at the very limit of species' tolerance to lack of resources can only be determined by more detailed experimentation.

These two experiments provide evidence for plasticity in stolon internode length and probability of branching, but not for plasticity in stolon branching angles, since the experimental arrangement for potting ramets did not allow natural expression of the two-dimensional structure of the clone. However, the limited data available for *G. hederacea* (Slade, unpublished data), *Ranunculus repens* and *Carex arenaria* (Sutherland & Stillman 1988), have not revealed any significant change in branching angle when these species have been exposed to different growing conditions.

There are very few published observations concerning the morphology of clonal herbs growing under field conditions which can be used to analyse foraging responses to resource supply. Most of the available information concerns the changes in morphology seen when species are subjected to different levels of competition, and in order to draw parallels between the experimental results and field observations, it must be assumed that greater competition reduces the availability of resources to the plant in the field. It must also be assumed in some cases that the observed differences in morphology are caused by plasticity, and not the result of genetical differences between clones. Despite the apparent corroboration of the experimental results by the few field observations, it must be recognized that there may have been other reasons for the morphological variation of species in the field. As this possibility cannot be completely discounted in the absence of detailed site descriptions and genetical analysis of sampled material, caution is advisable in interpreting field data. However, in competition with grasses, *Trifolium repens* has fewer stolon branches, and concentrates growth in linear extension to a greater extent than when it grows in isolation (Harper 1983). Comparison of *Ranunculus repens* clones from adjacent woodland and recently established park grassland showed that the woodland clones had significantly longer stolon internodes and higher dry matter allocation to stolons than the grassland clones (Lovett Doust 1981a, b). Under experimental conditions, Ginzo & Lovell (1973) reported that similar total lengths of primary stolons were produced by *Ranunculus repens* in high and low nutrient conditions, but that there was far less branching under low nutrient conditions. Data from *Solidago canadensis* are more contradictory and more difficult to interpret. Bradbury & Hofstra (1976) and Bradbury (1981) reported that rhizome lengths were greater in recently abandoned old fields than at later points in succession, when resources were probably scarcer and competition more intense. Hartnett & Bazzaz (1983) report changes in the opposite direction.

## MORPHOLOGICAL RESPONSES TO NEIGHBOURING RAMETS

There is little published information about changes in the morphology of clones of a given species when subjected to interactions with different neighbouring plant species. A preliminary attempt at analysing differences in clone structure displayed when (1) all ramet neighbours were parts of the same plant and (2) at least some ramet neighbours were parts of different plants was published by Bülow-Olsen, Sackville Hamilton & Hutchings (1984). This study involved analysis of the spatial structure of dense clones (monocultures) of *Trifolium repens* under field conditions. Stolon apices were mapped within areas of ground where two very dense, extensive and phenotypically different clones were abutting each other along a wide margin. Mapping was repeated on two dates one week apart, and the growth increment of each stolon apex during the interval was measured. The area of ground (or 'territory') nearest to each stolon apex was calculated by tessellating the two-dimensional plane with Voronoi polygons (Mead 1966; Liddle, Budd & Hutchings 1982). The technique provides a snapshot picture of the arrangement and packing density of stolon apices at one point in time, whereas in reality the positions of apices are continually changing because apices are being produced, growing and dying. However, given that these clones were believed to have been occupying the study site for several years (it had been sown with *Trifolium repens* and *Lolium perenne* 7 years before the study was conducted), the assumption that the technique provides a good indication of the stable density of stolon apices in each clone seems reasonable. Five pairs of clones were mapped in this way.

On the basis of the maps produced it was possible, by analysing the polygon size distributions, mean distance to nearest neighbours and mean growth rates of stolon apices, to designate one clone of each pair as a relative phalanx and the other as a relative guerilla. This interpretation of clone morphology was made from analysis of the hinterland region of the clones, i.e. that region where interclonal interactions were absent and ramets made contact only with other ramets belonging to the same clone. The *relative* densities of opposing clones can also be compared on the basis of their mean hinterland polygon areas (x-axis, Fig. 6.5). Pairs of clones with similar apex density each have values close to 1.0 on this axis whereas a clone which has a phalanx structure relative to its neighbour has a value < 1.0 and a clone with a guerilla structure relative to its neighbour has a value > 1.0. Further analysis revealed that whereas the mean areas of hinterland polygons of opposing clones differed significantly in all cases, the mean areas of their polygons in the border zones between them (i.e. the region where at least one neighbour of each apex is from the opposing clone) were not statistically different in any case (Bülow-Olsen, Sackville Hamilton & Hutchings 1984). The *relative* response of clone morphology to the presence of a neighbouring clone can be quantified, as on the y-axis in Fig. 6.5. In every case, the relative phalanx clone altered its morphology in a guerilla direction in the presence of its neighbour (i.e. its structure was less dense in its border region), whereas the guerilla clone of each pair displayed a more phalanx structure in its border zone. The best-fit curve through the data points passes almost exactly through the point 1, 1, although this is probably somewhat fortuitous. The indication is that, upon being faced with a conspecific clone of different packing density, the density of apices in a clone of *Trifolium repens* is adjusted towards that of its neighbour. The degree of adjustment is apparently greater for pairs of clones which start the interaction with greater differences in their density, and both clones take part in the response. This analysis of clone

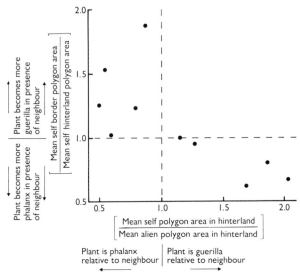

FIG. 6.5.    Relationship between the mean area of polygons in the hinterland region of a clone of *Trifolium repens* and that of its conspecific neighbour (*x*-axis), plotted against the extent to which mean polygon area alters between the hinterland and border regions of the clone (*y*-axis). Clones to the left of the dashed vertical line are more phalanx than their neighbour; clones to the right are more guerilla in structure. Clones above the horizontal dashed line become more guerilla in the presence of their neighbour whereas clones below the line become more phalanx. For further details see text.

behaviour in the presence of neighbours is clearly preliminary, however, and more data are needed to confirm these results.

## EXPLANATIONS FOR DIFFERENTIAL FORAGING BEHAVIOUR

What physiological explanations can be invoked for the morphological changes observed when clonal herbs are grown under different conditions? The most likely explanations rest on the fact that stolons and rhizomes are modified, horizontally oriented stems, and therefore their development and branching are probably under the same physiological controls as that of stems. Cook (1985), Lovell & Lovell (1985) and Mogie & Hutchings (unpublished data) have discussed the effects of environmental factors, particularly light quality and nutrient availability, upon clone development. These factors act mainly through their effects upon the concentrations of the plant growth substances which determine the degree of apical dominance, internode extension and branching displayed by stems. Although the interactions between the environmental factors and several growth substances which may be involved are very complicated, and still far from being fully elucidated, some speculation is possible about the ways in which differential foraging may be controlled in clonal herbs.

Light which is transmitted through a leaf canopy becomes spectrally biased, and has, compared with unfiltered light, a higher far red–red ratio and a lower proportion of blue light (Hutchings 1976; Holmes 1981). When the ratio of far red–red light incident on stem apices is increased, apical dominance is strengthened, lateral bud growth is inhibited and internodes elongate (Grime 1966, 1979; Child, Morgan & Smith 1981; Morgan 1981). When the proportion of the blue waveband in incident light is reduced, internode elongation is promoted (Thomas 1981; Quail 1983; Schafer & Haupt 1983). The relevance of these effects in the field is that these alterations in light quality would change the foraging behaviour of a clonal herb. There would be greater emphasis on linear expansion of the clone and on extensive foraging when light quality was spectrally biased by an overhanging canopy of leaves whereas in full light there would be more branching and more intensive foraging (see also Solangaarachchi & Harper 1987). Although differences in light quality, indicating altered availability of light, could cause the changes in morphology which have been described, these morphological changes would promote more efficient foraging when any essential resource changed in abundance because of differences in the density of competitors.

If the sensor for availability of essential resources from the environment is located in the spacing organs of a clonal herb, then a mechanism for controlling foraging behaviour which responds to light quality could only operate in stoloniferous species or species with superficial rhizomes, in which a degree of light sensing is possible. Another resource which could control foraging behaviour must be suggested for those species which forage by means of subterranean spacers. Nitrogen is one soil-based resource which could operate as a cue for the promotion of different morphologies as its abundance fluctuated. It is recognized that the metabolic consequences of changing nitrogen supply are immensely complex. However, its abundance could affect the development of stoloniferous species *via* sensors located in their roots. The synthesis of cytokinin, one of the growth regulators involved in controlling internode elongation and lateral bud development, is affected by soil nitrogen availability. When soil nitrogen availability is high, cytokinin production is high (Aung, De Hertogh & Staby 1969; Menhenett & Wareing 1975; McIntyre 1976; Radley 1976; Qureshi & McIntyre 1979; Menzel 1980). Cytokinin synthesis inhibits internode elongation and promotes lateral bud development; apical dominance declines and a highly branched structure with short internodes is produced. When soil nitrogen availability is low, cytokinin synthesis decreases. Apical dominance is reinforced, internode elongation is promoted, and lateral bud development is suppressed. Thus, under nitrogen-rich conditions, a relatively phalanx structure develops, whereas under nitrogen-deficient conditions a guerilla structure is produced.

Some experimental evidence which supports these theories is available in the literature. For example, a high far red–red light ratio reduced tillering in experimental plants of *Lolium multiflorum*, although many buds which could generate tillers were present (Deregibus, Sánchez & Casal 1983; Casal, Deregibus & Sánchez 1985). Greater apical dominance was suggested to be the reason for the lower tillering rate. Similarly, in *Paspalum dilatatum* and *Sporobolus indicus*, tillering rates increased in plants with red-light-emitting diodes placed at the base of their crowns, compared with controls (Deregibus *et al.* 1985), because the diodes reduced the ratio of far red–red light incident at tiller bases. Under pot conditions McIntyre (1967) demonstrated that *Agropyron repens* produced rhizomes from buds in its lowest leaf axils when grown under low nitrogen conditions, but that the same buds produced tillers when nitrogen levels were high. Similar changes in behaviour accompanied low and high water availability from the soil (McIntyre 1976). These results are consistent with the behavioural changes shown by clones of *Glechoma hederacea* under different levels of resource availability, and support the interpretations given above, concerning the control of foraging.

## THE EFFECTS OF PHYSIOLOGICAL INTEGRATION BETWEEN RAMETS ON FORAGING BEHAVIOUR

The ramets of a clonal plant may differ in their physiological states and in their functions. They may be producing daughter ramets directly by vegetative propagation, providing support for the establishment of new ramets which have been produced at other locations on the clone, or supporting older ramets which are growing under adverse conditions (Hartnett & Bazzaz 1983). Ramets may have purely vegetative functions or they may be involved in sexual reproduction. The states and functions of ramets also change through time. Organization of the roles of different ramets requires that as long as they remain connected to each other their behaviour is to some extent integrated. Physiological integration between connected ramets is not only necessary for coordination of these different activities, but it is also required to control the positioning and local density of ramets.

Those structures of clonal plants which have been referred to until now in this paper as spacers (Bell 1984) have an additional major function as routes along which resources can be moved within the growing plant. Several studies have been made of translocation patterns in clonal herbs, mainly using radioactive tracers (see Pitelka & Ashmun (1985) for a detailed review of the subject). Although the variety of methods involved makes generalizations about results difficult, investigations on many species have revealed considerable acropetal movement of photoassimilates (Hoshino 1972; Ginzo & Lovell 1973a, b; Newell 1982; Noble & Marshall 1983; Alpert & Mooney 1986) and nutrients (Guttridge 1959; Norton & Wittwer 1962; Grindey 1975; Noble & Marshall 1983) towards developmental sinks such as stolon and rhizome tips and actively growing leaves and roots. By contrast, there is little evidence of strong basipetal translocation in undisturbed clonal plants under normal growing conditions.

The consequences of this imbalance of acropetal and basipetal translocation of resources for the morphology and foraging behaviour of the growing plant can be examined by providing some ramets of experimental clones with good growing conditions, while other ramets are given poor conditions (Slade & Hutchings 1987c). *Glechoma hederacea* was again the experimental subject. The complete experimental design was complex, but for present purposes only a subset of the experimental treatments needs to be described (Fig. 6.6). Two sets of control clones were established receiving growing conditions which remained the same throughout the 12 weeks duration of

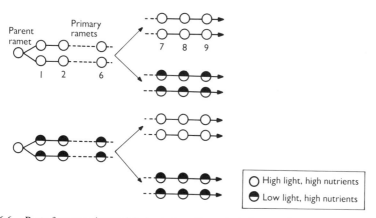

FIG. 6.6.  Part of an experimental design to investigate the influence of resource availability and translocation patterns upon morphology and foraging behaviour of *Glechoma hederacea*. Two sets of control clones and two sets of treatment clones were established. Control clones received the same growing conditions throughout the experiment. The growing conditions of all parts of treatment clones produced after the sixth primary ramets became established were different from those applied at the start of the experiment. The circles denote pots containing ramets.

the experiment. These conditions were (1) high light and high nutrients and (2) low light and high nutrients. The control clones provided base-line measurements of clone morphology against which the effects of changing the growing conditions during the experiment could be compared. In the case of treatment clones, growing conditions were changed after the sixth ramets along the primary stolons had rooted. Treatment clones grew (1) from high light and high nutrients into low light, high nutrient conditions (a treatment involving a change midway through the experiment from good to poor growing conditions), and (2) from low light and high nutrient conditions into high light, high nutrient conditions (a treatment involving a change from poor to good growing conditions). Comparisons were made between (1) the part of the treatment clones produced before the change in growing conditions and the corresponding part of those control clones which were receiving the same growing conditions in the first part of the experiment, (2) the part of the treatment clones produced after the change in growing conditions and the corresponding part of those control clones which were receiving the same growing conditions in the second part of the experiment (Table 6.3).

The change in growing conditions experienced by the treatment clones was accompanied by several significant morphological changes (Slade & Hutchings 1987c). Generally these corresponded with those already described as characteristic for clones foraging differentially in response to resource availability. In all respects the morphology of the part of treatment clones produced before the change in growing conditions matched that of the control clones receiving the same conditions (Table 6.3). However, the comparisons with the controls *after* the changes in growing conditions were applied, need to be analysed with great care. The part of the treatment clones produced after the change from unshaded to shaded conditions showed many highly significant differences from the corresponding part of the shaded control clones (Table 6.3). In contrast, the part of the treatment clones produced after the change from shaded to unshaded growing conditions, differed significantly in only a few respects (degree of branching, dry weight per unit length of stolon and clone dry weight) from the corresponding part of the resource-rich control clones. In many cases the level of significance of these differences was low (Table 6.3).

More understanding of the different impacts of these changing growing conditions upon subsequent clonal development can be obtained by analysing the characteristics of consecutive ramets along the stolons. When stolons grew from poor to good conditions, the morphological alterations indicating a changed foraging behaviour were virtually immediate, and completed by the time that the next internode had been produced (e.g. stolon internode lengths, Fig. 6.7). The clone ceased to forage like control clones growing in

TABLE 6.3. Mean (±S.E.) values of selected characteristics for the parts of control and treatment clones of *Glechoma hederacea* produced before and after a change in growing conditions was imposed upon the treatment clones. (For further details of the experimental arrangements see text). Clones were grown in (1) full light (+L) or under shading (−L) and (2) in nutrient-rich sand (+N) or nutrient-poor sand (−N). Statistical comparisons are made as follows: (i) *Treatment clones* +L+N/−L+N. The first parts of these clones are compared with the first part of +L+N/+L+N control clones. The second parts are compared with the second part of the −L+N/−L+N control clones. (ii) *Treatment clones* −L+N/+L+N. The first parts of these clones are compared with the first part of the −L+N/−L+N control clones and the second parts are compared with the second parts of the +L+N/+L+N control clones.

| | Control clones | | | | Treatment clones | | | |
|---|---|---|---|---|---|---|---|---|
| | +L+N/+L+N | | −L+N/−L+N | | +L+N/−L+N | | −L+N/+L+N | |
| Part of clone | First | Second | First | Second | First | Second | First | Second |
| Number of ramets | 53±3 | 55±6 | 16±2 | 7±1 | 55±4 N.S. | 32±5*** | 15±1 N.S. | 43±5 N.S. |
| Number of stolon branches | 14±1 | 24±2 | 4±1 | 0 | 13±1 N.S. | 7±1** | 3±1 N.S. | 14±2* |
| Stolon internode lengths (cm) | 5.8±0.2 | 6.6±0.2 | 9.5±0.3 | 10.0±0.3 | 6.7±0.2 N.S. | 9.1±0.2 N.S. | 8.9±0.2 N.S. | 7.4±0.2 N.S. |
| Σ Stolon length (cm) | 405±25 | 469±47 | 187±15 | 71±10 | 477±61 N.S. | 359±50*** | 175±7 N.S. | 420±50 N.S. |
| Dry weight per unit length of stolon (g.cm⁻¹ × 10⁻⁴) | 35.4±1.6 | 60.0±5.0 | 12.4±0.6 | 13.1±1.2 | 28.5±2.8 N.S. | 33.8±0.6*** | 13.8±0.4 N.S. | 39.6±3.4* |
| Clone dry weight (g) | 8.8±0.5 | 10.0±1.0 | 1.2±0.1 | 0.6±0.1 | 7.5±0.6 N.S. | 4.3±0.6*** | 1.6±0.1 N.S. | 5.7±1.1* |
| % allocation of dry weight to | | | | | | | | |
| Leaves | 38±1 | 40±2 | 41±1 | 39±2 | 43±1 N.S. | 34±1 N.S. | 39±1 N.S. | 42±2 N.S. |
| Petioles | 13±1 | 16±1 | 23±1 | 30±3 | 17±1 N.S. | 26±1* | 30±2 N.S. | 17±2*** |
| Stolons | 15±1 | 19±1 | 18±1 | 20±2 | 13±1 N.S. | 21±1 N.S. | 16±1 N.S. | 17±1 N.S. |
| Roots | 34±1 | 25±1 | 18±1 | 11±2 | 27±2 N.S. | 19±1** | 15±1 N.S. | 24±2** |
| Leaf area (cm²) | 450±46 | 582±83 | 209±16 | 99±35 | 600±67 N.S. | 437±78** | 210±13 N.S. | 436±68 N.S. |
| Leaf area per ramet (cm²) | 13.7±0.9 | 24.6±1.3 | 16.6±1.3 | 13.4±3.8 | 16.3±2.8 N.S. | 26.3±3.2* | 16.7±1.0 N.S. | 22.9±1.9 N.S. |
| Specific leaf area (cm² g⁻¹) | 143±19 | 167±19 | 461±59 | 436±82 | 205±18* | 337±15 N.S. | 461±44 N.S. | 171±17 N.S. |

N.S. = not significant; * = $P < 0.05$; ** = $P < 0.01$; *** = $P < 0.001$.

if its sister ramet was in non-saline soil. The benefit of support received by the ramets in saline soil in the treatment clones was measured as the difference between their weight and the weight of individual ramets in the saline control clones.

Those ramets of the treatment clones which were grown in non-saline soil had a significantly lower weight than ramets of control clones grown in non-saline soils. The cost, however, was comfortably exceeded ($P < 0.02$) by the benefit of the support provided to the ramets which were grown in saline soil in the treatment clones. These ramets were significantly heavier than ramets from the control clones grown under saline conditions. The results fulfil the essential requirement for the evolution of inter-ramet integration that, at least in the long term, the benefit accruing from integration must exceed the cost.

A similar experiment was performed upon *Glechoma hederacea* (Slade & Hutchings 1987d). Clones were grown under five different sets of growing conditions. Three sets of control clones were grown: (1) under full illumination, in sand watered with full strength nutrient solution, (2) under 75% shading, in sand watered with full strength nutrient solution, and (3) under full illumination, in sand watered with one-eighth strength nutrient solution. In the two remaining treatments, clones were grown under full daylight in sand watered with full strength nutrient solution until six ramets had rooted along each of the primary stolons. (This took approximately 7 weeks in all treatments.) Ramets produced after this were grown either (1) under shade, in sand watered with full strength nutrient solution or (2) in full light, in sand watered with one-eighth strength nutrient solution. Clones were harvested after 12 weeks and divided into (1) ramets and internodes one to six on the primary stolons, and their associated structures, and (2) primary ramets and internodes seven onwards and their associated structures. This enabled comparisons to be made of the growth of those parts of clones produced in different treatments before and after the date on which growing conditions were changed. The costs and benefits of inter-ramet integration could be analysed by using the same comparisons between control and treatment clones as in Salzman & Parker (1985).

The lack of significant differences between the weight of the first parts of clones subjected to a change in growing conditions, and that of the control clones receiving ample light and nutrients throughout the experiment (Fig. 6.8), indicates a lack of a cost to the ramets produced during the first part of the experiment incurred by being connected to ramets produced later, under either low light or low nutrient conditions. (The first parts of clones grown in all three of these sets of conditions had significantly greater weights than the first parts of either of the other control clones (Fig. 6.8)). However, the

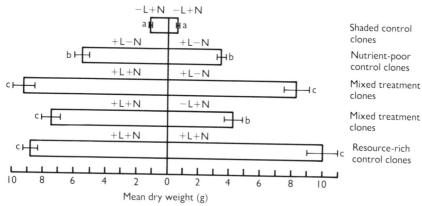

FIG. 6.8.  Mean total dry weight (±S.E.) of the first and second parts of clones of *Glechoma hederacea* subjected to five different sets of growing conditions. All parts of control clones received the same growing conditions throughout the experiment. The growing conditions of all parts of treatment clones produced after the sixth primary ramets became established, were different from those applied at the start of the experiment. Significant differences (at least $P < 0.05$) between the dry weights of the first and second parts of the clones growing in the five treatments are indicated by different letters. Statistical comparisons of dry weights *between* the two parts of the clones within treatments are not made (first produced part of clones on the left and second on the right). The growing conditions applying to each part of the experimental period are indicated ($+N$ = nutrient-rich; $-N$ = nutrient-poor; $+L$ = full light; $-L$ = shaded).

weights of the second parts of the treatment clones which were transferred to low light and to low nutrients during the experiment were significantly higher than the weights of the second parts of the shaded control clones and the nutrient-poor control clones respectively (both $P < 0.001$), indicating a marked benefit resulting from attachment to ramets growing under better conditions. These results only differ from those of Salzman & Parker (1985) in failing to reveal a significant cost caused by providing integrated support. It is suggested that the support provided in *G. hederacea* consists of resources surplus to the requirements of the providing ramets.

Both of these experiments allow straightforward analysis of the costs and benefits of inter-ramet integration. However, a further experiment with *Glechoma hederacea* reveals a need for more detailed analyses, and for greater consideration of the influence of clone structure upon the results obtained. The experiment commenced with clones consisting of two connected ramets which had been produced consecutively along the same stolon (Slade & Hutchings 1987d). Evidence has already been given showing that, on balance, more translocates pass from older to younger ramets than in the opposite direction along the same stolon. Because of this, the capacities of each ramet to support the other will not be equivalent: the costs and benefits of integration will differ according to the ability of the ramets in the treatments to supply, and to be supplied with, translocates. Therefore, for a full analysis of costs and benefits of inter-ramet integration in these much simplified two-ramet clones, two types of treatment clones need to be established in which connected ramets receive different growing conditions; in the first, the older ramet receives better conditions and in the second, the older ramet receives the poorer conditions.

The variable applied to the ramets in the experiment was nutrient supply. Control clones were established with both ramets receiving either full strength or one-eighth strength nutrient solution. Treatment clones were established with either the older ramet receiving full strength nutrient solution and the younger receiving one-eighth strength solution, or *vice versa*. The total weights of all the direct ramet progeny of each of the two parent ramets were measured after ten weeks of growth. Costs and benefits were calculated in the same way as in the experiment by Salzman & Parker (1985).

In the case of the treatment clones in which the older ramet was supplied with nutrient-poor conditions, and the younger ramet with nutrient-rich conditions, there was no evidence of any cost or benefit of integrated behaviour in comparison with the control clones. Thus, the part of these treatment clones produced by the older ramet did not differ significantly in any measured characteristic from the corresponding part of the nutrient-poor control clones (Table 6.4). Similarly, the part produced by the younger ramet did not differ significantly in any respect from the corresponding part of the nutrient-rich control clones.

In the treatment where the older ramet was in nutrient-rich soil and the younger in nutrient-poor soil, only very limited benefit to the younger ramet and its direct ramet progeny of being attached to ramets growing under favourable conditions could be demonstrated (Table 6.4). This initially surprising result is entirely a consequence of the structure of the clone which develops from the original two ramets. The only vascular route for acropetal translocation of resources to the part of the clone receiving a low nutrient supply was through the stolon internode connecting the two parent ramets. However, there are two other major routes for acropetal translocation from the older parent ramet, namely along the secondary stolons (Fig. 6.9). When the older parent ramet started to produce secondary stolons and daughter ramets, a considerable proportion of its acropetal translocation would have been towards them, rather than towards the younger parent ramet. In addition, the daughters of the older parent ramet would have exported translocates acropetally, towards their successors, rather than towards their parent ramets. Thus, the available supply of translocates to the younger, nutrient-poor part of the clone would have been severely limited because of

TABLE 6.4. Mean values ($\pm$ S.E.) of measured parameters for the first and second parts of *Glechoma hederacea* clones. Each part of the clone was grown either in nutrient-rich or nutrient-poor sand as indicated. For each parameter, significant differences (at least $P < 0.05$) between the same part of clones subjected to different experimental treatments, are demonstrated by different letters of the alphabet.

| | Nutrient-poor control clone | | Nutrient-poor nutrient-rich clone | | Nutrient-rich nutrient-poor clone | | Nutrient-rich control clone | |
|---|---|---|---|---|---|---|---|---|
| | First | Second | First | Second | First | Second | First | Second |
| Growing conditions | −N | −N | −N | +N | +N | −N | +N | +N |
| Mean number of ramets | 9±6 (a) | 41±4 (a) | 7±2 (a) | 66±8 (b) | 46±6 (b) | 48±3 (a) | 31±4 (b) | 57±2 (b) |
| Mean number of stolon branches | 3±1 (a) | 15±3 (a) | 2±1 (a) | 28±2 (b) | 19±3 (b) | 19±2 (a) | 13±2 (b) | 26±1 (b) |
| Mean stolon internode length (cm) | 7.7±0.5 (a) | 7.1±0.1 (a) | 8.1±0.6 (a) | 5.6±0.2 (b) | 6.0±0.2 (b) | 6.6±0.1 (c) | 6.1±0.1 (b) | 5.9±0.1 (b, c) |
| Total stolon length (cm) | 98±30 (a) | 704±82 (a) | 76±24 (a) | 584±80 (a, b) | 446±63 (b) | 452±35 (b) | 359±38 (b) | 573±12 (a) |
| Mean clone leaf area (cm$^2$) | 64±40 (a) | 632±88 (a) | 75±2 (a) | 949±134 (b) | 795±141 (b) | 510±73 (a) | 493±64 (b) | 935±45 (b) |

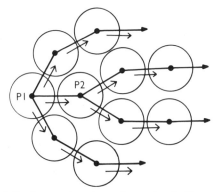

FIG. 6.9.   Directions of effective translocation pathways (marked by arrows) within experimental clones of *Glechoma hederacea*. Clones commenced growth as two adjacent parent ramets (P1, P2) joined by a connecting stolon. All stolons are growing towards the right hand side of the diagram. The circles denote pots containing ramets.

the structure of the clone. Its development would have been restricted, whereas the proliferation of those parts of the clone foraging in more favourable (nutrient-rich) conditions would have been promoted.

## CONCLUSION

For the plant population biologist interested in clonal herbs there are two major problems associated with the study of natural populations of these species in the field. Firstly there is the uncontrolled heterogeneity of the habitat, and secondly the variation in the genetical identity, and structure of the clones or fragments of clone under study. Simple experiments of the type described in this paper allow these problems to be largely avoided, and enable at least a start to be made on answering the questions posed in the introduction to this chapter. Although most of the experiments discussed take us some distance from the demographer's aim of documenting the statistics of births, deaths, diseases etc. in populations, they provide much needed help in enabling us to begin to understand and explain these statistics. Some of the results reveal greater subtlety in the behaviour of clonal herbs than was expected when the questions were originally formulated. They open up new insights into the complexity of these highly successful, but so far understudied organisms, and they may help us to devise new ways of approaching the investigation of their population biology under the more problematical conditions found in the field.

## ACKNOWLEDGMENTS

Part of this work was undertaken while A. J. Slade was in receipt of a Natural Environment Research Council studentship.

## REFERENCES

**Aarssen, L. W., Turkington, R. & Cavers, P. B. (1979).** Neighbour relations in grass-legume communities. II. Temporal stability and community evolution. *Canadian Journal of Botany*, **57**, 2695–703.

**Alpert, P. & Mooney, H. A. (1986).** Resource sharing among ramets in the clonal herb *Fragaria chiloensis*. *Oecologia, (Berlin)*, **70**, 227–33.

**Aung, L. H., De Hertogh, A. A. & Staby, G. (1969).** Temperature regulation of endogenous gibberellin activity and development in *Tulipa gesneriana* L. *Plant Physiology*, **44**, 403–6.

**Bell, A. D. (1974).** Rhizome organization in relation to vegetative spread in *Medeola virginiana*. *Journal of the Arnold Arboretum*, **55**, 458–68.

**Bell, A. D. (1979).** The hexagonal branching pattern of *Alpinia speciosa* L. (Zingiberaceae). *Annals of Botany*, **43**, 209–23.

**Bell, A. D. (1984).** Dynamic morphology: a contribution to population ecology. *Perspectives on Plant Population Ecology* (Ed. by R. Dirzo & J. Sarukhán), pp. 48–65. Sinauer, Massachusetts.

**Bell, A. D. & Tomlinson, P. B. (1980).** Adaptive architecture in rhizomatous plants. *Botanical Journal of the Linnaean Society*, **80**, 125–60.

**Bradbury, I. K. (1981).** Dynamics, structure and performance of shoot populations of the rhizomatous herb *Solidago canadensis* L. in abandoned pastures. *Oecologia (Berlin)*, **48**, 271–6.

**Bradbury, I. K. & Hofstra, G. (1976).** The partitioning of net energy resources in two populations of *Solidago canadensis* during a single developmental cycle in southern Ontario. *Canadian Journal of Botany*, **54**, 2449–56.

**Bülow-Olsen, A., Sackville Hamilton, N. R. & Hutchings, M. J. (1984).** A study of growth form in genets of *Trifolium repens* L. as affected by intra- and interplant contacts. *Oecologia (Berlin)*, **61**, 383–7.

**Casal, J. J., Deregibus, V. A. & Sánchez, R. A. (1985).** Variations in tiller dynamics and morphology in *Lolium multiflorum* Lam. vegetative and reproductive plants as affected by differences in red/far-red irradiation. *Annals of Botany*, **56**, 553–9.

**Child, R., Morgan, D. C. & Smith, H. (1981).** Morphogenesis and simulated shadelight quality. *Plants and the Daylight Spectrum* (Ed. by H. Smith), pp. 409–20. Academic Press, London.

**Clegg, L. (1978).** *The morphology of clonal growth and its relevance to the population dynamics of perennial plants.* Ph.D. thesis, University of Wales.

**Cook, R. E. (1985).** Growth and development in clonal plant populations. *Population Biology and Evolution of Clonal Organisms* (Ed. by J. B. C. Jackson, L. W. Buss & R. E. Cook), pp. 259–96. Yale University Press, New Haven.

**Deregibus, V. A., Sánchez, R. A. & Casal, J. J. (1983).** Effects of light quality on tiller production in *Lolium* spp. *Plant Physiology*, **72**, 900–2.

**Deregibus, V. A., Sánchez, R. A., Casal, J. A. & Trlica, M. J. (1985).** Tillering responses to enrichment of red light beneath the canopy in a humid natural grassland. *Journal of Applied Ecology*, **22**, 199–206.

**Ginzo, H. D. & Lovell, P. H. (1973a).** Aspects of the comparative physiology of *Ranunculus bulbosus* L. and *Ranunculus repens* L. I. Response to nitrogen. *Annals of Botany*, **37**, 753–64.

**Ginzo, H. D. & Lovell, P. H. (1973b).** Aspects of the comparative physiology of *Ranunculus*

# 7. PHYSIOLOGICAL AND DEMOGRAPHIC IMPLICATIONS OF MODULAR CONSTRUCTION IN COLD ENVIRONMENTS

## T. V. CALLAGHAN
*Institute of Terrestrial Ecology, Merlewood Research Station, Grange-over-Sands, Cumbria LA11 6JU, UK*

## SUMMARY

**1** The evolutionary selection pressures of cold environments are characterized by the limitation of resources and resource capture by the physical environment. In contrast, those of mesic temperate environments are dominated by competition for resources between opportunistic species.

**2** Deterministic growth in cold environments offers limited opportunities for locating and capturing resources, but conservative physiological strategies enhance the efficient use of water, and the efficient use and re-use of carbon and nutrients.

**3** The selection pressures of cold environments have resulted in the dominance of long-lived clonal and perennial plants.

**4** There has also been selection for close linkage between modular morphological form and the environment of growing points, achieved either through purely physical means or through physiological interdependence between modules.

## INTRODUCTION

Many plants, and perhaps the majority, are constructed from repeated morphological units, or modules (Prévost 1978). Such plants can be regarded as a metapopulation of modules (White 1979, 1984; Harper 1981), and an understanding of their overall structure, function and growth may be obtained from a knowledge of the physiological and demographic interrelationships between the component modules.

Modular plants may be aclonal or clonal. In aclonal plants, the entire plant functions as a single physiological unit and the death of part of the plant may result in the death of the whole plant or genet (Pitelka & Ashmun 1985). In contrast, clonal plants develop ramets which are capable of independent existence and they reproduce the genet when connections between them break. Clonal plants are an important component of global vegetation, particularly at high latitudes and altitudes (Cook 1983), but aclonal plants also occur in cold environments.

There are several comprehensive and up-to-date reviews on clonal plants and modular construction (Cook 1983, 1985; Jackson, Buss & Cook 1985; Harper, Rosen & White 1986) and on the physiological interrelationships within clones (Watson & Casper 1984; Pitelka & Ashmun 1985; Hutchings & Bradbury 1986; Hutchings & Slade 1988). Rather than attempt another comprehensive review, therefore, this chapter presents some examples in detail which illustrate a range of types of modular construction and their relationships with function. Emphasis is also placed on environmental conditions in the field, and in particular, on the severe conditions of cold regions. Here, resources are limiting to plant growth and their capture and utilization is constrained by the physical environment. Hitherto, most studies have concentrated on species, particularly crop species, of mesic temperate environments. In these environments, resources are relatively abundant but resource capture is constrained by competition between individuals.

Thus, in mesic temperate environments, plant competition and opportunistic growth seem to dominate plant responses to relatively abundant resources and strong biotic selection pressures. In some other environments, however, abiotic factors provide the major selection pressures and the dominant plant responses are to exist with minimal resources while surviving the extreme physical environment. In such environments, competitive exclusion may be replaced by commensalism or mutualism while slow, deterministic growth (Tomlinson 1982) tends to replace opportunism (Callaghan & Emanuelsson 1985).


## TUNDRA ENVIRONMENTS

One-quarter of the Earth's land surface is covered by the tundra, the climate of which is characterized by short, cool summers, long, cold winters, extremes of photoperiod and strong, desiccating winds. Tundra soils are generally primitive and infertile and experience extremes of temperature and hydration while often having a permanently frozen layer at some depth. Reviews of these conditions and the plants of these areas are given by Billings & Mooney (1968), Bliss (1971), Savile (1972), Lewis & Callaghan (1976), Aleksandrova (1980) Bliss, Heal & Moore (1981) and Callaghan & Emanuelsson (1985). There are great variations of habitat within the tundra, but two contrasting generalized habitats may be distinguished for the convenience of examining modular construction in relation to function. The polar semi-desert and fell-field environments are characterized by exposure, little winter snow cover, extremes of soil hydration and temperature, mineral soils and open vegetation with species aggregated into islands of vegetation with extensive areas of bare ground between them. In contrast, the forest-tundra, the ecotone between the treeless tundra and the taiga, is characterized by scattered trees of low stature with a complete vegetation cover below, smaller ranges of temperature and hydration and organic, rather than mineral, soils.

Clonal plants with modular construction occur in both types of habitat but the aggregation of modules and their interrelationships differ. At one extreme of habitat type, deterministic growth and tightly packed modules ameliorate the microclimate experienced by growing points. At the other extreme of habitat type, the production of modules with prolonged physiological interdependence tightly controls the internal environment of extended growing points. Examples of these two basic systems will be described.

## PHYSICAL ATTRIBUTES OF MODULAR CONSTRUCTION AND DETERMINISTIC GROWTH

Plants of the open habitats (and also those of deserts) are characterized by strongly deterministic growth, in contrast to the situation found in temperate species. Tussock and cushion growth forms are particularly important. Both reach their peaks of expression in the austral and sub-antarctic regions. On the Falkland Islands, for example, tussocks of *Poa flabellata* (Lam.) Hook. may reach a height of 2.5 m and a diameter of 1 m and cushions of *Bolax gummifera* (Lam.) Spreng. may be 1.3 m in height and 3 m in diameter (Moore 1968). In the Arctic, tussocks of *Eriophorum vaginatum* L. cover vast areas ($0.9 \times 10^6$ km$^2$, Miller 1982) while cushion plants abound in the High Arctic and are the primary colonizers in some alpine areas (Griggs 1956).

## THE CUSHION GROWTH FORM AND TEMPERATURE CONTROL

Each cushion develops from one tap root and the shoots, or modules, are packed together at extremely high density. The cushion may live for over 100 years (Schroeter 1926) and during this time the length of all of the internodes is strongly controlled. This deterministic growth and tight packing of modules produces a structure with a smooth and continuous surface. The control of internode length can only be broken with difficulty by applying hormones (gibberellic acid) or growing the cushions at high temperatures (Spomer 1964). The natural aerodynamic shape of the cushion creates a thin boundary layer and enables high temperature differentials to be maintained. This effect

tussock of *Eriophorum vaginatum* for the 100 years for which it may live (Mark *et al.* 1985). The type of modular construction shown by the tussock and tufted growth forms seems to be associated with an inherent inefficiency. The tussock or tuft extends to adjacent areas very slowly, compared with rhizomatous or stoloniferous species, and the exploitation of new space must be achieved by the efficient production and dissemination of seed.

It is obvious from the above examples that the growing points of plants, and flowering primordia, are particularly vulnerable to severe physical environments. In such environments, the physical protection of young modules is of great importance and leads to a rather 'sessile' growth form which would be vulnerable to competition for light from taller species, and competition for new space from more mobile species. Also, if adventitious rooting is impossible, as in many of the cushion plants, individual genets, supported by one tap root, are unlikely to grow indefinitely.

## PHYSIOLOGICAL ATTRIBUTES OF MODULAR CONSTRUCTION AND PLANT 'MOBILITY'

In less severe physical environments, particularly in closed vegetation, there is a shift from the control of the physical environment of growing points by the types of modular constructions discussed above, to the physiological control of their internal environments by the integration of modules. This overcomes the design limitations and finite life-spans of genets associated with the cushion and tufted/tussock growth forms by allowing the extension of young modules away from the physical protection of parent ancestral modules. This subsidized growth of young modules is analogous to the periods of postnatal care characteristic of many animals. Subsidized growth of growing points also allows foraging in clonal perennial plants (Hutchings & Slade, 1988). Foraging for resources is a major feature of the rhizomatous/stoloniferous growth form, but avoidance of interspecific competition can, perhaps, be an important aspect of foraging behaviour in some species. Consequently, it is interesting to investigate the mechanisms which support this growth form.

## SOURCE/SINK RELATIONSHIPS AND DIRECTION OF RESOURCE MOVEMENT

The basic principle of source/sink relationships and direction of resource movement is well illustrated by the work of Qureshi & Spanner (1971) on *Saxifraga sarmentosa* L. Under normal conditions, young modules, which are

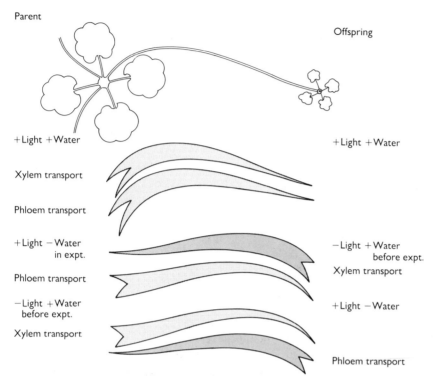

Parent

Offspring

+Light +Water

+Light +Water

Xylem transport

Phloem transport

+Light −Water
    in expt.

−Light +Water
    before expt.

Xylem transport

Phloem transport

−Light +Water
  before expt.

+Light −Water

Xylem transport

Phloem transport

FIG. 7.3. The effects of varying sink/source relationships between parent and attached offspring on the change in directions of transport in the phloem and xylem of *Saxifraga sarmentosa*. Arrows indicate directions of flow (data from Qureshi & Spanner 1971).

sinks for nutrients, water and carbon, import these in the xylem or phloem from the parent module which is a source (Fig. 7.3). As the young module ages, its dependence on the parent decreases until it becomes self-sufficient. In many species, self-sufficiency for photoassimilate precedes self-sufficiency for water and nutrients as leaves are produced before roots. If environmental conditions change when the young module has some degree of self-sufficiency, then the directions of both xylem and phloem transport can be reversed independently according to the changes from source to sink (Fig. 7.3). In addition to these changes in source/sink relationships between parent and offspring modules, the young module may again become dependent upon parental resources after it has achieved self-sufficiency, if environmental conditions change.

This phenomenon has been described by various authors in relation to clipping leaves in grasses (e.g. Marshall & Sagar 1965; Gifford & Marshall

1973) or shading young modules (Ong & Marshall 1979; Hartnett & Bazzaz 1983). However, few quantitative assessments have been made of the movement of resources and most studies of translocation in the field have been on crop plants or have been descriptive (Allessio & Tieszen 1975; Noble & Marshall 1983) with the notable exceptions of Ashmun, Thomas & Pitelka (1982) and Tietema (1980). It can be argued that there should be a 'cut-off point' beyond which a module in a favourable microsite should cease to subsidize a disadvantaged module lest the donor module becomes unviable. However, such 'cut-off points' do not yet appear to have been recorded.

Descriptive studies often show transport in different directions at the same time i.e., reciprocal translocation. Perhaps the best documented example of this was that reported by Svoboda & Bliss (1974) who painted leaves of *Dryas integrifolia* with $^{32}$P and then detected active roots and nodules from autoradiographs of the root system. The degree and period of module integration varies markedly from species to species (see Pitelka & Ashmun 1985 for a review), even within genera, e.g. *Viola* L. (Newell 1982).

Perhaps the one underlying basic principle is that the youngest modules, or growing points, require a more favourable internal environment in terms of supplies of resources, than modules at any other stage of development. If the growing points are self-sufficient for a resource, such as photoassimilates, then physiological interdependence between modules can be minimal. On the other hand, if there is a division of labour between modules, then interdependence is necessary, even at the expense of the older modules. For example, in conditions of nutrient stress, the growing points of *Lycopodium selago* L. are maintained with near optimal concentrations of phosphorus, which is essential in the energy transfer processes associated with growing points, while older modules have low levels. At high concentrations of phosphate supply, the older modules act as a buffer, accumulating excess phosphate (Headley 1986). An important application of this principle is that when assessing nutrient limitations on growth, organs with high physiological activity should not be sampled. Foliar analyses of tree leaves, for example, are far less indicative of phosphate limitation than are analyses of the roots (Harrison *et al.* 1987).

Pitelka & Ashmun (1985) have suggested that there are two types of modular integration, one associated with clear age or stage differences between modules and, often, directional growth, and another associated with interconnected modules, which are all at the same stage or age. The former is particularly characteristic of the more extreme environments where differences of age between modules can be great, and physical connections between modules are long lasting (Jónsdóttir & Callaghan 1988).

## GROWTH, LIFE-SPAN AND PHYSIOLOGICAL CONTINUITY BETWEEN MODULES IN SEDGES

Simple systems of modular organization and physiological integration are seen in the sedges *Carex bigelowii* Torr. and *Uncinia meridensis* Steyermark, and also in *Carex arenaria* L. (Tietema & van der Aa 1981; Noble & Marshall 1983). In *Uncinia meridensis*, guerilla modules (invasive modules with relatively long extension growth and infrequent branching; Clegg 1978) are non-photosynthetic in their first year and are subsidized in terms of water, nutrients and carbon by their parent modules (Callaghan 1984). Growth is exponential and probabilities of survival are high due to this 1 year period of 'postnatal care' (Table 7.1). As the leaf complement and roots develop in the young module, it becomes both a source and sink. When an inflorescence is produced, this acts as a new sink. If the inflorescence is photosynthetic, as in

TABLE 7.1. The generalized growth of sedge modules and the interrelationships between them (based mainly on Callaghan 1984 and Jónsdóttir & Callaghan 1988) C = carbon, N = nutrients, W = water).

| Age (years) | Leaf production | Growth | Allocation (dry matter) | Sink/source | Survival | Reproduction |
|---|---|---|---|---|---|---|
| 0 | 8 non-photo. bracts | Exponential | Rhizome only | Strong sink only (C, N, W) (strong source for hormones) | 57% | Initiation of buds |
| 1 | 4 leaves | 3 × increase | Leaves, roots *Not* rhizome | Weak sink, weak source (C, N, W) | 93% | Development of buds |
| 2 | 3 leaves (longest) | 2 × increase | Dead leaves, all living compartments | Mainly source. Export to younger (C, N, W) and older (C) modules | 70% | 4.5% probability of flowering |
| 3–4 | 3–4 leaves (maximum area) | 0 increase | Rhizome | Weak sink, weak source, almost independent | 47–18% | 8–23% probability of flowering |
| 5 | 4 leaves but all leaves dead by autumn | Decrease | Rhizome | Weak sink (C) weak source (N, W) | '0%' | 25% probability of flowering |
| 6 | none | Increase? | Rhizome? | Sink (C) weak source (N, W) | ? | None |

*Uncinia meridensis*, it may contribute to its own energy balance, as has been shown in wheat (Carr & Wardlaw 1965), but it may also export about 28% of its photoassimilate to interconnected young modules (Callaghan 1984). After the formation of an inflorescence, or when this eventually fails to occur, the module loses its leaves. There is then considerable uncertainty about the continuity between the old modules and younger ones and the physiological status of the old leafless modules. Little is known about the length of time over which physiological continuity between modules may persist and connections between modules are currently considered to be long-lived if they persist for more than 2 years (Pitelka & Ashmun 1985).

In the sedge *Carex bigelowii*, connections between modules may persist for up to twenty-seven generations representing a period of a similar number of years (Kershaw 1962; Jónsdóttir & Callaghan 1988). Translocation studies by Allessio & Tieszen (1975), Callaghan (1984) and Jónsdóttir & Callaghan (1988) show that the old modules survive for many years after they have lost their leaves and act as a transport system connecting younger generations of tillers. Recent work supplying nitrate to the roots of old (at least 7 years), apparently dead tillers has suggested that they take up and transport sufficient nitrate to switch on the enzyme nitrate reductase in young modules (Jónsdóttir & Callaghan, unpublished data).

By mechanisms such as this, the old modules contribute significantly to the survival of the youngest modules. Jónsdóttir & Callaghan (1988) have demonstrated the interdependence between modules of *Carex bigelowii* by cutting the connections between modules of different ages. The youngest modules, when separated from the rest of the plant, show poorer survival (Table 7.2) until, at one site, five generations of tillers remain connected, which may be taken as the minimum size of the physiological unit. However, the young modules control the proliferation of old, apparently dead tillers, presumably via the transport of auxins, in that breaking the connections

TABLE 7.2. Effect of severing various generations of modules of *Carex bigelowii* on the % survival of the youngest (T1) module (data from Jónsdóttir & Callaghan 1988).

| Treatment (see Fig. 7.4) | % Survival | |
|---|---|---|
| | in Iceland | in Lapland |
| Control (clone in tact) | 83.3 | 100.0 |
| Clone severed between T1 and T2 | 16.6 | 83.3 |
| Clone severed between T2 and T3 | 16.6 | 100.0 |
| Clone severed between T3 and T4 | 40.0 | 100.0 |
| Clone severed between T4 and T5 | 80.0 | 100.0 |
| Clone severed between T5 and T6 | | 100.0 |

between old and young modules reduces the fecundity of the young modules but stimulates old buds on old modules to break dormancy (Fig. 7.4). Similar experiments were carried out by Hartnett & Bazzaz (1983) and Lovett-Doust (1981a, b) with similar conclusions.

The net result of the processes which lead to the 'period of postnatal care' of young modules is a great increase in survival probabilities when compared with young plants produced from seed, a range of 0.43 to almost 1.0 compared with almost 0 to 0.07 respectively for tundra plants (Callaghan & Emanuelsson 1985).

The type of growth described above can be regarded as a rather moderate guerilla growth form with regular but limited extension growth subsidized by older modules. Phalanx and guerilla growth forms have been described in different species while Slade & Hutchings (1987a, b) have shown that the growth form displayed may be responsive to different nutrient and light regimes, even experienced within a clone: in poor nutrient conditions, *Glechoma hederacea* L. produces a few long internodes (the guerilla form) whereas in better nutrient regimes, proliferation of branches increases and internode length decreases (Slade & Hutchings 1987a; Hutchings & Slade 1988), thereby increasing the opportunity for exploitation of the resource. However, both growth forms may be produced by one parent module in one set of conditions. Species such as *Carex aquatilis* Wahlenb. (Shaver & Billings 1975), *Phragmites communis* Trin. (Haslam 1969) and the sedge *Rostkovia magellanica* (Lam.) Hook. (Callaghan 1977) produce two types of module. The guerilla module has a prolonged period of resource subsidy while the assimilatory phalanx module quickly achieves photosynthetic self-sufficiency (Fig. 7.5). The guerilla modules explore new distant ground while the phalanx modules expand the colony locally and, perhaps, subsidize the guerilla modules. One of the results of the heavy subsidization of the guerilla modules is that they may survive in unsafe microenvironments and extend beyond them to favourable microsites where phalanx modules can become established.

## SUBSIDIZED GROWTH AND THE EXPLOITATION OF SPACE

The rhizomatous and stoloniferous growth forms have been interpreted in the context of ramet survival and growth out of unfavourable microsites in patchy environments (e.g. Pitelka & Ashmun 1985), yet little evidence has been presented on the variations in microenvironment encountered by exploratory modules. Svensson (1987) has examined the complexity of variation in the microhabitats encountered by the horizontal branches, or

Kershaw, K. A. (1962). Quantitative ecological studies from Landmannahellir, Iceland. III. Variation in performance of *Carex bigelowii*. *Journal of Ecology*, **50**, 393–99.

Lewis, M. C. & Callaghan, T. V. (1976). Tundra. *Vegetation and the Atmosphere*. Vol. 2 (Ed. by J. L. Monteith), pp. 399–433, Academic Press, London.

Larcher, W. (1980). *Physiological Plant Ecology*. Springer, Berlin, Heidelberg, New York.

Lovett-Doust, L. (1981a). Intraclonal variation and competition in *Ranunculus repens*. *New Phytologist*, **89**, 495–502.

Lovett-Doust, L. (1981b). Population dynamics and local specialization in a clonal perennial *(Ranunculus repens)*. I. The dynamics of ramets in contrasting habitats. *Journal of Ecology*, **69**, 743–55.

Mark, A. F., Fetcher, N., Shaver, G. R. & Chapin III, F. S. (1985). Estimated ages of mature tussocks of *Eriophorum vaginatum* along a latitudinal gradient in central Alaska, U.S.A. *Arctic and Alpine Research*, **17**, 1–5.

Marshall, C. & Sagar, G. R. (1965). The influence of defoliation on the distribution of assimilates in *Lolium multiflorum* Lam. *Annals of Botany*, **29**, 365–70.

Miller, P. C. (1982). The availability and utilization of resources. *Tundra Ecosystems* (Ed. by P. C. Miller), Proceedings of a workshop held in San Diego, California 9–12 October 1978. *Holarctic Ecology*, **5**, 83.

Mølgaard, P. (1982). Temperature observations in high arctic plants in relation to microclimate in the vegetation of Peary Land, North Greenland, *Arctic and Alpine Research*, **14**, 105–15.

Moore, D. M. (1968). *The Vascular Flora of the Falkland Islands*. British Antarctic Survey Scientific Reports No. 60.

Newell, S. J. (1982). Translocation of 14-C assimilate in two stoloniferous *Viola* species. *Bulletin of the Torrey Botanical Club*, **109**, 306–17.

Noble, J. C. & Marshall, C. (1983). The population biology of plants with clonal growth. II. The nutrient strategy and modular physiology of *Carex arenaria*, *Journal of Ecology*, **71**, 865–77.

Oechel, W. C. & Van Cleve, K. (1986). The role of bryophytes in nutrient cycling in the taiga. *Forest Ecosystems in the Alaskan Taiga* (Ed. by K. Van Cleve, F. S. Chapin III, P. W. Flanagan, L. A. Viereck & C. T. Dyrness), pp. 121–37. Springer, New York, Berlin, Heidelberg & Tokyo.

Ong, C. K. & Marshall, C. (1979). The growth and survival of severely-shaded tillers in *Lolium perenne* L. *Annals of Botany*, **43**, 147–55.

Pitelka, L. F. & Ashmun, J. W. (1985). Physiology and integration of ramets in clonal plants. *Population Biology and Evolution of Clonal Organisms* (Ed. by J. B. C. Jackson, L. W. Buss & R. E. Cook), pp. 399–436. Yale University Press, New Haven.

Prévost, M. F. (1978). Modular construction and its distribution in tropical woody plants. *Tropical Trees as Living Systems* (Ed. by P. B. Tomlinson & H. Zimmermann), pp. 223–31. Cambridge University Press, Cambridge.

Qureshi, R. A. & Spanner, D. C. (1971). Unidirectional movements of tracers along the stolon of *Saxifraga sarmentosa*. *Planta*, **101**, 133–46.

Savile, D. B. O. (1972). Arctic adaptations in plants. *Canadian Department of Agriculture Monographs*, **6**.

Schroeter, C. (1926). *Das Pflanzenleben der Alpen*. Albert Baustein, Zurich.

Shaver, G. R. & Billings, W. D. (1975). Root production and root turnover in a wet tundra ecosystem, Barrow, Alaska. *Ecology*, **56**, 401–9.

Slade, A. J. & Hutchings, M. J. (1987a). The effects of nutrient availability on foraging in the clonal herb *Glechoma hederacea*. *Journal of Ecology*, **75**, 95–112.

Slade, A. J. & Hutchings, M. J. (1987b). The effects of light intensity on foraging in the clonal herb *Glechoma hederacea*. *Journal of Ecology*, **75**, 639–50.

Spomer, G. G. (1964). Physiological ecology studies of alpine cushion plants. *Physiologia Plantarum*, **17**, 717–24.

**Svensson, B. M. (1987).** *Studies of the metapopulation dynamics of* Lycopodium annotinum *and its microenvironment.* PhD. Thesis, University of Lund.

**Svensson, B. M. & Callaghan, T. V. (1988a).** Apical dominance and the simulation of metapopulation dynamics in *Lycopodium annotinum. Oikos*, **51**, 331–42.

**Svensson, B. M. & Callaghan, T. V. (1988b).** Small-scale vegetation analysis related to the growth of *Lycopodium annotinum* and variations in its microenvironment. *Vegetatio* (in press).

**Svoboda, J. & Hutchinson-Benson, E. (1986).** Arctic cushion plants as fallout 'monitors'. *Journal of Environmental Radioactivity*, **4**, 65–76.

**Svoboda, J. & Bliss, L. C. (1974).** The use of auto radiography in determining active and inactive roots in plant production studies. *Arctic and Alpine Research*, **6**, 257–60.

**Tamm, C. O. (1953).** Growth, yield and nutrition in carpets of a forest moss (*Hylocomium splendens*). *Meddelanden från Statens Skogsforskningsinstitut*, **43**, 1–140.

**Tietema, T. (1980).** Ecophysiology of the sand sedge, *Carex arenaria* L. II. The distribution of 14-C assimilates. *Acta Botanica Neerlandica*, **29**, 165–78.

**Tietema, T. & van der Aa, F. (1981).** Ecophysiology of the sand sedge *Carex arenaria* L. III. Xylem translocation and the occurrence of patches of vigorous growth within the continuum of a rhizomatous plant system. *Acta Botanica Neerlandica*, **30**, 183–89.

**Tomlinson, P. B. (1982).** Chance and design in the construction of plants. *Acta Biotheoretica*, **31A**, 162–83.

**Watson, M. A. & Casper, B. B. (1984).** Morphogenetic constraints on patterns of carbon distribution in plants. *Annual Review of Ecology and Systematics*, **15**, 233–58.

**Westoby, M. (1984).** The self thinning rule. *Advances in Ecological Research*, **14**, 167–225.

**White, J. (1979).** The plant as a metapopulation. *Annual Review of Ecology and Systematics,* **10**, 109–45.

**White, J. (1984).** Plant metamerism. *Perspectives on Plant Population Ecology* (Ed. by R. Dirzo & J. Sarukhán), pp. 15–47. Sinauer Associates, Sunderland, Massachusetts.

# 8. DISTRIBUTION LIMITS AND PHYSIOLOGICAL TOLERANCES WITH PARTICULAR REFERENCE TO THE SALT MARSH ENVIRONMENT

J. ROZEMA, M. C. T. SCHOLTEN[1], P. A. BLAAUW AND
J. DIGGELEN

*Department of Ecology and Ecotoxicology, Free University, PO Box 7161, 1007
MC Amsterdam, The Netherlands*

## SUMMARY

**1** Plants are generally considered to be indicators of specific environmental conditions (soil or climate). Classically, physiological response curves are obtained by experimental studies where one or few relevant environmental factors are varied at a limited number of levels, in order to analyze distribution and abundance of individuals in plant populations in the field. More generally, the relevance of physiological mechanisms in the explanation of ecological phenomena is considered.

**2** Taking the salt marsh environment as an example, differences in spatial and temporal distribution of plant populations are discussed in terms of: (1) physiological tolerance to environmental factors, (2) interspecific competition in the field and in the greenhouse, (3) relevance of plant parameters measured such as germination rate, relative growth rate, morphology, reproductive system, longevity and phenology, (4) (hemi)parasite plant relations, versus mycorrhizal interactions.

## ECOLOGY AND PHYSIOLOGY

The occurrence of plant species at a particular site has often been considered to indicate specific environmental conditions. Some plant species are abundant in calcareous areas and are absent in soils with acid soil conditions. Also halophytic and non-halophytic (glycophytic) plant species are distinguished. Nitrophytes are plant species characteristic of sites rich in inorganic nitrogen and heavy metal plants (metallophytes) occur more or less specifically

---

[1] Present address, Division of Technology for Society TNO Laboratory for Applied Marine Research, P.O. Box 57, 1780 AB Den Helder, The Netherlands.

on places with ore deposits near the earth's surface and on heavy metal mine waste heaps. The heavy metal content of the soil in mining areas exceeds that of adjacent places. In ecophysiological studies of heavy metal and other tolerances of plant species, it has often been noticed that a difference ('discrepancy') exists in the location of the 'physiological' and 'ecological optimum' (Ellenberg, 1981). The difference between the physiological and ecological optimum curve, or, when more environmental variables are considered, the physiological tolerance volumes (Etherington 1975) and the ecological tolerance volumes is usually ascribed to competitive exclusion.

Referring to the problem of the ecology of salt marsh plants, Ernst (1978) discusses the occurrence of *Suaeda maritima* in the field within a soil salinity range 400–700 mM NaCl, whilst in callus culture (von Hedenström & Breckle 1974) and sand culture biomass development is maximal at 100–150 mM NaCl. More generally, it has been noticed that *in vitro* activity of many cytoplasmic enzymes extracted from halophytes decreases in response to increased salinity in the same way as the response curve of enzymes from glycophytes (Flowers 1972; cf. Rozema *et al.* 1985). Cellular compartmentation is thought to represent a physiological adaptation both to increased soil salinity and excess uptake of heavy metals, where sensitive enzymes are separated from toxic levels of salt and heavy metals. The physiological costs for maintenance of such a compartmentation system has been called the physiological price for resistance (Yeo 1983) and may lead to decreased growth of adapted plants compared to non-adapted plants when cultivated in a medium without salt or heavy metals (Ernst 1986). Also, when organic solutes are involved in the resistance to heavy metals such as malate to complex with zinc for detoxification and transport, or compatible osmotic solutes (proline, glycinebetaine) in cellular osmoregulation in response to variation of external salinity, then correlated reduced growth is often interpreted as part of the physiological price for resistance (Yeo 1983). A low relative growth rate of plant species as a result of this could make these species weak competitors in comparison to non-adapted species under conditions without heavy metal and salinity stress.

Apart from the description of the obvious difference between the location of the optimum of the physiological and ecological response curve to environmental variables in many text books of plant ecology, convincing experimental studies that bridge the gap between vegetation studies and physiological research are rare. To relate the distribution of a plant species to one single environmental variable is always a simplification. Also often only vegetative growth is the parameter studied, and other plant parameters such as germination, flowering and seed production are neglected. When more environmental factors including their interactions and if possible more stages of the life-cycle of plants are taken into account, a more realistic approach is reached. Of course, the attractiveness of the general notion of a physiological potential which is restricted by interspecific competitive interference, is probably based on the apparent logic that is assumed behind this idea.

Controversy between physiological and ecological findings may arise from too easy or too early generalization or extrapolation from one discipline to another, which is illustrated below by studies of carbon dioxide fixation. In the last 20 years plant physiological information about photosynthetic carbon dioxide fixation schemes has made enormous progress and led among other things to the distinction of the C3, C4 and CAM types of $CO_2$ fixation, now treated in almost every textbook of plant ecology and plant physiology. The occurrence of C4 photosynthesis was correlated with environments with the highest light intensities, high temperatures for growth and photosynthesis and tolerance of water stress, and that of CAM with dry regions with a high light intensity (Larcher 1983). It has often been assumed that because of the high net photosynthetic rate of C4 plants, their productivity will be higher than that of C3-plants. As far as agricultural plant species are concerned and when maximal and average net assimilation rates are considered, this generalization often seems to be true (Larcher 1983). Caldwell (1974) reports studies on two adjacent cool salt desert communities dominated by *Atriplex concertifolia* (C4) and *Ceratoides lanata* (C3). The analyses demonstrate that neither of the above general features of C3- and C4-plant could simply be confirmed by extensive field measurements. Both communities are nearly monospecific and have an almost identical physical environment characterized by an annual precipitation of 200 mm, an annual temperature range of $-30$ to $+40°C$ and very saline and reasonably homogenous silty soils of lacustrine origin. Species with C4 photosynthesis are reputed as having high rates of $CO_2$ fixation. In Caldwell's field measurements however, the C3-species *Ceratoides* showed higher maximal rates of daily leaf carbon fixation than *Atriplex concertifolia* (C4). Nevertheless, the community dominated by the C4-*Atriplex* species exhibited greater annual carbon fixation, standing biomass and above- and below-ground productivity. This could be attributed to a greater quantity of foliage per unit ground area combined with the longer season of a positive carbon balance, lasting until the autumn months September and October; the carbon balance of *Ceratoides* (C3) was no longer positive later than July and early August. The higher productivity of the *Atriplex* dominated community could in this case not simply be related to the C4 photosynthetic pathway. This field study showed that generalizations from plant physiological and agricultural studies may turn out to be unproven speculations.

## DISTRIBUTION LIMITS OF PLANT SPECIES

The study of the pattern of distribution and abundance of plant species may be performed on different scales. The outcome of the study of the factors determining the distribution limits of plant species is dependent on the scale of the pattern of distribution. Geographically, the northern limit of Mediterranean plant species in Europe may be determined by the occurrence of winter frost that kills leaves lacking frost resistance mechanisms (Larcher 1983). At this level (macro) climatological factors are involved. Within such a geographical distribution area, populations of a species may be present in those environments (habitats) that have a characteristic set of edaphic and (micro) climatological conditions (e.g. soil type, acidity, humidity of the soil, temperature). The abundance of a particular plant species on acidic soils may not simply be interpreted as the plant species being acidophilous. Plant species occurring on acidic soils *resist* or *tolerate* the conditions of acidic soils better than other species do. In an extreme environment, like the desert, with diurnal temperature changes from 10–40°C, lichens may represent the only plant life form capable of maintaining a positive carbon balance by absorbing water from the humid atmosphere during the night and being photosynthetically active during the cool morning hours (Lange, Schulze & Koch 1970).

On a smaller scale within a site with acid soil conditions closer analysis of a vegetation with the abundance of one plant species will almost always show coexistence with other less abundant species. This may relate to spatial and temporal heterogeneity of a site and the existence of different microsites with different species each fitting in its own niche. Alternatively, coexistence of plant species may be explained in terms of interference both in the above-ground and below-ground environment.

In the case of salt marshes there is quite remarkable similarity in the composition of the vegetation of salt marshes from different parts of the world (Chapman 1960; Ranwell 1972). The genera *Salicornia* and *Spartina* occur in the vegetation of most salt marshes of the Northern hemisphere. In this respect salt marshes could, confusingly, be termed azonal vegetation (Mueller-Dombois & Ellenberg 1974) indicating that salt marshes are not restricted to particular geographic zones. Generally, it is assumed that this type of azonal vegetation implies that it is mainly edaphically controlled.

At this level, the local salt marsh environment, that is the area of coastal fringe between the 'high water mark spring tide' and the 'low water mark spring tide' (Ranwell 1972), the occurrence of many of the salt marsh plant species is almost completely limited to the salt marsh environment. Chapman (1960) stated that the occurrence of halophytes in their natural habitats is restricted to sites with a soil salinity no less than 1.0% of the dry weight of the soils; outside this range halophytes do not grow optimally, or do not compete well or do not complete their life-cycle. This value of soil salinity representing the limit between halophytes and glycophytes is to some extent arbitrary: rather than 1.5% it might also be 0.5%, or 2.0%, or, perhaps better, salinity of the soil moisture could be taken, such as 70 mM NaCl or other monovalent salts, or the water potential of the soil moisture solution, −3.3 bar or −0.33 MPa. (Chapman 1975; Flowers, Troke & Yeo 1977; Greenway & Munns, 1980; Munns, Greenway & Kirst, 1983). All in all, dependent on the scale of observation, for halophytes there is a clear relationship between the distribution limits of a plant species and the physiological tolerance of the species to the factor governing the salt marsh environment: soil salinity.

## PHYSIOLOGICAL TOLERANCES AND FACTORS LIMITING PLANT GROWTH

Since Blackman (1905) developed the 'law of the limiting factors' it has been widely accepted that a process like plant growth, depending on a number of different processes, is actually limited by the factor that is 'in the minimum'. This law does not only apply to factors that may be in short supply, e.g. water, temperature, light, atmospheric carbon dioxide, soil inorganic nitrogen, phosphorus, and every other essential mineral nutrient, but also to factors that are present in excess. In addition to high levels of the above factors, this law is also valid for salinity, heavy metals, and contaminants in the atmosphere and in the soil. Except perhaps in some agricultural systems, plants are not, as a rule, provided with a physiologically optimal supply of environmental resources in their natural habitat. Competition between plant species may lead to abundance of a plant species that successfully competes for light, water or available nutrients. The growth rate of plant species may be a most important plant characteristic in both the study of tolerance and in the study of the competitive interference between species. Yet, the rapidity with which plant species can adapt to the altered environmental conditions may relate to longevity, life form, and breeding system of plant species (cf. Grubb, Kelly & Mitchley 1982).

## THE SALT MARSH ENVIRONMENT

In the following section the salt marsh environment is used as an example to analyse the question of the importance of the study of physiological tolerances in the understanding of distribution limits of plant species. The reasons for this choice are:

**1**   The zoned salt marsh vegetation is a clear-cut case of a habitat where there is a sequence of abundance of different plant species in a seaside–landside transition that is well known and widely described for coastal sites all over the world.

**2**   There are numerous autecological and ecophysiological studies conducted with these salt marsh halophytes, so that in principle the physiological tolerance to prevailing environmental factors is known and can be used for a comparison between species in order to analyse the species distribution in the salt marsh.

**3**   The number of environmental factors governing salt marsh vegetation zonation is limited (Rozema *et al.* 1985) and the distribution pattern of plant species in the field seems to be experimentally testable.

Also, we presume that findings for this particular coastal habitat may also be valid in other natural habitats. The reason for this is, that although environmental (stress) factors will be peculiar to individual natural habitats, difference in growth, be it to a varying extent, is a common feature for all plants, and when competition between *plant* species occurs, it is always the same kind of limited resources which are involved, such as light, water, and mineral nutrients.

## HALOPHYTE SEED GERMINATION AND ZONATION OF SALT MARSH VEGETATION

Germination and establishment represents an important stage in the life-cycle of salt marsh halophytes. Not only are different parts of the salt marsh more or less frequently flooded with sea water, with saline ground water and soil moisture as a result, but in dry periods the salinity of the surface of the salt marsh soil may far exceed that of deeper layers (hypersalinity) (Mahall & Park 1976). On the other hand, variation of the soil texture composition (e.g. the content of sand versus clay, organic matter content) and climatic conditions like rainfall may cause great variation in surface run off, drainage and dilution of sea water. The result of this may be heterogeneity of soil moisture salinity in both space and time. Does the germination response of salt marsh halophytes correspond to their position in the salt marsh vegetation zonation? Or, in other words does seed from species from saline sites germinate better in response to increased salinity compared to species from less saline sites?

Ungar (1978) concluded that halophytes are generally more salt tolerant than glycophytes at the germination stage, while Rozema (1975) found definitely non-halophytic inland species like *Holcus lanatus*, *Stellaria media* and *Chamaenerion angustifolium* to have the same germination response to salinity as salt marsh species tested. In an analysis of life-form spectra of European salt marshes (Chapman 1960, 1964), it can be concluded that about 30% of the salt marsh species are annuals; the majority of the salt marsh vegetation is formed by perennial species. For short-lived and long-lived halophyte perennials it is not likely that salt tolerance at the germination stage and in another phase in the life-cycle of a plant is necessarily correlated. Woodell (1985) compared the seed germination response to salinity in coastal dune shingle and marsh plant species. Germination of seed of dune species was more reduced at increased salinity than of shingle, driftline or salt marsh species. For the salt marsh species the germination response to salinity could not easily be correlated to the position of plant species in the plant zonation. For all annual and perennial species tested, germination was found to be best at low salinity as was found for germination of two annual *Salicornia* species by Huiskes *et al.* (1985a). In field experiments Huiskes *et al.* (1985a) found almost complete germination of seed of both *Salicornia* species buried under 0.5 cm salt marsh soil in the lower and middle parts of the salt marsh during the months of February, March and April. During these months seedlings of the following species are abundant in the salt marsh: *Atriplex hastata, A. littoralis, Elymus pycnanthus* (driftline), *Aster tripolium, Triglochin maritima, Spergularia maritima*, and *Suaeda maritima* (Rozema unpublished). In western Europe the spring season is generally characterized by a marked surplus of precipitation over evaporation. Therefore, it seems that seeds of most salt marsh plant species germinate at a time that salinity of the soil moisture is low. Halophyte seed dormancy and survival of seed after immersion with sea water seem to be important characteristics of salt marsh species (Woodell 1985). Increased germination in fresh water after pretreatment in sea water has been related by Woodell (1985) to periods of rainfall that may be preceded by winter time spring tides flooding the seeds of dune and driftline species. Most salt marsh species survive immersion in sea water and show stimulation of germination after return to fresh water. This may relate to flooding and dispersal of halophyte seed with sea water inundation followed by rainfall in the winter time and spring in western Europe.

To conclude, it is unlikely that this dispersal of halophyte seed and subsequent germination will directly relate to salt marsh vegetation zonation. As a whole, there is a poor correlation between the halophyte seed germination response to salinity and the position of plant species in salt marsh vegetation. In the more particular case of winter and summer annuals in coastal ecosystems, the germination behaviour of the seed fits in with the timing of the onset of the short life-cycle ending in the setting of seed, and the dormant structure to survive unfavourable environmental conditions. In this regard, the temporal distribution pattern of these plant species is very much dependent on characteristics of the germination response. This strongly suggests that survival, growth and reproduction of seedlings and adult plants in the salt marsh will be more decisive in the position of plant species in salt marsh vegetation zonation and succession (Watkinson & Davy 1985). In a more general sense, dormancy may exist in the seed of annual species rather than in perennial plant species. The triggering of the dormancy mechanism relates to the optimum timing of the germination and development of winter annuals in the summer-dry dune ecosystem and salt marshes and of summer annuals occurring in so many different ecosystems (Janssen 1973; Watkinson 1978, 1981; Rozijn 1984).

## DISTRIBUTION OF SALT MARSH SPECIES AND PHYSIOLOGICAL TOLERANCES: A FIRST ATTEMPT

Salt marsh vegetation zonation represents a more or less constant sequence of plant species with variation that might relate to differences in frequency of sea water flooding, texture and relief of the salt marsh (see Fig. 8.1 with salt marsh vegetation zonation of a 'Wadden' type salt marsh). Despite many detailed phytosociological descriptions correlative analyses and many individual autecological studies, direct evidence demonstrating that sea water inundation governs salt marsh vegetation zonation was not available until a study by Cooper (1982) and later by Rozema *et al.* (1985) compared fifteen salt marsh species from all different zones in one experimental study. Such multispecies ecophysiological studies are time consuming and technically difficult in that greenhouse and cultivation conditions should be optimal for all species during a 3 month period. Diggelen (1988) repeated the multispecies experiment reported by Rozema *et al.* (1985) growing sixteen species under four combinations of non-saline versus saline and non-flooded versus flooded conditions. The mean relative growth rate was calculated and four different indices of flooding and salt tolerance (cf. Rozema *et al.* 1985). The outcome of Diggelen's experiments was that the ranking of salt marsh species according to their index of salt tolerance, obtained under non-flooded and flooded conditions, was significantly correlated with the ranking of salt marsh vegetation zonation in the field. This may lead us to conclude that sea water salinity will be an important environmental factor affecting the mean relative growth rate of the species and thereby their position along a salt marsh height gradient.

The above conclusion assumes that a species with a high index of salt tolerance may be expected to be a better competitor under saline conditions than a species with a lower index of salt tolerance. It should be noticed however, that the calculation of the above indices of tolerance obscures the

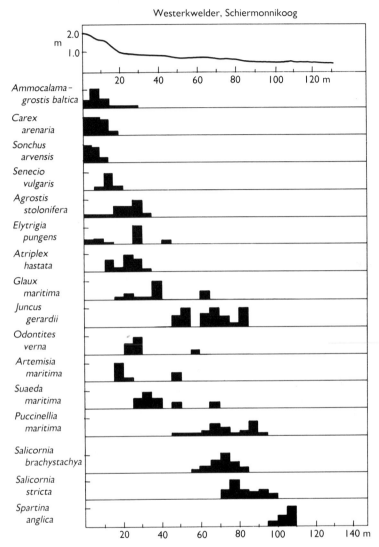

FIG. 8.1. Zonation of vegetation of the Westerkwelder salt marsh on the Wadden Island of Schiermonnikoog, along a height gradient. On the y-axis the biomass as a percentage of the maximum biomass for each species is indicated.

realized growth rates under a certain treatment: *Suaeda maritima* may have a higher index of salt tolerance under non-flooded and flooded conditions than *Salicornia dolichostachya*, but the latter species exhibits a higher relative growth rate under saline conditions (Diggelen 1988).

## THE ZONATION OF TWO MICROSPECIES OF SALICORNIA

*Salicornia dolichostachya* Moss ($2n = 36$) mainly inhabits lower zones in the salt marsh, below the mean high water (MHW) line, while the diploid *Salicornia brachystachya* König ($2n = 18$) (Heukels & van der Meyden 1983) occupies a zone in the salt marsh above the MHW line (Huiskes *et al.* 1985; Schat 1986; Rozema *et al.* 1987; Schat, List & Rozema 1987). The difference in distribution between the two microspecies may occur along a height gradient of only 10 cm. From detailed field observations it seems a likely explanation of the apparent difference in spatial distribution in terms of differential growth response to relevant environmental conditions of the lower salt marsh (Schat 1986). Among these factors are: salinity levels or the type of soil salinity variation both in space and time, variation of the oxygen content and redox potential of the salt marsh soil and the availability of nutrients in the soil, which might be dependent on the oxygen and redox status of the soil. All experimental studies performed so far seem to indicate that no significant differences exist between the two microspecies with respect to (1) seed dispersal and seed germination (Huiskes *et al.* 1985a), (2) seedling mortality (Huiskes, Schat & Elenbaas 1985), (3) response to factors related to anaerobic conditions (effects of reduced iron, manganese, and sulphide (Rozema *et al.* 1987), growth, mineral nutrition and carbon assimilation (Schat, List & Rozema 1987). These authors have concluded that it is possible to ascribe the zonation of the two *Salicornia* species to sensitivity of the root system to mechanical damage by tidal currents in *S. brachystachya*. Testing this hypothesis in greenhouse or in field studies, has proven difficult.

This case of zonation of microspecies of *Salicornia* illustrates how difficult it may be to find key factors from ecophysiological studies that convincingly explain observed differences in species distribution in the salt marsh. In a further series of field and greenhouse studies, the distribution of two salt marsh species occurring in the lower zones of salt marshes has been made in more detail.

## PHYSIOLOGICAL TOLERANCE AND COMPETITIVE RELATIONS OF *SPARTINA ANGLICA* AND *PUCCINELLIA MARITIMA*

*Spartina anglica* and *Puccinellia maritima* are both perennial grasses occurring in the lower parts of salt marshes. *Spartina anglica* occupies parts of the lower salt marsh mainly below the MHW level, and *Puccinellia maritima* inhabits sites around the MHW level and somewhat higher parts (Scholten & Rozema 1987). The difference and overlap of the distribution of these two species in salt marshes may depend on differences in their physiological tolerance to sea water flooding (Rozema *et al.* 1985) and to interspecific competitive relations, cf. Russell, Flowers & Hutchings (1985).

The well-developed aerenchyma within the root tissue of *Spartina anglica* represents an adaptation to anaerobic conditions. It has been indicated that excess reduced sulphur may be detoxified to dimethylsulphoniopropionate (DMSP) (Diggelen *et al.* 1986). A sodium chloride exclusion mechanism operates in the cells of the root tissue maintaining relatively low salt levels in the shoot and in addition to this the effective secretion of salt by the salt glands of the leaves of *Spartina anglica* (Rozema, Gude & Pollak 1981). *Puccinellia maritima* accumulates the compatible osmotic solute proline, while concentrations of methylated quaternary ammonium compounds reach high values in the shoot tissue of *Spartina anglica* (Stewart & Lee 1974; Wyn Jones & Storey 1978; Rozema *et al.* 1985; Diggelen *et al.* 1986).

In the greenhouse and experimental garden the growth of *Puccinellia maritima* and *Spartina anglica* has been compared (Table 8.1). Under non-saline conditions in clay and sand *Spartina* was found to have a relatively low mean relative growth rate compared to *Puccinellia maritima*. The index of salt tolerance defined as the ratio of the mean relative growth rate measured under saline and non-saline conditions, obtained for *Puccinellia maritima* at increased and low salinity was 0.80 under non-flooded and 0.83 under flooded

TABLE 8.1. Mean relative growth rate (mg g$^{-1}$ dry mass day$^{-1}$) of *Puccinellia maritima* and *Spartina anglica* in clay and sand culture on the experimental garden under drained and waterlogged conditions and under conditions of low soil water salinity (50 mM NaCl) and increased soil water salinity (350 mM NaCl). Average values with standard error of the mean.

| | *Puccinellia maritima* | | *Spartina anglica* | |
|---|---|---|---|---|
| | Drained | Waterlogged | Drained | Waterlogged |
| Clay | | | | |
| 50 mM | $56.9 \pm 1.2$ $n=6$ | $56.0 \pm 1.3$ $n=11$ | $9.6 \pm 2.0$ $n=3$ | $45.5 \pm 2.3$ $n=9$ |
| 350 mM | $45.6 \pm 3.8$ $n=12$ | $46.3 \pm 1.8$ $n=12$ | $23.3 \pm 3.6$ $n=9$ | $29.1 \pm 3.4$ $n=15$ |
| Sand | | | | |
| 50 mM | $47.2 \pm 2.2$ $n=6$ | $46.3 \pm 2.4$ $n=11$ | $3.4 \pm 0.9$ $n=3$ | $19.4 \pm 2.8$ $n=9$ |
| 350 mM | $43.4 \pm 2.2$ $n=12$ | $39.1 \pm 1.6$ $n=12$ | $7.3 \pm 3.3$ $n=7$ | $15.5 \pm 2.0$ $n=18$ |

conditions respectively (Table 8.2) in clay culture. Also, in sand culture *Puccinellia* maintains its mean relative growth rate and the indices of salt tolerance calculated do not differ markedly from data obtained for clay. Flooding tolerance indices calculated range from 0.90 to 1.02 for varied saline conditions in clay and sand. Growth results for *Spartina anglica* differ markedly. This species shows growth stimulation by salt, both in clay and sand culture but only under non-flooded conditions (Tables 8.2 and 8.3). The growth rate of *Spartina anglica* is strongly increased under waterlogged conditions when cultivated in clay and sand, but less strongly under saline conditions. It appeared that growth of *Spartina* is much more reduced on sand compared to clay than of *Puccinellia*. From these experiments it was concluded that *Spartina anglica* grows better than *Puccinellia maritima* under waterlogged and clayey conditions. On the other hand, the mean relative growth rate of *Puccinellia maritima* under saline waterlogged conditions (46.3 mg g$^{-1}$day$^{-1}$) exceeds that of *Spartina anglica* (29.1 mg g$^{-1}$ day$^{-1}$ $\pm$ 3.4). This leaves us with the question that the present experimental growth results cannot simply explain the position of these two species in salt marshes, with *Spartina* occupying the lower zones and *Puccinellia maritima* a somewhat higher position (Diggelen 1988).

In addition, the zonation of salt marsh halophytes and that of *Spartina* and *Puccinellia* in particular may show variance with the type of salt marsh (e.g. 'Wadden' type salt marsh *versus* Beach Plain type salt marsh; Beeftink 1977) and when salt marshes in the Wadden sea area are compared with estuarine salt marshes in the Delta region in south-west Netherlands. *Spartina anglica* is distributed as vast monocultures in the lower parts of muddy salt

TABLE 8.2.    Indices of salt tolerance and flooding tolerance of *Puccinellia maritima* and *Spartina anglica* calculated from data presented in Table 8.1, for drained ($-$F) and waterlogged ($+$F), low salt ($-$S) and high salt ($+$S) conditions.

|  |  | *Puccinellia* | *Spartina* |
|---|---|---|---|
| Clay |  |  |  |
| Salt tolerance | $-$F | 0.80 | 2.42 |
|  | $+$F | 0.83 | 0.63 |
| Flooding tolerance | $-$S | 0.98 | 4.74 |
|  | $+$S | 1.02 | 1.25 |
| Sand |  |  |  |
| Salt tolerance | $-$F | 0.92 | 2.15 |
|  | $+$F | 0.84 | 0.80 |
| Flooding tolerance | $-$S | 0.98 | 5.71 |
|  | $+$S | 0.90 | 2.12 |

TABLE 8.3. Mean relative growth rate (mg g$^{-1}$ day$^{-1}$) of *Puccinellia maritima* and *Spartina anglica* in the greenhouse in sand culture under drained and waterlogged conditions and under conditions of low soil water salinity (50 mM NaCl) and increased soil water salinity (350 mM) in the greenhouse (20°C, 70% RH, 300 μE m$^{-2}$ s$^{-1}$). Average value of 9 replicates with the standard error of the mean.

|  | Low salinity | Increased salinity |
|---|---|---|
| *Puccinellia maritima* | | |
| Drained | 66 ± 3.5 | 45 ± 4.0 |
| Waterlogged | 66 ± 5.3 | 32 ± 3.2 |
| *Spartina anglica* | | |
| Drained | 49 ± 1.5 | 41 ± 3.5 |
| Waterlogged | 46 ± 3.1 | 35 ± 2.1 |

marshes in the south-western part of the Netherlands. In salt marshes in the Wadden area, *Spartina* is less abundant and mostly develops more or less as isolated tussocks within stands of *Salicornia dolichostachya* with *Puccinellia maritima* present in the centre of these tussocks. Higher on the marsh, clumps of *Spartina* occur with *Puccinellia* growing within these *Spartina* clumps. At about 20 cm above MHW level there is a gradual transition towards a sward of *Puccinellia* with only few and small areas of *Spartina* (see Scholten & Rozema 1988). On the muddy estuarine salt marshes in the Delta area in south-west Netherlands, *Spartina* clones develop from 70 cm below the MHW level and fusion of these *Spartina* tussocks leads to an almost completely monospecific *Spartina* sward. Above the MHW line there is a sharp transition to the vegetation zone with dominance of *Puccinellia maritima*.

Based on these field observations some hypotheses and questions on the differences of distribution of these two species may be formulated to be tested in experimental greenhouse and field studies:

1 If *Puccinellia* is less tolerant to sea water flooding than *Spartina*, is the establishment of *Puccinellia* in the centre of *Spartina* tussocks then made possible due to the aerating effect of radial oxygen leakage from the *Spartina* roots?

2 The competitive advantage of *Spartina* over *Puccinellia* is less on sandy soils of the Wadden type salt marshes than on clayey marshes in the Delta region. In part the field observations on the distribution of the two species in the salt marshes can be explained on the basis of Diggelen's (1988) experimental studies showing improved growth of *Spartina* under waterlogged and clayey conditions.

## COMPETITIVE RELATIONS

In a Wadden type salt marsh on Schiermonnikoog at two sites differing only 4 cm in height, the biomass development of the two plant species was measured in plots where *Spartina* and *Puccinellia* occur together. In a homogeneous mixed stand, shoots of *Puccinellia* and *Spartina* were separately removed in some of the plots in early August 1984. ('Removal or deletion experiment'; Silander & Antonovics 1982). In November 1984, the aerial biomass of the two species in the plots was harvested and measured (Table 8.4).

The competitive influence of a species A on the growth of species B can be derived from the calculation of a ratio of interference (see Scholten *et al.* 1987; Scholten & Rozema 1987).

$$RI_{A \to B} = \frac{Ba - B}{Ba + B}$$

where Ba represents the biomass of species B in the mixed stand (where none of the two species has been removed) and B represents the biomass of species B in monoculture (the plot where species A has been removed). When there is no competitive influence of A on B the ratio is zero. When species A suppresses the growth of species B the ratio will be negative.

In the lower plot, where *Spartina anglica* is dominant (Table 8.4), the removal of *Spartina* resulted in a significantly higher biomass of *Puccinellia maritima*, growth of *Spartina* in the plot where *Puccinellia* had been removed did not significantly exceed growth of *Spartina* in the mixed plots. In other words, in the lower sites *Spartina* shoots suppress the growth of *Puccinellia* (ratio of competition—0.439), probably by light interception. At the same time in this lower zone there is no such competitive effect of *Puccinellia maritima* on *Spartina anglica* (ratio of competition—0.017).

TABLE 8.4. Biomass (dry mass, g m$^{-2}$) of aerial parts of *Puccinellia maritima* and of *Spartina anglica* in monoculture and mixed plots. Ratio of interference (R.I.), expressing competitive advantage of *Spartina* over *Puccinellia* (S→P) and vice versa (P→S) in lower and higher plots, respectively in the Westerkwelder salt marsh at Schiermonnikoog, The Netherlands.

|  | Lower plot | Higher plot |
|---|---|---|
| *Puccinellia* mono culture | 475 | 580 |
| *Puccinellia* mixed culture | 185 | 450 |
| *Spartina* mono culture | 445 | 310 |
| *Spartina* mixed culture | 430 | 70 |
| Ratio of interference |  |  |
| P→S | −0.017 | −0.632 |
| S→P | −0.439 | −0.126 |

In the higher plot with dominance of *Puccinellia maritima*, removal of *Puccinellia* was followed by significantly increased growth of shoot biomass of *Spartina anglica*, while no such effect was found in the lower plot. This result is a strong indication of competitive suppression of *Spartina* shoots by *Puccinellia* leaves (ratio of competition—0.632), while *Spartina* does not suppress *Puccinellia* (ratio of competition—0.126). More details of these field studies are given in Scholten & Rozema (1988) and Scholten, Blaauw, Stroetenga & Rozema (1987). Removal experiments also showed that the competitive ability of *Puccinellia* against *Spartina anglica* increased with increasing height on the salt marsh and that the competitive advantage of *Spartina* over *Puccinellia* increased with decreasing height.

The abundance of *Spartina anglica* in the clayey estuarine salt marshes in south-west Netherlands, with vast monospecific stands correlated not only with the growth stimulation of *Spartina anglica* under flooded conditions (Table 8.1) but also with the finding that the competitive ability of *Spartina* relative to *Puccinellia* is higher on clay than on sand (Scholten & Rozema 1988). Not only are the Wadden salt marshes more sandy than the estuarine salt marshes, there are also some climatological differences between them. The geographical distribution of *Puccinellia maritima* is in western and northern Europe, while the northern limit of *Spartina anglica* in Europe is at the Skallingen salt marsh in southern Denmark. Generally, *Spartina anglica* has a more southerly distribution than *Puccinellia maritima*.

Removal experiments carried out through the course of the growing season (Scholten *et al.*, unpublished data) revealed a number of findings. Growth of *Puccinellia* shoots may start earlier in the season, providing a competitive advantage relative to *Spartina anglica* on the higher parts of the sandy Wadden salt marshes, but not in south-west Netherlands in clayey salt marshes. In the south-west Netherlands, where salt marshes have temperatures about 3°C higher on average than at the Wadden Islands, the competitive ability of *Spartina* relative to *Puccinellia* is already high at the start of the growth period. In this area, sprouts of *Spartina* emerge earlier, allowing it to take the lead over *Puccinellia* in competition for light. This advantage is maintained or enlarged during the course of the summer and autumn. These field studies of the competitive relations of *Spartina* and *Puccinellia* have helped to explain the occurrence of the dense virtual monocultures of *Spartina* in the south-west Netherlands.

The occurrence of *Puccinellia maritima* within clumps of *Spartina anglica* could be explained physiologically in terms of *Puccinellia* requiring oxygen in the root environment which is provided by *Spartina anglica* through radial oxygen leakage. In field and greenhouse studies (Scholten & Schobben, unpublished data) no evidence was found for this hypothesis. Values of the redox potential value in pots with flooded mixed cultures of *Spartina anglica* and *Puccinellia maritima* did not differ significantly. In greenhouse experiments the only competitive interactions between the two species related to the capture of light and uptake of minerals. Transplantation experiments conducted at the Westerkwelder salt marsh at Schiermonnikoog indicated that *Puccinellia* plants within tussocks of *Spartina anglica* are protected against wind and wave action (Scholten & Rozema 1988). In the higher zones of the salt marsh vegetation, evidence has been found of radial oxygen leakage from the root system of *Juncus gerardii*, providing aerobic conditions required for growth of *Agrostis stolonifera* and *Festuca rubra* (Scholten & Blaauw 1987).

The competition between *Aster tripolium* and *Puccinellia maritima* was studied under different levels of inorganic nitrogen supply, in greenhouse conditions. The species were grown in mono and mixed cultures, following an 'addition' design (Scholten *et al.* 1987). During the first week above ground biomass growth of *Puccinellia* is repressed by *Aster*. This competitive influence of *Aster* on *Puccinellia* is reflected by a low, negative value of the ratio of interference; the repression of *Puccinellia* increases with decreasing supply of inorganic nitrogen (Scholten *et al.* 1987), indicating competition for nitrogen. After 2 to 4 weeks the competitive influence of *Aster* on *Puccinellia* is increased by shading of *Puccinellia* on the rich medium. Growth of *Puccinellia* during the experiment increases its small competitive influence on *Aster* in the pots with a medium inorganic nitrogen level. Competition for nitrogen between the species may also have taken place in the pots with the lowest nitrogen supply.

At the high nitrogen level, where there is competition for light, the number of shoots of *Puccinellia* increased with no change in shoot length. Under low levels of inorganic nitrogen in the soil, where competitive uptake of nitrogen is more likely to occur, shoot length is far more reduced (Scholten *et al.* 1987), while the number of shoots of *Puccinellia* is unaffected. In further experiments with *Puccinellia* and *Aster* growing in pots in the greenhouse, seedlings, young plants and old plants were used. It appeared that the competitive repression of *Puccinellia* on *Aster* plants was stronger with decreasing size of the *Aster* plants (Scholten *et al.* 1987).

Field studies of the competitive relations of *Agrostis stolonifera* and *Festuca rubra* ssp. *litoralis* using the same technique of removing one of two species confirmed the findings of the *Spartina–Puccinellia* case. In plots with a higher biomass of *Festuca* shoots relative to *Agrostis* removal of *Festuca* sprouts led to increased growth of *Agrostis* compared to the plot with both species present. Removal of *Agrostis* in plots where it was dominant was followed by increased growth of *Festuca*, indicating suppression of *Festuca* by *Agrostis* (Scholten *et al.* 1987). In greenhouse experiments it was found that *Agrostis* develops above-ground stolons that may occupy open space easily. On the contrary, *Festuca rubra* forms more new tillers close to the parental plants, producing a dense sward. *Agrostis stolonifera* obtains therefore a competitive advantage where rapid colonization and spread of open areas in salt marshes is possible. The competitive ability of *Festuca rubra* is limited in early successional stages with open spaces, but production of high density of shoot will steadily present competitive advantages which are relative to *Agrostis* and will contribute to successful competition with *Agrostis* for light interception and nutrient uptake (Scholten, unpublished data).

These findings indicate that the distribution of plant species along height gradients of salt marshes is also determined by the outcome of interspecific competition for light and nutrients. An early emergence during the season will help to acquire a competitive advantage of one species relative to another. Yet, plant species differ with respect to time of emergence. Plant species that develop aerial parts later during the season possibly avoid stress or damage due to unfavourable climatological conditions that may occur during this season.

In a study of a New England salt marsh community, Bertness & Ellison (1987) concluded that vegetation zones of *Spartina alterniflora*, *Spartina patens* and *Juncus gerardii* usually can be explained primarily in terms of the species responses to salinity and substrate redox potential. In addition to this, salt marsh vegetation zonation may also be explained as the outcome of competitive interference between the species as shown by reciprocal transplant studies. Two other salt-tolerant species, *Distichlis spicata* and *Salicornia europaea*, occur frequently in disturbed areas, such as places with deposition of dead plant material. *D. spicata* rapidly colonizes disturbed sites with vegetative runners while the annual *Salicornia europaea* invades such patches by seed germination.

# RELATIONSHIP OF AUTOTROPHIC HALOPHYTES WITH HETEROTROPHIC PLANTS ON SALT MARSHES

In competitive relationships species interact negatively with each other, albeit that the intensity may vary widely, depending on many environmental and plant characteristics. In the relationships between heterotrophic and autotrophic plants on salt marshes, the heterotrophic partner always benefits from inorganic and organic compounds obtained from the host plant.

The hemiparasite *Odontites verna* ssp. *serotina* (Scrophulariaceae) occurs on the salt marshes of the Wadden Islands in the upper zones, where it grows together with *Festuca rubra* ssp. *litoralis*, *Agrostis stolonifera*, *Juncus gerardii*, *Elymus pycnanthus* and *Trifolium repens* (see Fig. 8.1). Many hemiparasites of the Scrophulariaceae are not host-specific (Borg 1985). Govier, Brown and Pate (1968) have reported that plants of *Odontites* may reach a maximum size when one individual is connected by haustoria to several species of host plants. *Odontites verna* ssp. *serotina* is a summer annual hemiparasite producing seed in autumn. Dormancy of the seed is broken after a 12–14 weeks of moist–cold (4°C) treatment (Borg 1985). Roots of *Odontites* have xylem–xylem contact with the roots of host plants. The hemiparasite leaves contain chlorophyll and are capable of photosynthesis. Growth and reproduction is possible without a host plant, although the hemiparasite reaches only a minimum size. In the case of holoparasitism the reduction in yield of the host may cause economic problems, such as the yield of tobacco plants parasitized by *Orobanche*. In salt marshes where *Odontites* plants reach a high density, the size of the host plants may also be reduced, compared to sites without the hemiparasite (Borg 1985).

At the Westerkwelder salt marsh at Schiermonnikoog a study was made of the ecophysiological relationships between *Odontites* and its host plants. A comparison was made of the shoot tissue and the total water potential $\Psi_T$ of host plant species parasitized and non-parasitized by *Odontites* (Table 8.5). The total water potential of the shoot was measured at mid-day using a Scholander's pressure bomb. The root hemiparasite *Odontites* absorbs xylem sap with dissolved nutrients from its host by maintenance of a more negative water potential than that of the host plant species. There is a continuous flow of xylem water and dissolved nutrients which is related to a high transpiration rate of *Odontites* even during the dark period (Rozema *et al.* 1987a). The sodium and potassium content of the sap of the leaves of *Odontites* plants attached to *Juncus gerardii* or *Festuca rubra* ssp. *litoralis* exceeds that of the host plants. The sodium content of *Odontites* shoot tissue amounts to 1.128 mmol $g^{-1}$ dry mass compared to 0.175 mmol $g^{-1}$ dry mass Na in the tissue of *Juncus gerardii* leaves (Table 8.5). The sodium content of xylem sap of the possible host plants of *Odontites*, grown in a medium with 50 M NaCl, varied from 2.5–8 mM Na; the potassium content of the xylem sap was about 2.5 mM. These data show the enormous accumulation of sodium in the shoot of *Odontites* when attached to these monocotyledonous species.

Surprisingly, an analysis of the mineral and water relations of the holoparasite *Cuscuta salina* var *major* parasitizing halophytes in the lower zones in a San Francisco Bay salt marsh revealed (Rozema *et al.* 1986a) that the sodium and chloride content of the holoparasite was four to ten times less than that of the host plant tissue. The sodium levels in the stem holoparasite

TABLE 8.5. Mineral composition (mmol or μmol/g DW) of the shoot tissue and total water potential of the plant $\Psi_T$ (MPa) of two host plant species parasitized and non-parasitized by the hemiparasite *Odontites verna* ssp. *serotina* in the Westerkwelder salt marsh of Schiermonnikoog, The Netherlands, July 1984. Average values of four replicates.

| Species | Mineral element | | | | $\Psi_T$ | Na host | K |
|---|---|---|---|---|---|---|---|
| | mmol/g DW | | μmol/g DW | | | Na paras. | Ca |
| | K | Na | Ca | Mg | | | |
| Host *Festuca rubra* ssp. *litoralis* | 0.540 | 0.140 | 19.1 | 26.7 | −2.20 | 0.211 | |
| Hemiparasite *Odontites verna* | 0.416 | 0.665 | 50.5 | 105.3 | −2.75 | | 8.2 |
| Host *Festuca rubra* ssp. *litoralis* non-parasitized | 0.443 | 0.090 | 17.9 | 31.4 | −2.04 | | |
| Host *Juncus gerardii* | 0.508 | 0.175 | 10.3 | 23.0 | −2.10 | 0.136 | |
| Hemiparasite *Odontites verna* | 1.138 | 1.128 | 137.2 | 325.8 | −2.84 | | 8.3 |
| Host *Juncus gerardii* non-parasitized | 0.325 | 0.178 | 14.1 | 32.8 | −2.17 | | |
| L.S.D. $P=0.05$ | 0.214 | 0.085 | 11.3 | 19.9 | −0.20 | | |

*Cuscuta salina* may be less than that of the root hemiparasite *Odontites*. This is remarkable, since the sodium content of *Salicornia pacifica* when parasitized by *Cuscuta* (about 3–5 mmol g$^{-1}$ dry mass) was far higher than that of the host plants *Festuca rubra* ssp. *litoralis* and *Juncus gerardii* in the Schiermonnikoog salt marshes (0.14–0.18 mmol g$^{-1}$ dry mass). *Cuscuta* is a phloem feeder and this is apparently related to a low rate of uptake of Na and other minerals.

Wallace, Romney & Alexander (1978), studying the stem holoparasite *Cuscuta nevadensis* parasitizing desert halophytes, report similarly low Na concentrations in the parasite tissue. The shoot tissue concentrations of Na, K, Ca and Mg in the root holoparasite *Cistanche lutea*, a phloem feeder, on the salt-accumulating halophyte *Halimione portulacoides* are lower than the values assessed in the shoot of the host plant (Baumeister & Ernst 1978). It appears, that like in many other shoot hemiparasites (*Rhinanthus serotinus*, Klaren 1975; *Phoradendron villosum*, a mistletoe on *Quercus lobata*, Hollinger 1983) transpiration rates are high and may contribute to accumulation of ions in the hemiparasite tissue. Although hemiparasites of the Scrophulariaceae are considered to be host aspecific (Borg 1985), the xylem feeder *Odontites verna* ssp. *serotina* occurring on salt marshes is mostly found in upper parts of salt marshes attached to salt-excluding monocotyledons. *Cuscuta salina* and *Cistanche lutea* that feed on the phloem of host plant have

J. ROZEMA *et al.*

DeJong, T. M., Drake, B. G. & Pearcy, R. W. (1982). Gas exchange of Chesapeake tidal marsh species under field and laboratory conditions. *Oecologia (Berlin)*, **52**, 5–11.

Diggelen, J. van, Rozema, J., Dickson, D. M. J. & Broekman, R. (1986). *β*-3-Dimethyl-sulphoniopropionate, proline and quaternary ammonium compounds in *Spartina anglica* in relation to sodium chloride, nitrogen and sulphur. *New Phytologist*, **103**, 573–86.

Diggelen, J. van, Rozema, J. & Broekman, R. (1987). Mineral composition of and proline accumulation by *Zostera marina* L. in response to environmental salinity. *Aquatic Botany*, **27**, 169–76.

Diggelen, J. van (1988). *A comparative study on the ecophysiology of salt marsh halophytes.* Thesis, Free University, Amsterdam.

Diggelen, J. van, Rozema, J. & Broekman, R. A. (1987). Growth and mineral relations of salt marsh species on nutrient solutions containing various sodium sulfide concentrations. *Vegetation Between Land and Sea. Structure and Processes* (Ed. by A. H. L. Huiskes, C. W. P. M. Blom & J. Rozema), pp. 258–66. Junk, Dordrecht.

Ellenberg, H. (1953). Physiologisches und ökologisches Verhalten derselben Pflanzenarten. *Berichte der Deutsche Botanischen Gesellschaft*, **65**, 350–61.

Ellenberg, H. (1981). *Die Vegetation Mitteleuropas mit den Alpen.* Ulmer, Stuttgart.

Ernst, W. H. O. (1978). Chemical soil factors determining plant growth. Structure and functioning of plant populations. *Proceedings Nederlandse Academie van Wetenschappen*, **70**, 155–88.

Ernst, W. H. O. (1985). Impact of Mycorrhiza on metal uptake and translocation by forest plants. *Heavy metals in the environment.* (Ed. by T. D. Lekkas), pp. 596–599. CEP Consultants Edinburgh.

Ernst, W. H. O. (1986). Long term pollution and selection. *Proceedings Environmental Contamination*, pp. 10–15, CEP Consultants, Edinburgh.

Etherington, J. (1975). *Environment and Plant Ecology* 2nd ed. John Wiley, Chichester.

Firbas, F. (1967). Pflanzen geographie. *Lehrbuch der Botanik* (Ed. by D. von Denfer, K. Mägdefrau, W. Schumacher & F. Firbas), pp. 679–709. Gustav Fischer, Stuttgart.

Fitter, A. H. (1986). Acquisition and utilization of resources. *Plant Ecology* (Ed. by M. J. Crawley) pp. 375–405. Blackwell Scientific Publications, Oxford.

Flowers, T. J. (1972). Salt tolerance in *Suaeda maritima*. (L.) Dum. The effect of sodium chloride on growth, respiration, and soluble enzymes in a comparative study with *Pisum sativum* L. *Journal of Experimental Botany*, **23**, 310–21.

Flowers, T. J., Troke, P. F. & Yeo, A. R. (1977). The mechanism of salt tolerance in halophytes. *Annual Review of Plant Physiology*, **28**, 89–121.

Govier, R. N., Brown, J. G. S. & Pate, J. S. (1968). Hemiparasitic nutrition in Angiosperms. *New Phytologist*, **67**, 963–72.

Greenway, H. & Munns, R. (1980). Mechanisms of salt tolerance in monnalophytes. *Annual Review of Plant Physiology*, **31**, 149–90.

Grime, J. P. (1979). *Plant Strategies and Vegetation Processes.* John Wiley, Chichester.

Grubb, P. J., Kelly, D. & Mitchley, J. (1982). The control of relative abundance in communities of herbaceous plants. *The Plant Community as a Working Mechanism* (Ed. by E. I. Newman) pp. 79–97. Blackwell Scientific Publications, Oxford.

Heukels, H. & van der Meyden, R. (1983). Flora van Nederland. 20th ed. Wolters-Noordhoff, Groningen.

Hollinger, D. Y. (1983). Photosynthesis and water relations of the mistletoe. *Phoradendron villosum* and its host, the California valley oak, *Quercus lobata. Oecologia (Berlin)*, **60**, 396–400.

Huiskes, A. H. L., Stienstra, A. W., Koutstaal, B. P., Markusse, M. M. & van Soelen, J. (1985). Germination ecology of *Salicornia dolichostachya* and *Salicornia brachystachya. Acta Botanica Neerlandica*, **34**, 396–80.

**Huiskes, A. H. L., Schat, H. & Elenbaas, P. F. M. (1985).** Cytotaxonomic status and morphological characterisation of *Salicornia dolichostachya* and *Salicornia brachystachya*. *Acta Botanica Neerlandica*, **34**, 271–82.

**Janssen, J. G. M. (1973).** Effects of light, temperature and seed age on the germination of the winter annuals *Veronica arvensis* L. and *Myosotis ramosissima* Rochel ex Schult. *Oecologia (Berlin)*, **12**, 141–46.

**Klaren, C. M. (1975).** Physiological aspects of the hemiparasite Rhinantus serotinus. Ph.D. Thesis. Groningen.

**Lange, O. L., Schulze, E. D. & Koch, W. (1970).** Experimentell-ökologische Untersuchungen an Flechten der Negev-Wüste. II. $CO_2$-Gaswechsel und Wasserhaushalt von *Ramalina maciformis* (Del.). Bory am natürlichen Standort, während der sommerlichen Trockenperiode. *Flora*, **159**, 38–62.

**Larcher, W. (1983).** *Physiological Plant Ecology*. 2nd edn. Springer, Berlin.

**Mahall, B. E. & Park, R. B. (1976).** The outcome between *Spartina foliosa* Trin. and *Salicornia virginica* L. in salt marshes of Northern San Francisco Bay. II. soil water and salinity. Journal of Ecology, **64**, 793–809.

**Miles, J. (1979)** *Vegetation Dynamics*. Outline studies in ecology. Chapman & Hall, London.

**Mueller-Dombois, D. & Ellenberg, H. (1974).** *Aims and Methods of Vegetation Ecology*, Wiley, New York.

**Munns, R., Greenway, H. & Kirst, G. O. (1983).** Halotolerant eukaryotes. *Physiological Plant Ecology III* (Ed. by O. L. Lange, P. S. Nobel, C. B. Osmond & H. Ziegler) pp. 59–135. Springer-Verlag, Berlin.

**Pugh, G. J. F. (1979).** The distribution of fungi in coastal regions. *Ecological Processes in Coastal Environments* (Ed. by R. L. Jefferies & A. J. Davy) pp. 415–28. Blackwell Scientific Publications, Oxford.

**Ranwell, D. S. (1977).** *Ecology of salt-marshes and sand-dunes*. Chapman & Hall, London.

**Rozema, J., Gude, H. & Pollak, G. (1981).** An ecophysiological study of the salt secretion of four halophytes. *New Phytologist*, **89**, 201–17.

**Rozema, J. (1975).** The influence of salinity, inundation and temperature on the germination of some halophytes and non-halophytes. *Oecologia Plantarum*, **10**, 317–29.

**Rozema, J., Bijwaard, P., Prast, G. & Broekman, R. (1985).** Ecophysiological strategies of coastal halophytes from sand dunes and salt marshes. *Ecology of Coastal Vegetation* (Ed. by W. G. Beeftink, J. Rozema & A. H. L. Huiskes) pp. 499–522. Junk, Dordrecht.

**Rozema, J., Broekman, R., Arp, W., Letschert, J., van Esbroek, M. & Punte, H. (1986a).** A comparison of the mineral relation of a halophytic hemiparasite and holoparasite. *Acta Botanica Neerlandica*, **35**, 105–9.

**Rozema, J., Arp, W., Diggelen, J. van, van Esbroek, M., Broekman, R. & Punte, H. (1986b).** Occurrence and ecological significance of vesicular arbuscular mycorrhiza in the salt marsh environment. *Acta Botanica Neerlandica*, **35**, 457–67.

**Rozema, J., Arp, W., Diggelen, J. van, Kok, E. & Letschert, J. (1987a).** An ecophysiological comparison of measurements of the diurnal rhythm of the leaf elongation and changes of the leaf thickness of salt-resistant Dicotyledonae and Monocotyledonae. *Journal of Experimental Botany*, **38**, 442–53.

**Rozema, J., List, J. C. van der, Schat, H., Diggelen, J. van & Broekman, R. A. (1987b).** Ecophysiological response of *Salicornia dolichostachya* and *Salicornia brachystachya* to seawater inundation. *Vegetation Between Land and Sea. Structure and Processes*. (Ed. by A. H. L. Huiskes, C. W. P. M. Blom & J. Rozema) pp. 178–84. Junk, Dordrecht.

**Rozijn, N. A. M. G. (1984).** *Adaptive strategies of some dune annuals*. Ph.D. Thesis, Free University Amsterdam.

**Russell, P. J., Flowers, T. J. & Hutchings, M. J. (1985).** Comparison of niche breadth and overlaps of halophytes on salt marshes of differing diversity. *Ecology of Coastal Vegetation* (Ed. by W. G. Beeftink, J. Rozema & A. H. L. Huiskes) pp. 163–70. Junk, Dordrecht.

Schat, H. (1986). Niche differentiation within *Salicornia:* adaptations to seawater flooding. *Ecology of Estuarine Vegetation* (Ed. by J. Rozema) pp. 67–86 Amsterdam-Yerseke (in dutch).

Schat, H., List, J. C. van der & Rozema, J. (1987). Ecological differentiation of the microspecies *Salicornia dolichostachya* Moss and *Salicornia brachystachya:* growth, mineral nutrition, carbon assimilation and development of the root system in anoxic and hypoxic culture solution. *Vegetation Between Land and Sea. Structure and Processes* (Ed. by A. H. L. Huiskes, C. W. P. M. Blom & J. Rozema) pp. 162–76. Junk, Dordrecht.

Scholten, M. C. Th., Blaauw, P. A., Stroetenga, M. & Rozema, J. (1987). The impact of competitive interactions on the growth and distribution of plant species in salt marshes. *Vegetation Between Land and Sea. Structure and Processes.* (Ed. by A. H. L. Huiskes, C. W. P. M. Blom & J. Rozema) pp. 268–79. Junk, Dordrecht.

Scholten, M. C. Th. & Rozema, J. (1988). The competitive ability of *Spartina anglica* on Dutch salt marshes. *Spartina anglica–a Review of Current Research* (Ed. by A. J. Gray) NERC, ITE Furzebrook, in press.

Silander, J. A. & Antonovics, J. (1982). Analysis of interspecific interactions in a coastal community—a pertubation approach. *Nature (London)*, **298**, 557–60.

Stewart, G. R., Lee, J. A. (1974). The role of proline accumulation in halophytes. *Planta*, **120**, 279–89.

Ungar, I. A. (1978). Halophyte seed germination. *The Botanical Review*, **44**, 233–64.

von Hedenström, H. & Breckle, S. W. (1974). Obligate halophytes? A test with tissue culture methods. *Zeitschrift zur Pflanzenphysiologie*, **74**, 183–5.

Wallace, A., Romney, E. M. & Alexander, G. V. (1978). Mineral composition of *Cuscuta nevadensis* Johnston (Dodder) in relation to its hosts. *Plant and Soil*, **50**, 227–31.

Watkinson, A. R. (1978). The demography of a sand dune annual. II. The dynamics of seed populations. *Journal of Ecology*, **66**, 35–44.

Watkinson, A. R. (1981). The population ecology of winter annuals. *The Biological Aspects of Rare Plant Conservation.* (Ed. by H. Synge) pp. 253–64. Wiley, Chichester.

Watkinson, A. R. & Davy, A. J. (1985). Population biology of salt marsh and sand dune annuals. *Ecology of Coastal Vegetation* (Ed. by W. G. Beeftink, J. Rozema & A. H. L. Huiskes) pp. 487–97. Junk, Dordrecht.

Woodell, S. R. J. (1985). Salinity and seed germination patterns in coastal plants. *Ecology of Coastal Vegetation* (Ed. by W. G. Beeftink, J. Rozema & A. H. L. Huiskes) pp. 223–29. Junk, Dordrecht.

Wyn Jones, R. G., Storey, R. (1978). Salt stress and comparative physiology in the gramineae. II Glycine-betaine and proline accumulation in two salt and waterstressed barley cultivars. *Australian Journal of Plant Physiology*, **5**, 817–29.

Yeo, A. R. (1983). Salinity resistance: physiologies and prices. *Physiologia Plantarum*, **58**, 214–22.

# 9. DISTRIBUTION LIMITS FROM A DEMOGRAPHIC VIEWPOINT

R. N. CARTER AND S. D. PRINCE

*School of Biological Sciences, Queen Mary College, Mile End Road,*
*London E1 4NS, UK*

## SUMMARY

**1** A demographic approach to distribution limits implies a focus upon the ways in which species change in abundance towards their geographical limits. There is little evidence to support the traditional view that populations become smaller towards limits, but there is evidence that they become less frequent.

**2** Species may exhibit clumped patterns of distribution towards their limits. This is discussed in relation to work on *Lactuca serriola*. Stable patterns of distribution are the result of dynamic processes of colony turnover. The concept of populations of populations—here called suprapopulations— assists in understanding these processes.

**3** An epidemic model of suprapopulation dynamics suggests that small changes in climatic factors eliciting small plant responses may cause limits that are abrupt relative to climatic gradients.

**4** Fugitive species are likely to show this behaviour at their limits, while long-lived species of stable habitats are not. The model further suggests that the availability of colonizable sites for a species is a parameter in the equilibrium controlling its climatic limit. Dispersal effectiveness is also such a factor.

**5** The assumptions underlying the model and the scope of the model are discussed. The model has value for strategic, not predictive, purposes.

## INTRODUCTION

This chapter deals with the geographical distribution limits of species and the role that demographic studies might play in explaining them. Distribution limits are simply lines drawn around species ranges on a map, and as such they are abstractions from reality. They do however relate to the ecology of real plants in dividing areas where they grow from areas where they do not. Species are universally believed to occur where environmental conditions favour them, and limits are therefore explained in terms of the physiological responses of plants to favourable conditions within their ranges, and, more importantly, to unfavourable conditions elsewhere. It is worthy of note that geographical distribution patterns depend on the scale of the maps on which they are observed, and so do the reasons adduced to explain them. Geographical limits could be observed at many scales, from the world scale to that of an English county, for example, and different types of explanation might accordingly be required (Fig. 9.1). The arguments presented here apply to limits at the scale of the British Isles, as shown in the maps (Fig. 9.2) in Perring & Walters (1962).

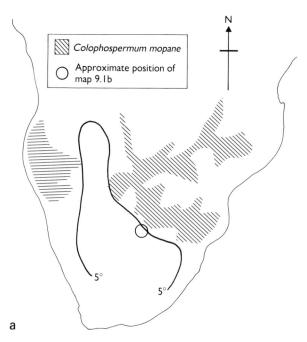

FIG. 9.1a.    The distribution limit of *Colophospermum mopane* in (a) southern Africa at large (adapted from Henning & White 1976), and (b) eastern Botswana (R. W. Madams, unpublished data). At the larger scale it correlates with the mean 5°C July isotherm. At the smaller scale it correlates with rocky hills, drainage channels, coarse grained sands and cultivated areas, from which it is absent. At the larger scale the correlation with soils is obscure; suitable soils occur on both sides of the limit, but they vary at too fine a scale to present systematic pattern at this scale. Conversely, at the smaller scale the correlation with climate is obscure, because climate varies at too coarse a scale. Thus, in the field the limit always appears to correlate with features of the terrain and the critical responses of the plant to climate are masked by larger responses to local factors. If the importance of scale were not recognized the limit at the larger scale might wrongly be attributed to soils. In (b) solid lines outline rocky hills.

Tswapong Hills

Tswapong Hills

▤ *Colophospermum*
  *mopane*

FIG. 9.1b.

*annua, Melampyrum arvense* and *Kickxia elatine* (Pigott 1970), *Kickxia spuria, Lathyrus nissolia* and *Lactuca serriola* (Carter & Prince 1981). The distribution limit of the winter annual ruderal *Lactuca serriola* has been extensively investigated in Britain. This species has been known in south-eastern England since the earliest days of botanical recording (Carter & Prince 1982). Towards its limit it shows no marked decline in any aspect of fecundity, either in natural sites at its limit, or in transplant sites beyond it (Carter & Prince 1985; Prince & Carter 1985), it shows no tendency to fail in establishment (Prince & Carter 1985), it regenerates indefinitely in transplant sites beyond its limit (Prince & Carter 1985) and it does not become noticeably less frequent towards its limit in the uniform environment of the M5 motorway (Prince, Carter & Dancy 1985). Small changes in performance that can be detected, either towards the limit or in transplant sites, include a slight delay in flowering (Dancy 1984; Carter & Prince 1985), a slight restriction in the range of habitats occupied towards the limit (Carter & Prince 1985), small (but doubtfully disadvantageous) changes in the habit of winter growth and very small changes in the numbers of viable achenes produced, detectable only by lumping sites in regions (Dancy 1984). These could scarcely account for the distribution limit under the tolerance limit paradigm.

The validity of the model is also supported by the fact that it provides explanations for the seemingly unrelated observations concerning limits discussed in the previous section. If the utility of the suprapopulation concept is accepted, the model can be viewed as a null hypothesis (in the sense of Strong 1980) for distribution limit control. It describes what happens at limits when nothing special happens, because it is based on minimal assumptions. It provides a better null hypothesis than the tolerance limit paradigm, which is poorly defined and makes many hidden assumptions, as for example that environmental factors have effects on plants in many sites over many generations that correspond on a one-to-one basis with their effects on individuals in one generation.

# REFERENCES

Bailey, N. T. J. (1975). *The Mathematical Theory of Infectious Diseases and its Applications.* Griffin, London.

Becker, N. (1977). On a general epidemic model. *Theoretical Population Biology,* **1**, 22–36.

Becker, N. (1979). The uses of epidemic models. *Biometrics,* **35**, 295–305.

Cadbury, D. A., Hawkes, J. G. & Readett, R. C. (1971). *A Computer-mapped Flora: a Study of the County of Warwickshire.* Academic Press, London.

Carter, R. N. & Prince, S. D. (1981). Epidemic models used to explain biogeographical distribution limits. *Nature, London,* **293**, 644–45.

Carter, R. N. & Prince, S. D. (1982). A history of the taxonomic treatment of unlobed-leaved prickly lettuce, *Lactuca serriola* L., in Britain. *Watsonia*, 14, 59–62.

Carter, R. N. & Prince, S. D. (1985). The geographical distribution of prickly lettuce (*Lactuca serriola*) I. A general survey of its habitats and performance in Britain. *Journal of Ecology*, 73, 27–38.

Clymo, R. S. (1962). An experimental approach to part of the calcicole problem. *Journal of Ecology*, 50, 707–31.

Dancy, K. J. (1984). *Studies on the distribution limit of Lactuca serriola L. in Britain.* Ph.D. thesis, University of London.

Davison, A. W. (1970). The ecology of *Hordeum murinum* L. I. Analysis of the distribution in Britain. *Journal of Ecology*, 58, 453–66.

Davison, A. W. (1977). The ecology of *Hordeum murinum* L. III. Some effects of adverse climate. *Journal of Ecology*, 65, 523–30.

Diamond, J. M. (1979). Community structure: is it random, or is it shaped by species differences and competition? *Population Dynamics* (Ed. by R. M. Anderson, B. D. Turner & L. R. Taylor), pp. 165–81. Blackwell Scientific Publications, Oxford.

Doignon, P. (1954). Persistance d'une fougère Atlantique sous le climat de Fontainebleau. *Cahiers des Naturalistes*, 9, 51–52.

Erickson, R. O. (1943). Population size and geographical distribution of *Clematis fremontii* var. *riehlii*. *Annals of the Missouri Botanical Garden*, 30, 63–68.

Erickson, R. O. (1945). The *Clematis fremontii* var. *riehlii* population in the Ozarks. *Annals of the Missouri Botanical Garden*, 32, 413–60.

Ford, M. J. (1982). *The Changing Climate: Responses of the Natural Flora and Fauna.* George Allen & Unwin, London.

Gadgil, M. (1971). Dispersal: population consequences and evolution. *Ecology*, 52, 253–61.

Good, R. (1931). A theory of plant geography. *New Phytologist*, 30, 149–71.

Griggs, R. F. (1914). Observations on the behaviour of some species at the edges of their ranges. *Bulletin of the Torrey Botanical Club*, 41, 25–49.

Harper, J. L. (1981). The meanings of rarity. *The Biological Aspects of Rare Plant Conservation* (Ed. by H. Synge), pp. 189–203. Wiley, Chichester.

Hengeveld, R. & Haeck, J. (1982). The distribution of abundance. I. Measurements. *Journal of Biogeography*, 9, 303–16.

Henning, A. C. & White, R. E. (1974). A study of the distribution of *Colophospermum mopane* (Kirk ex Benth.) Kirk ex J. Leon: the interactions of nitrogen, phosphorus and soil moisture stress. *Proceedings of the Grassland Society of Southern Africa*, 9, 53–60.

Iversen, J. (1944). *Viscum, Hedera* and *Ilex* as climate indicators. *Geologiske Forense Stockholm Forhandlung*, 66, 463–83.

Kermack, W. O. & McKendrick, A. G. (1927). Contributions to the mathematical theory of epidemics. I. *Proceedings of the Royal Society, Series A*, 115, 700–21.

Levins, R. (1970). Extinction. *Some Mathematical Questions in Biology* (Ed. by M. Gerstenhaber), *Lectures on Mathematics in the Life Sciences*, 2, 77–107. American Mathematical Society, Providence.

MacArthur, R. H. & Wilson, E. O. (1967). *The Theory of Island Biogeography.* Princeton University Press, Princeton.

Matthews, J. R. (1937). Geographical relationships of the British Flora. *Journal of Ecology*, 25, 1–90.

May, R. (1974). *Stability and Complexity in Model Ecosystems.* Princeton University Press, Princeton.

Perring, F. H. & Walters, S. M. (1962). *Atlas of the British Flora.* Nelson, London.

Pigott, C. D. (1970). The response of plants to climate and climatic change. *The Flora of a Changing Britain* (Ed. by F. H. Perring), pp. 32–44. Classey, Faringdon.

Pigott, C. D. (1981). Nature of seed sterility and natural regeneration of *Tilia cordata* near its northern limit in Finland. *Annales Botanici Fennici*, **18**, 255–63.

Pigott, C. D. & Huntley, J. P. (1981). Factors controlling the distribution of *Tilia cordata* at the northern limit of its range. III. Nature and causes of seed sterility. *New Phytologist*, **87**, 817–39.

Prince, S. D. & Carter, R. N. (1985). The geographical distribution of prickly lettuce (*Lactuca serriola*), III. Its performance in transplant sites beyond its distribution limit in Britain. *Journal of Ecology*, **73**, 49–64.

Prince, S. D., Carter, R. N. & Dancy, K. J. (1985). The geographical distribution of prickly lettuce (*Lactuca serriola*) II. Characteristics of populations near its distribution limit in Britain. *Journal of Ecology*, **73**, 39–48.

Rabinowitz, D. (1981). Seven forms of rarity. *The Biological Aspects of Rare Plant Conservation* (Ed. by H. Synge), pp. 205–17. Wiley, Chichester.

Richards, P. W. & Evans, G. B. (1972). Biological flora of the British Isles: *Hymenophyllum* Sm. *Journal of Ecology*, **60**, 245–68.

Salisbury, E. J. (1926). The geographical distribution of plants in relation to climatic factors. *Geographical Journal*, **57**, 312–35.

Salisbury, E. J. (1932). The East-Anglian flora. *Transactions of the Norfolk and Norwich Naturalists Society*, **13**, 191–263.

Salisbury, E. J. (1939). Ecological aspects of meteorology. *Quarterly Journal of the Royal Meteorological Society*, **65**, 337–58.

Salisbury, E. J. (1953). A changing flora as shown in the study of weeds of arable land and waste places. *The Changing Flora of Britain* (Ed. by J. E. Lousley), pp. 130–39. Botanical Society of the British Isles, Oxford.

Salisbury, E. J. (1961). *Weeds and Aliens*. Collins, London.

Sheail, J. (1976). *Nature in Trust*. Blackie, Glasgow.

Sheldon, J. C. & Burrows, F. M. (1973). The dispersal effectiveness of the achene-pappus units of selected Compositae in steady winds with convection. *New Phytologist*, **72**, 665–75.

Strong, D. R. Jr (1980). Null hypotheses in ecology. *Synthese*, **43**, 271–85.

Wallace, B. (1981). *Basic Population Genetics*. Columbia University Press, New York.

White, J. (1979). The plant as a metapopulation. *Annual Review of Ecology and Systematics*, **10**, 109–45.

Whittle, P. (1955). The outcome of a stochastic epidemic—a note on Bailey's paper. *Biometrika*, **42**, 116–22.

Wilson, E. O. (1976). *Sociobiology: The New Synthesis*. Belknap Press, Cambridge, Massachusetts.

# 10. GENETIC INFLUENCES ON THE DISTRIBUTION AND ABUNDANCE OF PLANTS

## JANIS ANTONOVICS[1] AND SARA VIA[2]

[1]*Botany Department, Duke University, Durham, North Carolina 27706, USA and*
[2]*Entomology Department and Section of Ecology and Systematics, Cornell University, Ithaca, New York, 14850 USA*

## SUMMARY

**1** The frequent observation of genetic differentiation amongst ecologically contrasting populations has led to the inference that the abundance and distribution of organisms has a 'genetic component'.

**2** This chapter uses theoretical models to explore the conditions under which the genetic component may or may not affect (a) population distribution along an environmental gradient, and (b) numerical abundance within a heterogeneous environment.

**3** Under the logistic model of population growth the intrinsic rate of population increase, $r$, influences population size if the population is genetically polymorphic, if density dependence acts only during one phase of the life-cycle, or if birth-rates and death-rates are affected separately by density-dependent versus independent factors.

**4** To understand empirically the effects of genetic differentiation on abundance there is a need to assess the forces of population regulation on contrasting genotypes by using reciprocal transplant experiments at a range of densities.

**5** In a heterogeneous environment, intrinsic growth-rate, carrying capacity, and the fraction of individuals entering each sub-habitat determine both gene frequency and population size.

**6** If population parameters in one sub-habitat are genetically correlated with those in a second sub-habitat, then strong evolutionary forces in one sub-habitat may dominate responses in the other. This may have large consequences for numerical abundance, with the possibility that evolutionary change results in a decline in overall population size.

## INTRODUCTION

It has become a commonplace observation that plant populations from diverse habitats and different geographical regions are genetically differentiated, and form 'ecotypes' adapted to those regions. Recently abundant evidence has also been gathered for adaptive microdifferentiation over very short distances. Thus it appears that genetic differentiation is in some measure responsible for the habitat and geographical range of a species. The idea that the distribution and abundance of a species has a genetic component is important because it not only brings into question the use of 'the species' (or its Latin binomial) as an ecologically useful unit (Harper 1982) but also extends our inquiry on the determinants of species distribution beyond simple proximal physiological properties of species as sufficient explanations, into a consideration of forces responsible for evolutionary limits to those properties (Pickett 1976; Antonovics 1976a).

While in extreme cases the role of genetic differentiation in determining species distribution cannot be denied (such as metal tolerance on toxic waste spoil, or where geographically diverse ecotypes die when reciprocally transplanted to alien sites), it is important to remember that although differential survival and reproduction may lead to adaptive genetic change, such change may not necessarily affect population size. Indeed, the result of adaptation may be a *decrease* rather than increase in population size (as in cases where selection favours larger body size or larger territories). In order to gain a full understanding of how genetic change may affect distribution and abundance of a species we need to explore models that simultaneously consider genetic and numerical change, and to use these models as a framework for empirical studies. Only in this way will we gain a realistic sense of 'where population genetics begins and ecology ends'. The relatively recent fusion of population genetics and population ecology theory has given us a basis from which to approach this question (Anderson 1971; Charlesworth & Giesel 1972; Roughgarden 1972). This chapter explores the conditions under which genetic change will affect population distribution along a gradient and numerical abundance within a heterogeneous environment. It also aims to use particular examples to illustrate both the dynamical nature of gene frequency/population size interactions, as well as the great need for experiments which depart from simple demonstrations of 'adaptive differentiation'.

## DISTRIBUTION ALONG AN ENVIRONMENTAL GRADIENT

### A simplistic scenario

Consider a continuous environmental gradient along which there are a number of discrete populations (Fig. 10.1). Along this gradient, as the environment becomes 'harsher', birth-rates decline and death-rates increase.

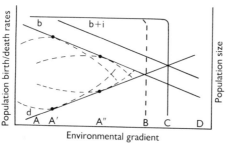

FIG. 10.1. Schematic diagram of demographic processes along an environmental gradient. b = birth-rate, d = death-rate, i = net immigration-rate. Curved lines associated with positions A′ and A″ along the environmental gradient represent values of b and d for genotypes sampled from these positions and transplanted to other positions along the gradient. The horizontal line shows the predicted population size assuming continuous population growth and equal density dependent regulation along the gradient. Beyond C population size is zero. See text for further explanation.

There is also some immigration and emigration, giving a positive net migration rate. Then the population will increase in regions A–C, but decline in regions C–D. Within A–C, the region B–C would have negative population growth if it were not for immigration from less harsh regions, circumstances referred to previously as a 'demographically marginal' population (Antonovics 1976b). Given that we have continuous population growth which follows the logistic equation, $dN/dt = rN(1-N/K)$, the population would be expected to reach some carrying capacity, $K$. In this simple deterministic situation, at equilibrium, the abundance of a species *within* the distribution range of the species (A–C) will be determined solely by $K$ (i.e. forces of density-dependent regulation) while the distributional range of the species will be determined by the intrinsic growth rate $r > 0$, i.e. where birth- and immigration-rates exceed death-rates. Actual distributional range will be affected if genotypes can be produced that have a greater birth-rate and/or lower death-rate, in the regions just beyond C.

Elsewhere, (Antonovics 1976b) I have emphasized that there are a substantial number of evolutionary forces that may retard the evolution of populations at the margins of their ecological ranges. When considered singly, the limits imposed by gene flow, small population size, coevolutionary responses of associated species, or by difficulty of generating adaptive character combinations may not be absolute. But combinations of these forces may result in relatively static species boundaries. Thus at an ecotone between a field and woodland, we could infer that evolutionary response of *Anthoxanthum* to the woodland habitat was probably limited by the difficulty of evolving unique character combinations given a small population size and

correlated loss of adaptation to other sub-habitats. This maladaptation leads
to decreases in population abundance in some sub-habitats. In this model,
the rate of adaptation to the $i$th niche depends on (1) the difference between
the phenotypic mean and the optimum phenotype, (2) the width of the fitness
function, (3) the frequency of each niche, and (4) the magnitude and signs of
the genetic variances and covariances of the character states. In general,
when constraining genetic covariances lead to loss of adaptation in one or
more of the niches, it will be the one(s) in which evolution is occurring most
slowly due to asymmetries in any of the four aspects just discussed.

The effects of genetic correlations across environment on mean fitness
(and thus on $e^r$) will be shown for a two-environment case when one
environment is rare ($q_2 = 0.3$), and when there is moderate genetic variance
and weak selection in each environment. If the character states expressed in
the two environments are uncorrelated (Fig. 10.6a), then mean fitness in both
sub-habitats increases fairly rapidly. However, if the character states in the
two sub-habitats are selected in the same direction but are negatively
genetically correlated (in Fig. 10.6b), then the mean fitness in the rare
environment can decrease dramatically as a correlated response to adaptation
in the more common environment where evolution is occurring more rapidly.
In this example, mean fitness in the rare environment drops to such a low
state that it is effectively a population sink, slowing the rate of overall
population growth ($\bar{W} = e^r$) considerably compared to what is attained in the
absence of the constraining genetic correlation.

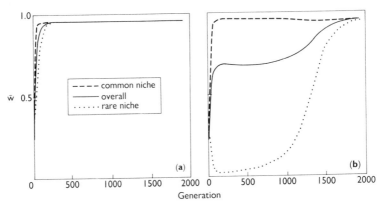

FIG. 10.6.    Effects of genetic correlation across two environments in a quantitative character
on the mean fitness ($W$) both within each sub-habitat. Phenotypic optima are $\theta_1 = 15, \theta_2 = 27$.
Relative frequencies of environments 1 and 2 are 0.3 and 0.7. (a) Genetic correlation between
character states in different environments is zero, so that adaptation can proceed independently
in each niche. (b) Genetic correlation across environments is $-0.78$.

These examples show that even for the density-independent case, the abundance of plants in different sub-habitats can critically depend on the genetic structure of plant populations. Although we have not yet integrated models of joint evolution of population parameters with models of density-dependent population regulation in each of the subhabitats, the potential consequences of such evolution can be illustrated by some examples. Thus consider that $\lambda_{AA}$ is initially 1.25 (i.e. $r_{AA}=0.25$) in both habitats; it then evolves to 4.75 in niche 2, but this evolution of high $\lambda$ in niche 2 is accompanied by a decline in $\lambda$ to 0.25 in niche 1. This approximates the situation shown in Fig. 10.6. Although the intrinsic rate of population increase ($\lambda$ or $r$) averaged over both habitats is increasing, the overall population size declines (Fig. 10.7). This is because in niche 2, the population is more severely limited by its carrying capacity and so the net increase in $\lambda$ in niche 2 does not fully compensate for the decrease in $\lambda$ in niche 1. Therefore in heterogeneous habitats, net increases in $r$ do not necessarily result in increased population size.

While the above example represents a quantum shift in the population parameters, we can study this shift in a more continuous way, by incrementing $r$ in the two habitats in each generation. When we do this, large shifts in population size may occur (Fig. 10.8a) sometimes in a very abrupt manner. This is because in the models, $\lambda$ below 1 (i.e. negative $r$) leads to shifts in patterns of density dependence, so resulting in substantial shifts in population dynamics. Thus in Fig. 10.8a, the model specification is such that when $r$ becomes negative then if the population is above carrying capacity, density effects result in zero population growth. However, change this assumption, and consider that when $r$ becomes negative, then there is no density

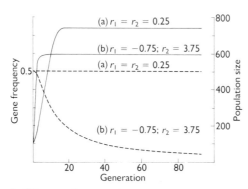

FIG. 10.7.   Effect of a difference of $r$ on population size and gene frequency; niche 1: $K_{AA}=800$, $K_{AB}=800$, $K_{BB}=700$; niche 2: $K_{AA}=700$, $K_{AB}=800$, $K_{BB}=800$). (a) $r$ in both niches $=0.25$ (b) in niche 1, $r=-0.75$; in niche 2, $r=3.75$.

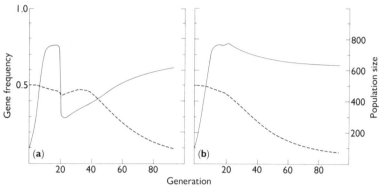

FIG. 10.8.    Effect of a gradual change in r in the two niches on population size and gene frequency. Parameters as in Fig. 10.7, except that r changes such that linearly over 100 generations, in niche 1 0.25→0.75, and in niche 2 from 0.25→3.75. (a) Density dependence in r logistic when $r > 0$. When $r < 0$, $r = -1$ when $N > K$, and $r = r(N/K - 1)$ when $N < K$. (b) Density dependence in r logistic when $r > 0$. When $r < 0$, no density dependence occurs.

regulation at all. The result is that now populations size increases (Fig. 10.8b), sometimes abruptly so. Unfortunately, it is not clear to what extent these abrupt changes are model dependent, because given the mathematical form of the model, it is difficult to maintain a functional form of the density dependence for negative values of r similar to that for positive values.

Nevertheless, these models illustrate how populations may possibly go extinct by virtue of 'over-specialization'. Thus consider a population in an area, a small proportion of which is a sub-habitat that has a very high carrying capacity, but low population growth-rate, and another much larger portion of that area which is a sub-habitat that has a low growth-rate, but also a low carrying capacity. Then in the small sub-habitat there will be strong selection to increase r. If this is negatively correlated with r in the abundant habitat, then the population will evolve as before, decreasing in the larger area, and increasing in the smaller area. However it is possible that while this is occurring, total population size actually decreases (Fig. 10.9). This is because most of the migrants enter the abundant sub-habitat, where they now have a low r, while the increased r in the rare habitat is insufficient to maintain the population in view of the large amount of emigration to the more abundant habitat. (As r continues to increase in the rare habitat, the population eventually 'recovers', but in the meantime, it has gone through a very severe bottleneck.) It is actually rather unlikely, given the unequal amount of gene flow, that selection in the rarer habitat would be sufficient to result in negatively correlated responses in the more common habitat. However, the example illustrates how relatively simple scenarios can be

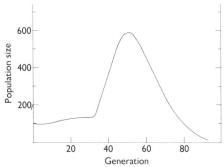

FIG. 10.9. Effect of a gradual change in *r* in the two niches on population size; niche 1: $K = 200$, initial $r = 0.4$; niche 2: $K = 800$, initial $r = -0.5$; *r* is decremented by 0.01 in niche 1 and increased by 0.05 in niche 2 each generation. From the mating pool, a fraction 0.85 enters niche 1, and a fraction 0.15 enters niche 2.

where $N_t$ is the seed pop
the net rate of increase f
cohort, and $R'_o$ is the c
proportion of every seed
germinating in Cohort 2
and under the influence

Now the effect of se
population over a perio
in $R_o$ and $R'_o$ is known
for two cohorts for seve
for a range of values of
dynamics of *Bromus* t
several successive coho
sites through three g
Quintanilla *et al.* (198
growing in an arable fi
iterated for three gene
and in each generati
obtained from the fiel
environmental condi
disease, herbivory and
$R_o$ and $R'_o$. $N_{t+3}$ wa
particular combinatic

In order to apply
actually varied in nu
were lumped into two
and one 'spring' coho
autumn and $R'_o$ for s
fitnesses in each p
recruitment. Fitness
years.

The effect on 1
germinating in sprir
to be typical, a mixe
dry site, whilst sprii
mesic and moist site
is difficult to compa
timing of germinat
Emergence in 1977
correlation with s
$r_s = 0.964$ and 0.9

constructed which show dramatic effects of genetic change on population abundance. Assessment of whether these scenarios are at all realistic will require considerably more empirical evidence than is currently available.

## CONCLUSION

The models presented in this chapter have been used to illustrate the various ways in which genetic variation in population parameters can affect numerical abundance. Examples have been chosen to emphasize that a basic knowledge of mechanisms limiting population size are essential before one can begin to translate any experimental observation of genetic differentiation into a statement about the numerical effects of such differentiation. Because knowledge of regulatory mechanisms is unavailable for any natural plant population (with few exceptions), analytical experimental approaches should involve adjusting densities of not only extant populations, but also of reciprocally transplanted populations. Input–output studies of contrasting genotypes sown at a range of densities into natural habitats can be used to predict equilibrium population sizes in the simplest situation of spatially separated, yet ecotypically differentiated populations. In more complex situations, such as heterogeneous environments, analysis of the component genetic and ecological factors determining abundance may prove difficult or impossible.

Nevertheless, other approaches are feasible. These approaches do not attempt to analyse the component phenomena, but simply ask about rates of population growth and spread given different input levels of genetic variance. Input levels of genetic variance can be controlled initially by varying the

this chapter a
demographic st
population-dyna
demographic ar
able environme

*A*

First consider
cohorts, an aut
and a spring c
This life-cycle
difference equ

FIG. 11.2 M

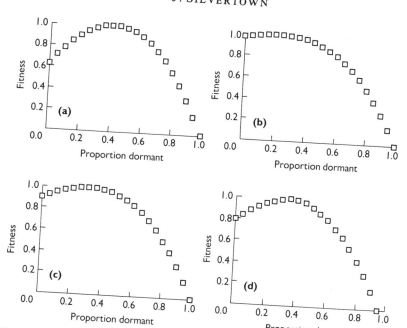

FIG. 11.7. Changes in parental fitness with changes in the proportion of dormant seeds for (a–c) *Emex australis* and (d) *Carrichtera annua*. Calculations made using Model 2, equation 2. See Table 11.1 for sources of data; (a) Ungarie A; (b) Ungarie B; (c) Two Wells.

The values of fitness plotted in Fig. 11.8 were calculated this way. The procedure did not alter the qualitative predictions of Model 2 for any species.

## DISCUSSION

Even though these results have been obtained with as few as 3 years' data, the two simple difference equation models appear to be useful in analysing the effect of germination delays upon parental fitness, and make several predictions which fit the field situation. In particular, Model 2 shows that the environmental variation measured in natural habitats is sufficient to favour the seed dormancy observed in the populations of some annual species found in them. This is very strong evidence that generational seed dormancy in *Sinapis, Emex* and *Carrichtera* is adaptive. Though the models and data are inadequate to apply a quantitative test of optimality, these results are a qualitative confirmation of the results of theory (e.g. Cohen 1966; Venable 1985; Brown & Venable 1986). It is significant that all three incorrect predictions made from Model 2 predicted that there should be no dormancy in species which in fact show it. This kind of error could result either from

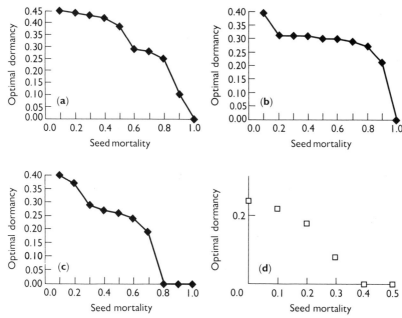

FIG. 11.8. The optimal dormant fraction for seed mortality *d* in (a–c) *Emex australis* and (d) *Carrichtera annua*. Calculations made using Model 2, equation 3. (a) Ungarie A; (b) Ungarie B; (c) Two Wells.

unrepresentativeness of the demographic data collected for the field populations in question, or from ignoring density dependence in the model used on these data.

## The effect of density dependence

The majority of studies reported density-dependent regulation occurring in the field, though this is not put explicitly into the models. A model with density dependence would favour seed dormancy more than Model 2 does because the latter model allows geometric population increase. Any life-history strategy involving a reproductive delay (such as dormancy) is at an automatic disadvantage in such a population. Model 1 also lacks density dependence but since Cohorts 1 and 2 reproduce at the same time in this model (though their germination is staggered), inserting density dependence may not alter the predictions of Model 1.

Although it is possible to see intuitively what effect density dependence would have upon the dormancy predictions of Model 2, it is not actually possible to modify the model to check this because, as already mentioned, $R_o$

incorporates all those effects, both physical and biotic, density-dependent and density-independent, that determine individual fitness. $R_o$ measures the final outcome of all density-dependent and density-independent processes (and their interactions) in a population after they have occurred. It is difficult, *post facto*, to disentangle the contribution of density-dependent and density-independent factors to the value of $R_o$ in a particular year and to the variance of $R_o$ in a series of years in the way which would be necessary to build a version of Model 2 which explicitly incorporated density dependence. One way might be to use a modified $k$-factor analysis which takes account of both temporal and spatial variation in mortality factors (Hassell 1985).

## *Dormancy without environmental variation*

Density dependence may be important in the evolution of seed dormancy for quite another reason too. It has recently been suggested by Ellner (1986) that density dependence *per se*, in the absence of any density-independent environmental variation, could favour the evolution of dormancy heteromorphism in seeds. The fitness of a plant is lowered by sibling competition among its progeny, so that a plant which spreads the germination of its seeds over several generations by making some of them dormant should have higher fitness. This should apply particularly to species with a high fecundity and poor seed dispersal because sibling competition will be strongest in these.

## *Seed dormancy and population dynamics*

The two simple population models used to explore the effect of dormancy upon fitness clearly imply that the time lags which dormancy introduces into a plant population can affect its dynamics. MacDonald & Watkinson (1981), Schmidt & Lawlor (1983) and Pacala (1986) explore this with a variety of models. As might be expected, seed dormancy is capable of damping population fluctuations. MacDonald & Watkinson's (1981) model also predicts that dormancy may sometimes be destabilizing though Pacala's (1986) model does not.

MacArthur (1972) suggested that seed dormancy (or in terms of Model 2, the value of $P$) could affect the spatial distribution of annuals. Compared with species for which $P < 1$, species with no dormancy ($P = 1$) should be patchy in distribution because, in the absence of dispersal, populations can only persist where local conditions are always good enough to maintain $R_o > 0$ and the average value of $R_o > 1$. Because of the latter condition, where these plants do occur they should be very abundant. It is interesting to note that

Andrew (1986) reports that *Sorghum intrans*, which has no dormancy, has a very patchy local distribution.

### *The effect of seed dormancy on the evolution of other traits*

This chapter concentrates upon the evolution of seed dormancy itself, but it should be noted that seed dormancy may have consequences for the evolution of other traits. Firstly, as several authors have pointed out, the existence of a pool of dormant seed in the soil acts as a 'memory' of past selection which may buffer the influence of contemporary selection and of genetic drift (Templeton & Levin 1979; Brown & Venable 1986). This is still a subject largely unexplored by fieldworkers (but see Epling, Lewis & Ball 1960 and Bosbach, Hurka & Haase 1982). Secondly, seed dormancy is a trait frequently correlated with other life-history characters (e.g. Silvertown 1980) and one which may substitute for them in its effect upon fitness. Seed dormancy and iteroparity may be related in this way. Venable & Lawlor (1980) and Klinkhamer *et al.* (1987) regard seed dormancy as dispersal in time and as an evolutionary alternative to dispersal in space. Venable & Brown (1988) model seed size, seed dispersal and seed dormancy as traits with correlated effects upon fitness. Ritland (1983) modelled the joint evolution of seed dormancy and flowering time. In conclusion it is clear that seed dormancy is a plant trait with many important consequences which are relatively easy to investigate through a combination of tractable population models and existing field techniques.

## ACKNOWLEDGMENTS

I thank the Section of Ecology & Systematics, Cornell University for their hospitality and support during the initiation of this study and I am grateful to Carlos Cordero, Mike Hutchings, Bridget Smith and Larry Venable who commented upon the manuscript.

## REFERENCES

**Andrew, M. H. (1986).** Population dynamics of the tropical annual grass *Sorghum intrans* in relation to local patchiness in its abundance. *Australian Journal of Ecology,* **11**, 209–17.

**Baskin, J. M. & Baskin, C. C. (1985).** Life cycle ecology of annual plant species of cedar glades of Southeastern United States. *The Population Structure of Vegetation* (Ed. by J. White), pp. 371–98. Junk, Dordrecht.

**Bosbach, K., Hurka, H. & Haase, R. (1982).** The soil seed bank of *Capsella bursa-pastoris* (Cruciferae)—its influence on population variability. *Flora,* **172**, 47–56.

**Brown, J. S. & Venable, D. L. (1986).** Evolutionary ecology of seed-bank annuals in temporally varying environments. *American Naturalist,* **127**, 31–47.

**Cohen, D. (1966).** Optimising reproduction in a randomly varying environment. *Journal of Theoretical Biology*, **12**, 119–29.

**Edwards, M. (1980).** Aspects of the population ecology of charlock. *Journal of Applied Ecology*, **17**, 151–71.

**Ellner, S. (1985a).** ESS germination strategies in randomly varying environments I. Logistic-type models. *Theoretical Population Biology*, **28**, 50–79.

**Ellner, S. (1985b).** ESS germination strategies in randomly varying environments II. Reciprocal yield-law models. *Theoretical Population Biology*, **28**, 80–116.

**Ellner, S. (1986).** Germination dimorphisms and parent–offspring conflict in seed germination. *Journal of Theoretical Biology*, **123**, 173–85.

**Epling, C., Lewis, H. & Ball, E. M. (1960).** The breeding group and seed storage: a study in population dynamics. *Evolution*, **14**, 238–55.

**Fernandez-Quintanilla, C., Navarrette, L., Andujar, J. L. G., Fernandez, A. & Sanchez, M. J. (1986).** Seedling recruitment and age-specific survivorship and reproduction in populations of *Avena sterilis* L. ssp. *ludoviciana* (Durieu) Nyman. *Journal of Applied Ecology*, **23**, 945–55.

**Garwood, N. C. (1982).** Seasonal rhythm of seed germination in a semideciduous tropical forest. *Tropical Forest: Seasonal Rhythms and Long-term Changes* (Ed. by C. G. Leigh, Jr., A. S. Rand & D. M. Windsor), pp. 173–85. Smithsonian Institution Press, Washington D.C.

**Garwood, N. C. (1983).** Seed germination in a seasonal tropical forest in Panama: a community study. *Ecological Monographs*, **53**, 159–81.

**Grime, J. P. (1979).** *Plant Strategies and Vegetation Processes*. Wiley, Chichester.

**Hassell, M. P. (1985).** Insect natural enemies as regulating factors. *Journal of Animal Ecology*, **54**, 323–34.

**Jain, S. K. (1976).** Patterns of survival and microevolution in plant populations. *Population Genetics and Ecology* (Ed. by S. Karlin & E. Nevo), pp. 49–89. Academic Press, New York.

**Klinkhamer, P. G. L., de Jong, T. J., Metz, H. A. J. & Val, J. (1987).** Life history tactics of annual organisms: the joint effects of dispersal and delayed germination. *Theoretical Population Biology*, **32**, 127–56.

**Loria, M. & Noy-Meir, I. (1979).** Dynamics of some annual populations in a desert loess plain. *Israel Journal of Botany*, **28**, 211–25.

**MacArthur, R. H. (1972).** *Geographical Ecology*. Harper & Row, New York.

**MacDonald, N. & Watkinson, A. R. (1981).** Models of an annual plant population with a seed bank. *Journal of Theoretical Biology*, **93**, 643–53.

**Mack, R. N. & Pyke, D. A. (1983).** The demography of *Bromus tectorum*: variation in time and space. *Journal of Ecology*, **71**, 69–93.

**Mack, R. N. & Pyke, D. A. (1984).** The demography of *Bromus tectorum*: the role of microclimate, grazing and disease. *Journal of Ecology*, **72**, 731–48.

**Pacala, S. W. (1986).** Neighborhood models of plant population dynamics 4. Single-species and multispecies models of annuals with dormant seeds. *American Naturalist*, **128**, 859–78.

**Ritland, K. (1983).** The joint evolution of seed dormancy and time of flowering in annual plants living in a variable environment. *Theoretical Population Biology*, **24**, 213–43.

**Schmidt, K. P. & Lawlor, L. R. (1983).** Growth rate projection and life history sensitivity for annual plants with a seed bank. *American Naturalist*, **121**, 525–39.

**Silvertown, J. (1980).** Seed size, life span, and germination date as coadapted features of plant life history. *American Naturalist*, **118**, 860–64.

**Silvertown, J. (1984).** Phenotypic variety in seed germination behavior: the ontogeny and evolution of somatic polymorphism in seeds. *American Naturalist*, **124**, 1–16.

**Symonides, E. (1974).** Populations of *Spergula vernalis* Willd. on dunes in the Torun Basin. *Ekologia Polska*, **22**, 379–416.

**Symonides, E. (1977).** Mortality of seedlings in natural psammophyte populations. *Ekologia Polska*, **25**, 635–51.

Symonides, E. (1979). The structure and population dynamics of psammophytes on inland dunes II. Loose-sod populations. *Ekologia Polska*, **27**, 191–234.

Symonides, E. (1983a). Population size regulation as a result of intra-population interactions. I. Effect of density on the survival and development of individuals of *Erophila verna* (L.) C.A.M. *Ekologia Polska*, **31**, 839–81.

Symonides, E. (1983b). Population size regulation as a result of intra-population interactions. II. Effect of density on the growth rate, morphological diversity and fecundity of *Erophila verna* (L.) C.A.M. individuals. *Ekologia Polska*, **31**, 888–912.

Templeton, A. R. & Levin, D. A. (1979). Evolutionary consequences of seed pools. *American Naturalist*, **114**, 232–49.

Venable, D. L. (1985). The evolutionary ecology of seed heteromorphism. *American Naturalist*, **126**, 577–95.

Venable, D. L. & Brown, J. S. (1988). The selective interactions of dispersal, dormancy, and seed size as adaptations for reducing risk in variable environments. *American Naturalist*, **131**, 360–84.

Venable, D. L. & Lawlor, L. (1980). Delayed germination and dispersal in desert annuals: escape in space and time. *Oecologia*, **46**, 272–82.

Weiss, P. W. (1981). Spatial distribution and dynamics of populations of the introduced annual *Emex australis* in south-eastern Australia. *Journal of Applied Ecology*, **18**, 849–64.

# 12. POPULATION DYNAMICS OF ANNUAL PLANTS

## E. SYMONIDES

*Institute of Botany, Warsaw University, Al. Ujazdowskie 4, 00478 Warsaw, Poland*

## SUMMARY

**1** The aim of this paper is to survey the life-cycle characteristics of annual plants, together with the factors determining their abundance and dynamics. It is demonstrated, using a range of examples, that annual plants display considerable variation in their life-histories and dynamics. The only common feature to all annual plants is that the vegetative individuals live for less than 12 months.

**2** Some annuals are truly semelparous, whilst others are better thought of as uniseasonally iteroparous. Depending on the species and environment, annual plants may produce only a few or many seeds. These may or may not be capable of long distance dispersal and may or may not possess long term dormancy capabilities. Seeds may either germinate simultaneously or over an extended period.

**3** Populations of seedlings may consist of either discrete or overlapping cohorts, and display a wide variety of survivorship curves ranging from extreme Deevey type I to type III.

**4** Some demographic and life-history characters seem to be characteristic of species, whilst others may vary with the environment. Among annuals there are many exceptions to the often quoted correlations between individual fecundity, seed longevity and juvenile mortality, predicted in theoretical models.

**5** Population regulation operates via density-dependent mortality, density-dependent fecundity or both processes. No uniform pattern of seasonal or long-term dynamics is common to all species. The main difference between the dynamics of annuals, biennials and perennials lies in the different time scale of events.

## INTRODUCTION

'Plants stand still and wait to be counted . . .' Harper (1977). Simple counts of the number of individuals, however, fail to reveal the dynamic nature of plant populations: new recruits appear, individuals may die, and others survive to produce offspring. These events are particularly well defined in populations of annual plants, in which the development of individuals from seedlings to generative plants occurs within a year and often within only a few months or even weeks. It might at first appear then that the population dynamics of all annuals would be rather similar. Detailed studies on seed dispersal, dormancy and germination, seedling emergence, plant survival and reproduction, conducted over the last few years, however, show that this is not the case, and provide an understanding of the range of population dynamics exhibited by annual plant species.

This chapter is concerned with a survey of crucial events in the life-cycle of annual plants. The main aims of this review are to show that (a) only a few demographic and life-history characteristics are common to most species, (b) there is no uniform pattern in either the seasonal or annual dynamics of plants, and that (c) basic differences between the dynamics of annuals and perennials lie only in the time scale of life-history events. Of necessity only a selection of examples can be used to illustrate these various points. The article consists of two parts: the first is on the dynamics of seeds and the second on the dynamics of growing plants. In both the emphasis is on what determines the fate of individuals. The question of why population size is as large as it is and why, in a single community, the population size of one species is higher or lower than that of another have generally been neglected, but see Watkinson (1985) and Grubb, Kelly & Mitchley (1982) respectively.

## DYNAMICS OF SEEDS

### Dispersal in space

The evolution of dispersal can be expected to be favoured where there is the chance of colonizing a site more favourable than the one that is presently inhabited (Gadgil 1971). Theoretically, a tendency to disperse should be common among annuals, as most species live in unfavourable environments, very often unstable and unpredictable. In fact, both dispersal behaviour and its effectiveness are strongly diversified in annuals, and in general, three groups of species may be distinguished among them.

The first group of species lack specialized structures for dispersal and consequently possess very poor seed dispersal. For these plants the distance that seeds fall is determined primarily by their weight, the height of seed release, environmental conditions (e.g. wind direction and velocity) and vegetation cover. Poor seed dispersal has been found in annuals from a range of habitats including forests—*Floerkea proserpinacoides* (Smith 1983a), closed grasslands—*Euphrasia stricta* (During *et al.* 1985), open grasslands—*Phlox drummondii* (Levin & Clay 1984), coastal dunes—*Vulpia fasciculata* (Watkinson 1978a), inland dunes—*Erophila verna* (Symonides 1984) and deserts—*Lappula redowskii* (Freas & Kemp 1983). The seeds of some of these species are relatively heavy (*Floerkea, Vulpia*), whilst those of others are exceptionally light (*Erophila, Euphrasia*). Presumably poor seed dispersal in all of these species is associated with the fact that suitable sites occur reliably in the immediate environment of the parent plant (Watkinson 1978a).

Conversely, seeds of annuals of the second group either possess special structures for dispersal (pappus, wings, elaiosomes), or are light enough to be easily carried by wind or water. As a consequence they may be dispersed over long distances. For example the diaspores of *Spergula vernalis* (= *S.morisonii*) (Symonides 1974), *Taeniatherum asperum* (Bartolome 1979), *Pectis angustifolia* (Freas & Kemp 1983), *Lactuca serriola* (Carter & Prince 1985) and *Salicornia europaea* (Watkinson & Davy 1985) are all known to be capable of being dispersed over long distances. All of these species are characteristic of temporary or severely disturbed habitats. Yet even amongst such species the fate of the large majority of seeds is probably to be dispersed not far from the parent plant. In general though, far less is known about the long-distance than short-distance dispersal of seeds. Undoubtedly long distance dispersal between local populations can provide an important source of genetic novelty. Also, long-distance dispersal will enable species to colonize new habitats rapidly and escape from the pressures of overcrowding and herbivore and pathogen attack.

Annual plants of the third group produce both dispersing and non-dispersing diaspores. The difference between seed types, as in the case of *Chenopodium album* and many other members of the Chenopodiaceae, Compositae, Cruciferae and Gramineae, usually results from the production of seeds with different weights. This is a maternal effect and is not associated with genetic differences between seeds. Sometimes seed dimorphism is associated with differences in the structure of diaspores. For example, *Heterotheca latifolia* produces two types of achene: some are light with a thin coat and have a pappus, others are heavy, with a thick coat and have no pappus (Venable & Levin 1985). Fruits of *Cakile edentula* consist of two segments: one of them is easily broken off from the parent plant and provides a long-distance dispersal unit, whilst the other remains attached to the parent plant and is likely to be buried in a suitable habitat around the base of the parent plant (Keddy 1982). However, it is the geocarpic annuals such as *Gymnarrhena micrantha* and *Emex spinosa* that perhaps possess the most extreme dispersal dimorphism. Besides producing heavy, subterranean diaspores, plants may also produce many light, easily dispersed, aerial seeds under favourable conditions (Koller & Roth 1964; Evenari, Kadouri & Gutterman 1977). The roles of dispersing and non-dispersing diaspores are different: the former are to colonize new habitats whilst the latter are to exploit safe sites close to the dead parent plant. Usually the stay-at-home seed is produced from flowers below ground or from unopened, cleistogamous flowers, whilst the seeds that are dispersed more are often from chasmogamous flowers. Thus seeds with recombinant and potentially novel genotypes have a greater tendency to disperse.

The pattern of seed dispersal and dormancy both determine the potential population in a given habitat. For plant populations in which there is no carry over of dormant seeds from one generation to the next, the abundance of each subsequent generation is determined, to a large extent, by the number and spatial distribution of seeds produced in the previous generation (Fig. 12.1). That is not to say that the number of reproducing plants will be directly proportional to the number of seeds. The aggregation of seeds may result in both density-dependent mortality and fecundity (Symonides 1983a,b, 1984) affecting both population size in the current year and future generations. In a study of the factors determining the abundance of *Cakile edentula* on a sand dune gradient Watkinson (1985) showed that the population as a whole was regulated by the density-dependent control of fecundity on the seaward end

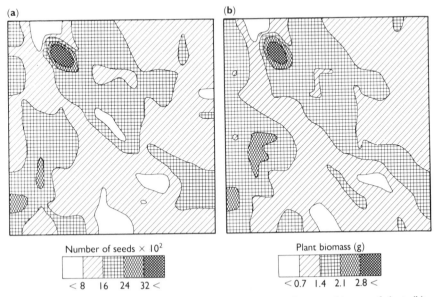

FIG. 12.1. The relationship between the seed rain (a) and the subsequent biomass of plants (b) in a population of *Salicornia patula* (redrawn with permission from Wilkoń-Michalska 1976).

of the gradient, but that the actual level of abundance along the gradient was determined primarily by the pattern of seed dispersal. Indeed without the landward migration of seeds, populations would not persist in those areas where the species is most abundant in the field.

## *Dormancy: dispersal in time*

Seed dormancy (enforced, induced or innate) is a widespread phenomenon amongst annuals. Until recently, prolonged dormancy and a persistent seed bank were considered to be characteristic of all annual plants. Nevertheless, it may be inferred from a number of recent studies that, as in the case of dispersal in space, three groups of annuals may be distinguished in respect of dispersal in time.

Seeds of the first group of species are characterized by periodic germination over many years. This group includes many weeds (e.g. Maillette 1985; Maxwell *et al.* 1986), but also desert plants (Friedman & Elberse 1976), some dune annuals (Symonides 1978a), and plants of ruderal (Arthur, Gale & Lawrence 1973) and some saline habitats (Watkinson & Davy 1985). All of these species are characteristic of hazardous environments in which the probability of successful reproduction is low. The presence of a seed bank can then be regarded as a bet-hedging tactic in an unpredictable environment (Cohen 1966; Leon 1985). The majority of persistent seed banks consist of buried seeds (Grime 1979) but in the case of the desert annual *Anastatica hierochuntica* the dry, hardened and curled-up skeletons of the dead parent plants serve to protect the seeds, which are released gradually (Friedman & Stein 1980).

Another large group of annuals contrasts strongly with the first in that they possess only a transient seed bank and have no carry-over of seed from one year to the next. Again these species are characteristic of various habitats but they can all be considered as largely predictable in that there is a high annual probability of successful reproduction. Thus there is no need for bet-hedging. Annuals without a seed bank include *Cerastium atrovirens* (Mack 1976), *Euphrasia stricta* (During *et al.* 1985), *Impatiens capensis* (Winsor 1983), *Lactuca serriola* (Marks & Prince 1981) and *Phlox drummondii* (Leverich & Levin 1979). In *Vulpia fasciculata* (Watkinson 1978b) less than 0.5% of the seeds survive in soil for more than a year, and in *Floerkea proserpinacoides* less than 0.3% (Smith 1983a).

The species of the third group have both dormant and non-dormant seeds, the proportion depending upon the species and habitat. These two types of seed may be morphologically identical, as in the case of *Erucastrum gallicum* (Klemow & Raynal 1983) and *Papaver dubium* (Arthur, Gale &

Lawrence 1973). However, more often they are dimorphic. In the case of *Phleum arenarium* it is only the small seeds that enter the seed bank (Ernst 1981), whilst in *Heterotheca latifolia* only the ray achenes are viable for longer than a year (Venable & Levin 1985).

It has already been stated that long-term seed dormancy can be seen in terms of bet-hedging by the individual plant in habitats where there is a high probability of seedling failure in any given year. This idea is largely discussed by Cohen (1966). But the presence of a large reservoir of seeds in the soil also has the consequence of stabilizing population fluctuations from year to year and preventing rapid directional selection (Epling, Lewis & Bell 1960). It has also been suggested by Venable & Lawlor (1980) that there will be a negative correlation between dispersal in space and dispersal in time. That is, plants with high dispersal will have a high germination fraction. This correlation appears to hold for species producing two dispersal morphs on each plant, but as might be expected does not always hold where comparisons are made between species. For example, the dispersing seeds of *Spergula vernalis* form a persistent bank, while non-dispersing seeds of its neighbour, *Erophila verna*, are short-lived (Symonides 1974, 1984).

### Seed and seedling mortality

Understanding what determines the fates of seeds is central to understanding what determines the potential size of populations. But whilst it is known that seed losses in some plant populations are small and in others big, the reasons are rarely clear. Low seed mortality has been found mainly in species with low fecundity, poor dispersal and heavy seeds. This group includes geocarpic annuals (approximately 100% of their subterranean seeds survive), and also *Floerkea proserpinacoides* and *Vulpia fasciculata* (usually less than 20% of seeds die). However, seed losses are, as a rule, much higher. In many species 80–99% of seeds die before entering the soil, as in *Koenigia islandica* (Reynolds 1984), *Sedum smallii* and *Minuartia uniflora* (Sharitz & McCormick 1975), *Phlox drummondii* (Leverich & Levin 1979) and *Erophila verna* (Symonides 1984). A large fraction of seeds may also die in the soil, if they are, for example, washed down too deeply (Maxwell *et al.* 1986) or as a result of seed predation on the soil surface if seed predators are particularly abundant (Borchert & Jain 1978).

From, unfortunately, very scarce data it may also be inferred that the mortality of individuals at the time of germination may also be very variable amongst species. Approximately 90% of individuals in populations of *Salicornia patula* die at this stage (Wilkoń-Michalska 1976) whilst the corresponding figures are 40% in *Floerkea proserpinacoides* (Smith 1983a),

12–28% in *Vulpia fasciculata* (Watkinson 1978b), and less than 1% for the subterranean seeds of geocarpic annuals (Evenari, Kadouri & Gutterman 1977).

## DYNAMICS OF GROWING PLANTS

### *Seed germination and seedling emergence*

The fraction of seeds germinating and the phenology of seedling emergence vary considerably between species, even amongst the annuals within a single community (cf. Bartolome 1979; Baskin & Baskin 1985). For some annuals the fraction of seeds germinating is very high. For example, 90–100% of the seeds of the following species germinate under favourable conditions: *Floerkea proserpinacoides* (Smith 1983a), *Pectis angustifolia* (Freas & Kemp 1983), *Salicornia patula* (Wilkoń-Michalska 1976) and *Vulpia fasciculata* (Watkinson 1978b). In others this fraction is considerably lower. Under natural conditions the proportion of seeds germinating in populations of *Phlox drummondii* does not exceed 10–20% (Levin & Clay 1984). The corresponding figure is 5–6% in *Erophila verna* (Symonides 1984) and 1.6% in *Spergula vernalis* (Symonides 1974). The fraction of germinating seeds is neither correlated with species fecundity, nor with seed weight or longevity: *V. fasciculata* differs in respect of these characters from *Pectis angustifolia*, and *Phlox drummondii* from *E. verna*. Among species producing dimorphic seeds the smaller seeds generally exhibit a remarkably lower germination capacity than the larger ones. For example, the respective germination percentages are 6.4% and 68.5% in *Ludwigia leptocarpa* (Dolan 1984) and 4.1% and 30.5% in *Heterotheca latifolia* (Venable & Levin 1985).

The fraction of seeds germinating determines the initial size of a population, but the subsequent seasonal dynamics are governed mainly by the phenology of seedling emergence. Both survival and fecundity depend on the timing of this event. Usually, the time of germination is determined by physical conditions, although it sometimes depends on the time of seed maturation, as in the case of *Geranium carolinianum* (Roach 1986). A major distinction can be made between those annuals that germinate in either the autumn or the spring, the so called winter and summer annuals. The former are common in a range of community types (cf. Schaal & Leverich 1982; Baskin & Baskin 1985; Watkinson & Davy 1985). Silvertown (1981), in an analysis of germination time in calcareous grasslands, suggested that the prevalence of autumn germination among annuals is a consequence of their characteristically small seed size. However, this cannot be considered as a general rule for all habitats as there is no such correlation amongst arable

TABLE 12.1. A comparison between the type of survivorship curve and individual fecundity in some annual species from various environments.

| Species | Survivorship Deevey type | Seeds per plant | Environment | References |
|---|---|---|---|---|
| *Vulpia fasciculata* | I | 1.7 | Coastal dunes | Watkinson & Harper 1978 |
| *Cerastium atrovirens* | I | 7.3 | Coastal dunes | Mack 1976 |
| *Floerkea proserpinacoides* | I | <23 | Deciduous forests | Smith 1983b,c |
| *Raphanus raphanistrum* | I | 1000–2240 | Wetlands | Stanton 1984 |
| *Bromus tectorum* | I–III | 250–5500 | Grasslands | Mack & Pyke 1983 Young & Evans 1985 |
| *Erophila verna* | I–III | 35–1280 | Grasslands | Symonides 1983a,b |
| *Polygonum confertiflorum* | II | 2.5 | High-montane zone | Reynolds 1984 |
| *Polygonum douglasii* | II | 5.8 | High-montane zone | Reynolds 1984 |
| *Koenigia islandica* | II | 8.2 | High-montane zone | Reynolds 1984 |
| *Erigeron canadensis* | II–III | 371–74 700 | Old-fields | Regher & Bazzaz 1979 |
| *Ludwigia leptocarpa* | II–III | 1200–3180 | Stream-banks | Dolan & Sharitz 1984 |
| *Spergula vernalis* | II–III | 91–653 | Inland dunes | Symonides 1974 |
| *Cerastium semidecandrum* | III | 28–320 | Grasslands | Symonides 1979b |
| *Androsace septentrionalis* | III | 952–1301 | Grasslands | Symonides 1979a |
| *Sedum smallii* | III | 114 | Granite out-crops | Sharitz & McCormick 1973 |
| *Minuartia uniflora* | III | 305 | Granite out-crops | Sharitz & McCormick 1973 |

conditions but also with seed type. During long drought, populations of the geocarpic annuals *Emex spinosa* and *Gymnarrhena micrantha* consist exclusively, or nearly exclusively, of cleistogamous individuals, populations of which conform to Deevey type I survivorship. In years with fairly abundant rainfall, however, type II or type III survivorship occurs, depending on the proportion of chasmogamous individuals in the population. Many of the latter usually die as seedlings (Zeide 1978). The survival of the geocarpic annual *Amphicarpum purshii*, has also been shown experimentally by Cheplick & Quinn (1982) to vary with soil moisture and seed type (Fig. 12.3b).

It is not only the weather that affects the pattern of survivorship in populations of annuals. Under similar weather conditions, local populations of a single species may differ in their survivorship type, depending on the habitat conditions. For example, the survivorship pattern of *Spergula vernalis* in the early successional stages of a dune system corresponds to a type II curve, whereas on the fixed dunes it accords to a type III (Symonides 1974). Conversely in *Veronica dillenii*, fewer juvenile individuals survive in open than closed grasslands (Fig. 12.3c). Studies on *Erophila verna* have also shown that the shape of a survivorship curve for a population depends also upon its

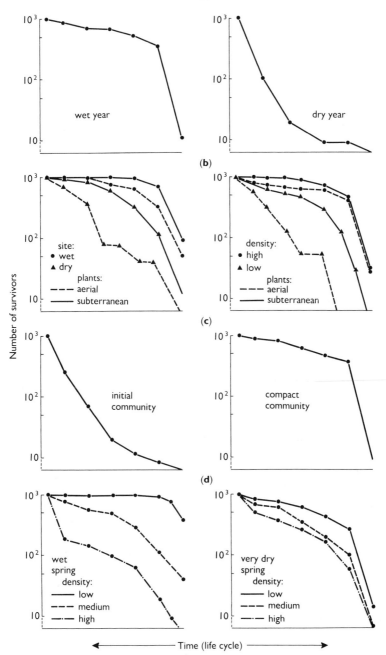

FIG. 12.3. The effect of various factors on survivorship in populations of (a) *Erucastrum gallicum* (drawn from data in Klemow & Raynal 1983); (b) *Amphicarpum purshii* (drawn from data in Cheplick & Quinn 1982); (c) *Veronica dillenii* (unpublished data); (d) *Erophila verna* (drawn from data in Symonides 1983a).

density and that the influence of density on survival is also affected by the weather (Fig. 12.3d). Density and soil moisture also interact to determine the survival of *Amphicarpum purshii* (Cheplick & Quinn 1982).

Where several distinct cohorts are produced within a growing season, their survivorship may either be similar as in the case of *Ludwigia leptocarpa* (Dolan & Sharitz 1984) or vary considerably depending upon the weather conditions. For example, in *Bromus tectorum* high survival (Deevey type I) characterizes cohorts of seedlings that emerge in spring, regardless of soil moisture, whereas survival of the autumn emerging cohorts is much more variable ranging from Deevey type I to type III depending upon the weather conditions (Mack & Pyke 1983).

### Seed production

The total seed production in populations of annuals varies between over a million per $m^2$ in *Salicornia patula* to just a few seeds in *Emex spinosa* and *Gymnarrhena micrantha*. The size of the seed crop depends upon the number of plants surviving to the reproductive period and their fecundity.

The fraction of individuals that survive to the flowering period and then produce seed is extremely variable in populations of individual species. Irrespective of the population size, this value ranges from 100% in *Androsace septentrionalis* (Symonides 1979a) and *Euphrasia stricta* (During *et al.* 1985), to merely a few per cent in *Ludwigia leptocarpa* (Dolan & Shartiz 1984) and *Soliva pterosperma* (Maxwell *et al.* 1986). The total number of seeds produced by individual plants may also be extremely variable, ranging from one to three subterranean diaspores in geocarpic species (Koller & Roth 1964) or a similar number of seeds in *Vulpia fasciculata* (Watkinson & Harper 1978) to hundreds of thousands of seeds in *Salicornia patula* (Wilkoń-Michalska 1976) and *Chenopodium album* (Harper 1977). Both parameters can vary between years, between successive cohorts and also within a single cohort.

Theoretically, high fecundity should be typical of annual plants. Optimal life-history models make some important predictions about the life-history characteristics that might be expected to maximize individual fitness in annual organisms. These include four elements (1) a rapid switch from vegetative to reproductive growth late in the life-cycle (Vincent & Pulliam 1980). Whilst such a switch has been found in some annuals such as *Echinocystis lobata* (Silvertown 1985) and *Aira caryophyllea* (Rozijn & van der Werf 1986), others like *Aira praecox* (Rozijn & van der Werf 1986) continue vegetative growth during the reproductive phase; and (2) translocation of nutrients from vegetative to reproductive organs (Chapin 1980). Again this occurs in some annuals such as *Phleum arenarium* (Ernst

1983), but not in others, e.g. *Cassia fasciculata* (Kelly 1986); (3) high reproductive effort (King & Rougharden 1982). In fact this is extremely variable among annuals and may range from over 60% as in *Spergula arvensis* (Maillette 1985), to as low as 5% in *Medicago laciniata* (Friedman & Elberse 1976). Moreover, reproductive effort may vary significantly among local populations of a single species, even in successive generations (e.g. Symonides 1974, 1979a; Hickman 1975). It should be noted, however, that the above determinations of reproductive effort are based on biomass or energy allocation. It might be that nutrient allocation is more appropriate; (4) maximization of seed number within the available resources, resulting in low individual seed weight (Baker 1974). Again this is true for many annuals but others, such as *Floerkea proserpinacoides, Vulpia fasciculata* and geocarpic annuals under unfavourable conditions, produce just a few relatively heavy seeds.

Annual plants are typically regarded as semelparous in that reproduction is followed by death within a single season. Such a categorization of annuals, however, hides a diversity of reproductive schedules. Some annuals are indeed truly semelparous in that they cluster their progeny in one, condensed reproductive phase, but others produce seed gradually over an extended period or in several discrete flushes as a result of either the production of a succession of monocarpic shoots or the persistent reproductive activity of polycarpic shoots (Watkinson & White 1986). In a sense they are uniseasonally iteroparous (Kirkendall & Stenseth 1984).

## Seasonal changes in population size

The pattern of seasonal change in population size varies among species with the number and relative abundance of emerging cohorts and with the death rate of seedlings from those cohorts. On the basis of the data in the literature on seedling emergence and survival it is possible to construct ten model curves for the seasonal population dynamics of annual plants (Fig. 12.4). The first four characterize populations composed only of one cohort, the next two of populations with two cohorts, and the last four of populations with more than two cohorts. Their description is given below:

1 The number of individuals rises rapidly to a seasonal maximum and then decreases (Fig. 12.4a). The greater the synchrony in germination and the higher the juvenile mortality, the more rapid are the changes in population abundance. Examples include *Salicornia patula* (Wilkoń-Michalska 1976), *Erophila verna* (Symonides 1983a), *Impatiens capensis* (Winsor 1983) and *Euphrasia stricta* (During *et al.* 1985).

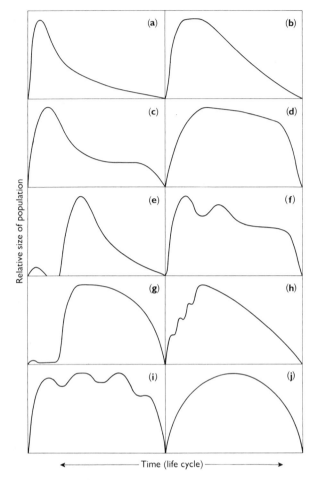

FIG. 12.4. Model curves of seasonal population dynamics. Populations composed of (a–d) one cohort, (e–f) two cohorts and (g–j) more than two cohorts. For explanations see text.

**2** A short phase of relatively constant population size occurs between the periods of population increase and decrease (Fig. 12.4b). This occurs during winter, or more rarely during spring to summer, when the small, juvenile plants are dormant. Examples include the winter annuals *Spergula vernalis* (Symonides 1974) and *Bromus mollis* (Wu & Jain 1979), and the summer annual *Trichostema lanceolata* (Heisey & Delwiche 1985).

**3** The period of rapid population increase is followed by a fairly rapid decrease in size and then stasis during flowering and fruiting (Fig. 12.4c).

Examples include *Trifolium arvense* (Symonides 1978b) and *Impatiens noli-tangere* (Falencka 1983).

**4** Following germination the number of individuals then declines only slowly over the rest of the life-cycle until death follows reproduction for the rest of the plants (Fig. 12.4d). Such a pattern results from high juvenile survival, often associated with winter dormancy in winter annuals. It is an extremely common pattern of seasonal population dynamics. Examples include *Cerastium atrovirens* (Mack 1976), *Floerkea proserpinacoides* (Smith 1983b), *Phlox drummondii* (Leverich & Levin 1979), *Vulpia fasciculata* (Watkinson & Harper 1978) and probably also geocarpic annuals if their populations are dominated by cleistogamous individuals.

**5** A bimodal pattern of abundance can be expected if the first cohort of plants produced in autumn is small (Fig. 12.4e) or if few of the autumn cohort of seedlings survive. The essential pattern of population dynamics can then be expected to be governed primarily by the fate of the second, spring cohort. Examples include both winter and summer annuals and include *Androsace septentrionalis* and *Cerastium semidecandrum* (Symonides 1979a, b).

**6** The seasonal change in the pattern of abundance of this group of plants is clearly bimodal (Fig. 12.4f). The first peak is only slightly higher than the second, despite the larger size of the early cohort. The first peak results from the high mortality of seedlings and virginile plants in the first cohort. Examples include the summer annual *Leavenworthia stylosa* (Baskin & Baskin 1972) and the winter annual *Lactuca serriola* (Marks & Prince 1981).

**7** In spite of three seedling cohorts only one distinct peak of abundance occurs in this group of plants (Fig. 12.4g). The first, small cohort dies almost entirely. The seasonal population dynamics is thus governed by the fate of the second cohort. Seedlings emerge gradually, but the considerable decrease in their abundance is then moderated by the emergence of the third cohort, which is relatively small. Examples include *Erucastrum gallicum* (Klemow & Raynal 1983).

**8** Abundance increases in a series of waves with the emergence and death of subsequent seedling cohorts and is then followed by a long decrease in the population size (Fig. 12.4h). Examples include *Ludwigia leptocarpa* (Dolan & Sharitz 1984), and probably also *Taeniatherum asperum* (Bartolome 1979) and *Erigeron canadensis* (Regher & Bazzaz 1979).

**9** and **10** For plants that produce many cohorts of seedlings the seasonal changes in population size are characterized, either by irregular fluctuations (Fig. 12.4i), or a steady increase and then decline in numbers (Fig. 12.4j). Examples include *Bromus tectorum* (Mack & Pyke 1983) and *Poa annua* (Law 1981).

Irrespective of the fundamental type of seasonal dynamics exhibited by a particular species variations in the basic pattern can be expected from year to year depending on the conditions (Fig. 12.5).

## Population regulation

Why a particular number of individuals occurs in a population can be explained by studying the numbers of births, deaths, immigrants and

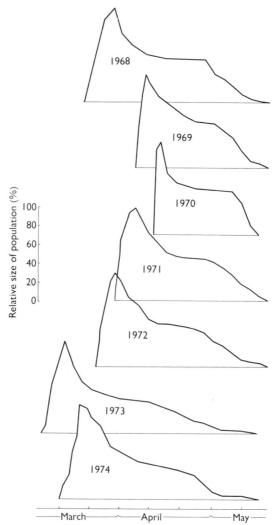

FIG. 12.5. Seasonal changes in the relative size of *Erophila verna* populations in 7 successive years (drawn and modified from Symonides 1983a).

emigrants and the way that these various parameters are affected by both density-independent and density-dependent processes (Watkinson 1985). It is, however, only density-dependent processes that can regulate population size and be invoked to explain why population densities persist within narrow bounds and why the growth-rate of populations decrease as population size increases. Observations to date indicate that population regulation operates primarily through density-dependent processes that affect mortality and fecundity. It is possible though that density could regulate population size through effects on seed dispersal.

Density-dependent mortality has been recorded in annuals from a range of habitats in both the seed and vegetative phases of the life-cycle. Sometimes, density-dependent mortality may reduce a wide range of seedling densities to a much smaller range of adult densities. For example, at a density of 5000 seedlings m$^{-2}$ in a population of *Soliva pterosperma* about 50% of individuals died in 2 weeks, whereas none died at a density of 100 m$^{-2}$ (Maxwell *et al.* 1986). Similarly in *Erophila verna*, 50% of seedlings died in a population of 6000 seedlings m$^{-2}$ in a period of 2 to 3 weeks, whereas only a few percent died where the density was 100–200 m$^{-2}$ (Symonides 1983a). Mortality as a result of crowding may result from factors both internal and external to the population. Thus intraspecific competition for limited resources may result in density-dependent mortality through self-thinning, whilst the level of crowding can also be expected to affect mortality through the level of herbivore and pathogen activity.

Density-dependent mortality regulates population size with varying degrees of effectiveness in different species and local populations of a single species. It has been found to be the only agent regulating population size in populations of *Cerastium semidecandrum* (Symonides 1979b) and *Lactuca serriola* (Marks & Prince 1981). But crowding may not only affect mortality, it may also affect the proportion of plants that flower and also the number of seeds produced per plant (see Figs 12.3 and 12.6). The negatively density-dependent control of reproduction has been recorded in a wide range of annual plants (Watkinson 1985) including *Floerkea proserpinacoides* (Smith 1983c), *Polygonum confertiflorum* (Reynolds 1984) and *Vulpia fasciculata* (Watkinson & Harper 1978). In all such species the plastic response of plants to density, as a result of intraspecific competition, produces a relatively constant seed yield, irrespective of the number of fruiting individuals.

Both density-dependent mortality and fecundity have been observed in populations of *Erophila verna* (Symonides 1983a, b), *Impatiens capensis* (Winsor 1983), *Phlox drummondii* (Leverich & Levin 1979) and *Spergula arvensis* (Maillette 1985). The relative importance of these different density-dependent processes appears to depend upon the species and upon the range of densities being considered. Moreover the degree of population regulation can be expected to vary among species and with the prevailing environmental conditions. Undoubtedly, a very tight regulation of population size rarely occurs, but it is observed under extremely unfavourable conditions in geocarpic annuals. For example, one individual of *Emex spinosa* produces three to four subterranean seeds, from which only one seedling usually emerges. Hence, the parent plant is replaced by one offspring, which occupies exactly the same site as the parent (as outlined by Evenari, Kadouri & Gutterman 1977).

Whilst understanding the various density-dependent and density-independent processes that operate in populations is crucial to understanding what determines population size, this is often difficult in practice. For example, compensatory growth may lead to complex interactions between these various processes. Thus, in populations of *Vulpia fasciculata* the time at which individuals die affects the way in which fecundity responds to density, and consequently population size (Watkinson 1982). The importance of manipulating population densities to unravel the significance of various density-dependent and density-independent processes is illustrated by the study of Borchert & Jain (1978). They found that seed predators reduced the seed pool in populations of *Avena fatua* by approximately 75%. As a result the plants grow in less crowded aggregations than in areas from which seed predators have been excluded and seedling mortality is reduced. Here the level of seedling mortality is dependent upon the level of seed predation. Studies carried out over one year are also crucial to understanding population regulation in certain cases. This is well illustrated by the study of Wilkoń-Michalska (1976) on *Salicornia patula*. If the density of plants is high one year, then the numerous seeds produced have difficulty entering the soil, because their dead parents occupy the site for many months. Hence, the size of the next generation of plants is considerably reduced. However, the plants at this much lower density produce an enormous number of seeds which are able to enter the soil. Thus the population increases in size again in the next year, producing an oscillation in population numbers.

The relationship between density-dependent processes and the spatial structure of populations has been neglected in most population studies, in spite of its undoubted importance. It has already been argued that both density-dependent mortality and density-dependent fecundity are common in natural populations of annual plants. These density-dependent processes, however, are likely to vary under different environmental conditions. It might be expected, therefore, that since natural populations consist usually of, more or less, discrete aggregations of different density and that since different sites are, more or less, favourable for the growth of plants, that the temporal and spatial dynamics of annual plants will be complex. This is well illustrated by *Erophila verna* (Symonides 1983a, b, 1984).

Each year seedlings of *E.verna* colonize small gaps in a compact grassland. The seedling densities in these gaps are remarkably variable, but the differences gradually diminish over the life-cycle due to density-dependent mortality, so that the differences are fairly small at the fruiting stage. Nevertheless, the fraction of individuals surviving to reproduction and their fecundity depend upon seedling density (Fig. 12.6). Hence, the number of

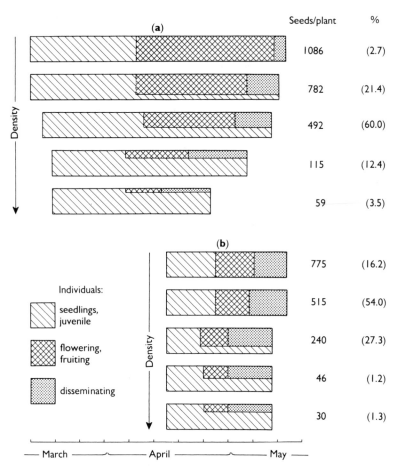

FIG. 12.6. The effects of density and longevity on the proportion of fruiting individuals and their fecundity in populations of *Erophila verna*. Long-lived (a) and short-lived (b) individuals. The densities range from 200 to 6000 m$^{-2}$. The contribution of individuals growing at each density to the total seed production of the population is given in brackets (drawn from data in Symonides 1983a, b).

seeds produced by the low-density aggregations of plants is considerably higher than the number produced by the high-density aggregations. Moreover, the seeds produced in the plots occupied by low-density aggregations of seedlings enter the soil easier and germinate better than the seeds produced in the gaps occupied by high densities of plants. In the next generation high densities of seedlings appear in those gaps that were scarcely inhabited in the previous year and vice versa (Fig. 12.7). Oscillations in population size are, therefore, produced within a single gap, but because different seed densities are found in different gaps the size of the whole population changes only slightly. For example, the number of seedlings in two 0.01 m$^2$ gaps in 4 successive years were 2, 76, 1, 58 and 62, 255, 1 respectively, whereas the total number of seedlings in an area of 16 m$^2$ varied only slightly from 5040, 4854, 4879 to 5144 (Symonides 1983a). From these data it can be seen that the apparent overall population stability hides a very dynamic population flux associated with the diverse spatial structure of the population.

Plant responses to crowding affect not only population regulation but also the genetic structure of populations. Density affects the relative fitness of individuals in a population through both differential mortality and fecundity. This has been clearly shown for *Salicornia europaea* (Davy & Smith 1985). Genetic changes during self-thinning have also been reported in artificial populations of *Phlox drummondii* (Bazzaz, Levin & Schmierbach 1982), suggesting differential mortality responses of genotypes to density. Some studies also show that density affects the distance of pollen dispersal in both wind and insect-pollinated species (Antonovics & Levin 1980). At high densities, wind speed is usually reduced, and, as plants are smaller, pollen travels less far than in low-density populations. In the case of insect-pollinated plants, the search distances of pollinators are smaller as a result of behavioural adjustment of insects, whose search patterns are determined by resource availability (Levin & Kerster 1974). Given that near neighbours may be closely related the level of inbreeding can be expected to rise in high-density populations of plants (Ellstrand, Torres & Levin 1978). However, in species with both cross- and self-pollination very low densities may also result in higher levels of inbreeding as a result of self-pollination, e.g. *Bromus tectorum* (Young & Evans 1985). All of these problems are thoroughly discussed by Antonovics & Levin (1980).

### Long-term population dynamics

Each year the size of the population of vegetative plants is affected by (a) the number of seeds produced in the previous generation, (b) seed dispersal in

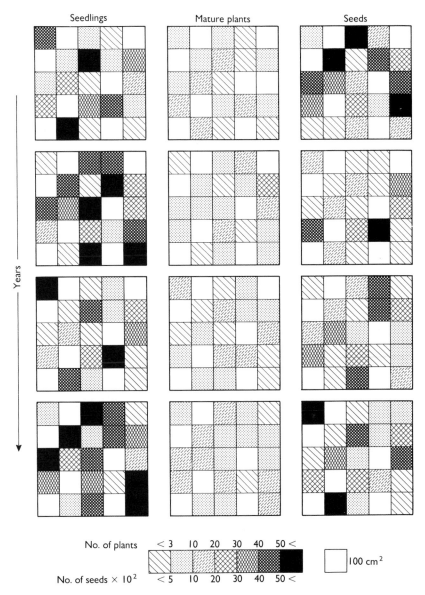

Fig. 12.7. Changes in the spatial structure and seed production of a population of *Erophila verna* in 4 successive years (Symonides unpublished data).

space and time, (c) seed survival and germinability, (d) the number of micro-sites suitable for seed germination and seedling emergence, and (e) the weather conditions during seedling emergence. All of these factors may vary from year to year and hence affect the long-term population dynamics of a species.

It is possible to recognize three main types of long-term population dynamics. The first concerns ephemeral species. Populations of such plants occur primarily as a seed bank, with germination occurring on the rare, unpredictable occasions when conditions are favourable. Some species may appear in most but not all years, as in the case of *Eriophyllum wallacei* (Juhren, Went & Phillips 1956), whilst others such as *Anastatica hierochuntica* may grow only once in several years (Friedman & Stein 1980). Ephemeral species are particularly characteristic of deserts but many weeds behave similarly, e.g. *Ipomoea hederacea* (Whigham 1984). The second group of species includes those 'wandering' plants, populations of which are very short-lived at a particular site, and which rely on long-distance dispersal for the colonization of new sites, e.g. *Lactuca serriola* (Prince, Carter & Dancy 1985). The third group encompasses all those species that are present on a site for many years. All of these plants show fluctuations or oscillations in abundance from year to year, some large and some small.

Irregular fluctuations in population size are particularly associated with unstable and unpredictable environments where they are a consequence of the variable biotic and physical environment. Such populations are not strongly regulated (Fig. 12.8a). Sometimes fluctuations in population size are extremely rapid. For example, a grassland fire reduced populations of *Bromus tectorum* to ten individuals $m^{-2}$, but then 3 years later they had increased to over 10000 (Young & Evans 1985). In general, changes in the number of individuals from year to year result from changes in the level of various density-independent processes that are particularly associated with external factors such as weather.

Regular increases and decreases in population size from year to year (oscillations) are poorly documented. However, such oscillations occur in *Erophila verna* (Symonides, Silvertown & Andreasen 1986) and *Salicornia patula* (Wilkoń-Michalska 1976). They appear to be characteristic of species populations which have high fecundity and germinability, and which occur in predictable habitats. The size oscillations are a consequence of the nature of the density-dependent regulation of population size.

There are two reasons why some populations maintain relatively stable population sizes over successive years: stable environmental conditions and strong density-dependent regulation. Small changes in population size from year to year often occur in closed communities of late successional stages (Fig. 12.8b). It may be that the presence of a seed bank also tends to buffer changes in population size but fairly stable population sizes are also typical of some populations of annual species which lack a seed bank (cf. Symonides 1979b).

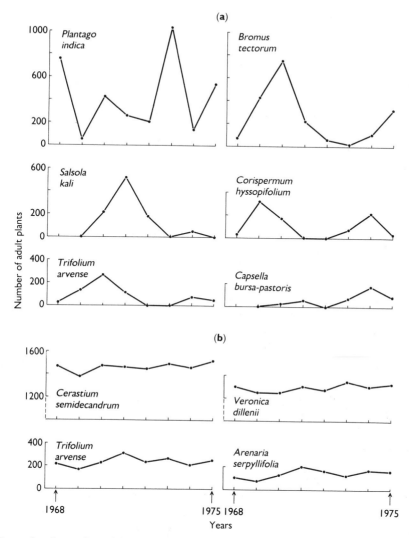

FIG. 12.8. Comparison of the long-term dynamics of populations of plants in (a) ruderal and (b) grassland communities. The communities are located close to each other (drawn from data in Symonides in 1979a, b).

## CONCLUSIONS

Rabotnov (1978) suggested that 'annual plants seem to have originated in arid regions with an open vegetation and a poor water supply'. On the basis of data presented here, however, this thesis must be disputed. Annuals occupy a large diversity of environments, they belong to various taxa and exhibit significant variability in their life-history characteristics. It is more probable that the short, annual life-cycle has evolved from perennial progenitors in various environments where disturbance results in the chance of an adult surviving from year to year being very low (Schaffer & Gadgil 1975).

It is beyond doubt that the annual life-cycle represents an arbitrary division among the continuum of life-history strategies (Harper 1977). Short longevity is the only feature that distinguishes annuals from biennials and perennials. Indeed local populations of some species are either annual, biennial or even perennial (e.g. *Lavatera arborea*), depending upon the conditions. Even within a single population where most individuals complete their life-cycle within 12 months, there are those that live longer.

This chapter documents, above all, the significant variation that is found in the population dynamics of annual plants. It also demonstrates the considerable similarity in demographic and life-history characteristics that are found in annuals from extremely different environments (cf. *Floerkea proserpinacoides* and *Vulpia fasciculata*), and also some of the crucial differences that occur between annuals, even from a single community (cf. *Erophila verna* and *Spergula vernalis*). Moreover, the examples described in this chapter indicate the lack of correlation between life-history characteristics. Low fecundity is not always associated with high germinability or high seedling survival. Neither are a high reproductive effort, nor a rapid switch from vegetative growth to reproduction the rule in annuals. As the number of studies on annual plants grows, so many of the gross generalizations about annual plants seem less certain.

## ACKNOWLEDGMENTS

The helpful comments of Drs A. R. Watkinson and A. J. Davy, as well as those of Mrs Anna Mazur are gratefully acknowledged.

## REFERENCES

Antonovics, J. & Levin, D. A. (1980). The ecological and genetic consequences of density-dependent regulation in plants. *Annual Review of Ecology and Systematics*, 11, 411–52.

Arthur, A. E., Gale, J. S. & Lawrence, K. J. (1973). Variation in wild populations of *Papaver dubium*. VII. Germination time. *Heredity*, 30, 189–97.

Baker, H. G. (1974). The evolution of weeds. *Annual Review of Ecology and Systematics*, 5, 1–24.

Bartolome, J. W. (1979). Germination and seedling establishment in California annual grassland. *Journal of Ecology*, 67, 273–81.

Baskin, J. M. & Baskin, C. C. (1972). Influence of germination date on survival and seed production of *Leavenworthia stylosa*. *American Midland Naturalist*, 88, 318–23.

**Baskin, J. M. & Baskin, C. C. (1985).** Life cycle ecology of annual plant species of cedar glades of southeastern United States. *Handbook of Vegetation Science. Vol. 3. The population structure of vegetation* (Ed. by J. White), pp. 371–98, Junk, Dordrecht.

**Bazzaz, F. A., Levin, D. A. & Schmierbach, M. R. (1982).** Differential survival of genetic variants in crowded population of *Phlox. Journal of Applied Ecology*, **19**, 891–900.

**Borchert, M. I. & Jain, S. K. (1978).** The effect of rodent seed predation on four species of California annual grasses. *Oecologia, Berlin*, **33**, 101–13.

**Carter, R. N. & Prince, S. D. (1985).** The geographical distribution of prickly lettuce (*Lactuca serriola*). I. A general survey of its habitats and performance in Britain. *Journal of Ecology*, **73**, 27–38.

**Chapin, F. S. (1980).** The mineral nutrition of wild plants. *Annual Review of Ecology and Systematics*, **11**, 233–60.

**Cheplick, G. P. & Quinn, J. A. (1982).** *Amphicarpum purshii* and the 'pessimistic strategy' in amphicarpic annuals with subterranean fruit. *Oecologia, Berlin*, **52**, 327–32.

**Cohen, D. (1966).** Optimizing reproduction in a randomly varying environment. *Journal of Theoretical Biology*, **12**, 119–29.

**Davy, A. J. & Smith, H. (1985).** Population differentiation in the life-history characteristics of salt-marsh annuals. *Vegetatio*, **61**, 117–25.

**Dolan, R. W. (1984).** The effect of seed size and maternal source on individual size in a population of *Ludwigia leptocarpa* (Onagraceae). *American Journal of Botany*, **71**, 1302–7.

**Dolan, R. W. & Sharitz, R. R. (1984).** Population dynamics of *Ludwigia leptocarpa (Onagraceae)* and some factors affecting size hierarchies in a natural population. *Journal of Ecology*, **72**, 1031–41.

**During, H. J., Schenkeveld, A. J., Verkaar, H. J. & Willems, J. H. (1985).** Demography of short-lived forbs in chalk grassland in relation to vegetation structure. *Handbook of Vegetation Science. Vol. 3. The population structure of vegetation* (Ed. by J. White), pp. 341–70, Junk, Dordrecht.

**Ellstrand, N. C., Torres, A. M. & Levin, D. A. (1978).** Density and the rate of apparent outcrossing in *Helianthus annuus (Asteraceae)*. *Systematic Botany*, **3**, 403–7.

**Epling, C., Lewis, H. & Bell, F. M. (1960).** The breeding group and seed storage: a study in population dynamics. *Evolution*, **14**, 238–55.

**Ernst, W. H. O. (1981).** Ecological implication of fruit variability in *Phleum arenarium* L., an annual dune grass. *Flora*, **171**, 387–98.

**Ernst, W. H. O. (1983).** Mineral nutrition of two contrasting dune annuals, *Phleum arenarium* and *Erodium glutinosum*. *Journal of Ecology*, **71**, 197–209.

**Evenari, M., Kadouri, A. & Gutterman, Y. (1977).** Eco-physiological investigations on the amphicarpy of *Emex spinosa* (L.) Campd. *Flora*, **166**, 223–38.

**Falencka, M. (1983).** *Struktura i dynamika populacji* Impatiens noli-tangere *w różnych warunkach ekologicznych*. Ph. D. thesis, Warsaw University.

**Freas, K. E. & Kemp, P. R. (1983).** Some relationships between environmental reliability and seed dormancy in desert annual plants. *Journal of Ecology*, **71**, 211–17.

**Friedman, J. & Elberse, W. T. (1976).** Competition between two desert varieties of *Medicago laciniata* (L.) Mill. under controlled conditions. *Oecologia, Berlin*, **22**, 321–39.

**Friedman, J. & Stein, Z. (1980).** The influence of seed-dispersal of *Anastatica hierochuntica* (Cruciferae) in the Negev Desert, Israel. *Journal of Ecology*, **68**, 43–50.

**Gadgil, M. (1971).** Dispersal: population consequences and evolution. *Ecology*, **52**, 253–61.

**Grime, J. P. (1979).** *Plant strategies and vegetation processes*. Wiley and Sons, New York.

**Grubb, P. J., Kelly, D. & Mitchley, J. (1982).** The control of relative abundance in communities of herbaceous plants. *The Plant Community as a Working Mechanism* (Ed. by E. I. Newman), pp. 79–97, Blackwell Scientific Publications, Oxford.

**Harper, J. L. (1977).** *Population biology of plants*. Academic Press, London.

**Heisey, R. M. & Delwiche, C. C. (1985).** Allelopathic effects of *Trichostema lanceolatum* (Labiatae) in the California annual grassland. *Journal of Ecology*, **73**, 729–42.

**Hickman, J. C. (1975).** Environmental unpredictability and plastic energy allocation strategies in the annual *Polygonum cascadense* (Polygonaceae). *Journal of Ecology*, **63**, 689–701.

**Inouye, R. S., Byers, G. S. & Brown, J. H. (1980).** Effects of predation and competition on survivorship, fecundity and community structure of desert annuals. *Ecology*, **61**, 1344–51.

**Juhren, M., Went, F. W. & Phillips, E. (1956).** Ecology of desert plants. IV. Combined field and laboratory work on germination of annuals in the Joshua Tree National Monument, California. *Ecology*, **37**, 318–30.

**Keddy, P. A. (1982).** Population ecology on an environmental gradient: *Cakile edentula* on a sand dune. *Oecologia, Berlin*, **52**, 348–55.

**Kelly, C. A. (1986).** Extrafloral nectaries: ants, herbivores and fecundity in *Cassia fasciculata*. *Oecologia, Berlin*, **69**, 600–5.

**King, D. & Roughgarden, J. (1982).** Multiple switches between vegetative and reproductive growth in annual plants. *Theoretical Population Biology*, **21**, 194–204.

**Kirkendall, L. R. & Stenseth, N. C. (1985).** On defining 'breeding once'. *American Naturalist*, **125**, 189–204.

**Klemow, K. M. & Raynal, D. J. (1983).** Population biology of an annual plants in a temporally variable habitat. *Journal of Ecology*, **71**, 691–703.

**Koller, D. & Roth, N. (1964).** Studies on the ecological and physiological significance of amphicarpy in *Gymnarrhena micrantha* (Compositae). *American Journal of Botany*, **51**, 26–35.

**Law, R. (1981).** The dynamics of a colonizing population of *Poa annua*. *Ecology*, **62**, 1267–77.

**Leon, J. A. (1985).** Germination strategies. *Evolution: essays in honour of John Maynard Smith* (Ed. by P. J. Greenwood, P. H. Harvey & M. Slatkin), pp. 129–42. Cambridge University Press, Cambridge.

**Leverich, W. J. & Levin, D. A. (1979).** Age-specific survivorship and reproduction in *Phlox drummondii*. *American Naturalist*, **113**, 881–903.

**Levin, D. A. & Clay, K. (1984).** Dynamics of synthetic *Phlox drummondii* populations at the species margin. *American Journal of Botany*, **71**, 1040–50.

**Levin, D. A. & Kerster, H. W. (1974).** Gene flow in seed plants. *Evolutionary Biology*, **7**, 139–220.

**Mack, R. N. (1976).** Survivorship of *Cerastium atrovirens* at Aberffraw, Anglesey. *Journal of Ecology*, **64**, 309–12.

**Mack, R. N. & Pyke, D. A. (1983).** The demography of *Bromus tectorum*: variation in time and space. *Journal of Ecology*, **71**, 69–93.

**Maillette, L. (1985).** Modular demography and growth patterns of two annual weeds (*Chenopodium album* L. and *Spergula arvensis* L.) in relation to flowering. *Studies on Plant Demography. A Festschrift for John L. Harper* (Ed. by J. White), pp. 239–255, Academic Press, London.

**Marks, M. & Prince, S. (1981).** Influence of germination date on survival and fecundity in wild lettuce *Lactuca serriola*. *Oikos*, **36**, 326–30.

**Maxwell, C. P., Jacob, N., Bollard, S. & Lovell, P. (1986).** Factors affecting establishment and survival of *Soliva* (Onehung weed) at Auckland, New Zealand. *New Zealand Journal of Botany*, **24**, 79–87.

**Miller, T. E. (1987).** Effects of emergence time on survival and growth in an early old-field plant community. *Oecologia, Berlin*, **72**, 272–78.

**Prince, S. D., Carter, R. N. & Dancy, K. J. (1985).** The geographical distribution of prickly lettuce *(Lactuca serriola)*. II. Characteristics of populations near its distribution limit in Britain. *Journal of Ecology*, **73**, 39–48.

**Rabotnov, T. A. (1978).** On coenopopulations of plants. *Structure and functioning of plant* *populations* (Ed. by A. H. J. Freysen & J. W. Woldendorp), pp. 1–26, North Holland Publishing Company, Amsterdam.

**Regher, D. L. & Bazzaz, F. A. (1979).** The population dynamics of *Erigeron canadensis*, a successional winter annual. *Journal of Ecology*, **67**, 923–33.

**Reynolds, D. N. (1984).** Population dynamics of three annual species of alpine plants in the Rocky Mountains. *Oecologia, Berlin*, **62**, 250–55.

**Roach, D. A. (1986).** Timing of seed production and dispersal in *Geranium carolinianum*-effect on fitness. *Ecology*, **67**, 572–76.

**Rozijn, N. A. M. G. & van der Werf, D. C. (1986).** Effect of drought during different stages in the life-cycle on the growth and biomass allocation of two *Aira* species. *Journal of Ecology*, **74**, 507–23.

**Schaal, B. A. & Leverich, W. J. (1982).** Survivorship patterns in an annual plant community. *Oecologia, Berlin*, **54**, 149–51.

**Schaffer, W. M. & Gadgil, M. D. (1975).** Selection for optimal life histories in plants. *The Ecology and Evolution of Communities* (Ed. by M. Cody & J. Diamond), pp. 142–57, Harvard University Press, Cambridge, Massachusetts.

**Sharitz, R. R. & McCormick, J. F. (1973).** Population dynamics of two competing annual plant species. *Ecology*, **54**, 723–40.

**Silvertown, J. W. (1981).** Seed size, life span, and germination date as coadapted features of plant life history. *American Naturalist*, **118**, 860–64.

**Silvertown, J. W. (1982).** *Introduction to plant population ecology.* Longman, London.

**Silvertown, J. W. (1985).** Survival, fecundity and growth of wild cucumber, *Echinocystis lobata*. *Journal of Ecology*, **73**, 841–49.

**Smith, B. H. (1983a).** Demography of *Floerkea proserpinacoides*, a forest-floor annual. III. Dynamics of seed and seedling populations. *Journal of Ecology*, **71**, 413–25.

**Smith, B. H. (1983b).** Demography of *Floerkea proserpinacoides*, a forest-floor annual. I. Density-dependent growth and mortality. *Journal of Ecology*, **71**, 391–404.

**Smith, B. H. (1983c).** Demography of *Floerkea proserpinacoides*, a forest-floor annual. II. Density-dependent reproduction. *Journal of Ecology*, **71**, 405–12.

**Stanton, M. L. (1984).** Developmental and genetic sources of seed weight variation in *Raphanus raphanistrum* L. (Brassicaceae). *American Journal of Botany*, **71**, 1090–98.

**Symonides, E. (1974).** Populations of *Spergula vernalis* Willd. from different dune biotopes of the Toruń Basin. *Ekologia Polska*, **22**, 417–40.

**Symonides, E. (1978a).** Number, distribution and specific composition of diaspores in the soil of the plant association Spergulo-Corynephoretum. *Ekologia Polska*, **26**, 111–22.

**Symonides, E. (1978b).** Effect of population density on the phenological development of individuals of annual plant species. *Ekologia Polska*, **26**, 273–86.

**Symonides, E. (1979a).** The structure and population dynamics of psammophytes on inland dunes. II. Loose-sod populations. *Ekologia Polska*, **27**, 191–234.

**Symonides, E. (1979b).** The structure and population dynamics of psammophytes on inland dunes. III. Populations of compact psammophyte communities. *Ekologia Polska*, **27**, 235–57.

**Symonides, E. (1983a).** Population size regulation as a result of intra-population interactions. I. Effect of density on survival and development of individuals of *Erophila verna* (L.) C. A. M. *Ekologia Polska*, **31**, 839–82.

**Symonides, E. (1983b).** Population size regulation as a result of intra-population interactions. II. Effect of density on the growth rate, morphological diversity and fecundity of *Erophila verna* (L.) C. A. M. individuals. *Ekologia Polska*, **31**, 883–912.

**Symonides, E. (1984).** Population size regulation as a result of intra-population interactions. III. Effect of *Erophila verna* (L.) C. A. M. population density on the abundance of the new generation of seedlings. Summing-up conclusions. *Ekologia Polska*, **32**, 557–80.

Symonides, E., Silvertown, J. W. & Andreasen, V. (1986). Population cycles caused by overcompensating density-dependence in an annual plant. *Oecologia, Berlin*, **71**, 156–58.

Venable, D. L. & Lawlor, L. (1980). Delayed germination and dispersal in desert annuals: escape in space and time. *Oecologia, Berlin*, **46**, 272–82.

Venable, D. L. & Levin, D. A. (1985). Ecology of achene dimorphism in *Heterotheca latifolia*. I. Achene structure, germination and dispersal. *Journal of Ecology*, **73**, 133–45.

Vincent, T. L. & Pulliam, H. R. (1980). Evolution of life history strategies for an asexual plant model. *Theoretical Population Biology*, **17**, 215–31.

Watkinson, A. R. (1978a). The demography of a sand dune annual: *Vulpia fasciculata*. III. The dispersal of seeds. *Journal of Ecology*, **66**, 483–98.

Watkinson, A. R. (1978b). The demography of a sand dune annual: *Vulpia fasciculata*. II. The dynamics of seed populations. *Journal of Ecology*, **66**, 35–44.

Watkinson, A. R. (1982). Factors affecting the density response of *Vulpia fasciculata*. *Journal of Ecology*, **70**, 149–61.

Watkinson, A. R. (1985). On the abundance of plants along an environmental gradient. *Journal of Ecology*, **73**, 569–78.

Watkinson, A. R. & Davy, A. J. (1985). Population biology of salt marsh and sand dune annuals. *Vegetatio*, **62**, 487–97.

Watkinson, A. R. & Harper, J. L. (1978). The demography of a sand dune annual: *Vulpia fasciculata*. I. The natural regulation of populations. *Journal of Ecology*, **66**, 15–33.

Watkinson, A. R. & White, J. (1986). Some life-history consequences of modular construction in plants. *Philosophical Transactions of The Royal Society London B*, **313**, 31–51.

Whigham, D. F. (1984). The effect of competition and nutrient availability on the growth and reproduction of *Ipomoea hederacea* in an abandoned old field. *Journal of Ecology*, **72**, 721–30.

Wilkoń-Michalska, J. (1976). Struktura i dynamika populacji *Salicornia patula* Duval-Jouve. *Rozprawy UMK, Toruń*.

Winsor, J. (1983). Persistence by habitat dominance in the annual *Impatiens capensis* (Balsaminaceae). *Journal of Ecology*, **71**, 451–66.

Wu, K. K. & Jain, S. K. (1979). Population regulation in *Bromus rubens* and *B. mollis*: life cycle components and competition. *Oecologia, Berlin*, **39**, 337–57.

Young, J. A. & Evans, R. A. (1985). Demography of *Bromus tectorum* in *Artemisia* communities. *Handbook of Vegetation Science. Vol. 3. The population structure of vegetation* (Ed. by J. White), pp. 489–502, Junk, Dordrecht.

Zeide, B. (1978). Reproductive behavior of plants in time. *American Naturalist*, **112**, 636–39.

# 13. THE EFFECTS OF ENVIRONMENTAL HETEROGENEITY IN SPACE AND TIME ON THE REGULATION OF POPULATIONS AND COMMUNITIES

## NORMA FOWLER

*Department of Botany, University of Texas, Austin, Texas 78713, USA*

## SUMMARY

**1** The regulation of plant community structure is closely related to the regulation of population size. It is therefore an example of an area in which the methods of population ecology, especially demographic methods, may be profitably applied to a problem in community ecology. The role of environmental variation in space and time is central to the regulation of both plant populations and plant communities.

**2** Temporal and spatial variation in the environment may obscure the action of density- and frequency-dependent regulating factors; this is widely appreciated in the case of year-to-year variations but often overlooked in the case of small-scale environmental variation in space. In descriptive studies, spatial heterogeneity in the environment can obscure a negative relationship between plant performance and measures of density (e.g. nearest neighbour distances), resulting in no apparent relationship, or even producing a positive relationship.

**3** Most of the mechanisms by which plant species coexist are dependent upon environmental variation in space or in time, or, in the case of disturbance, in both.

**4** Temporal variation in the environment may make density-dependent population regulation, and also community regulation, an infrequent event in many plant communities. Environmental variation can perhaps delay the competitive exclusion of species from a community, and simultaneously weaken the action of factors that promote coexistence.

## INTRODUCTION

The field of plant demography has been well-established for many years and continues to progress rapidly, as evidenced by the contributions to this and other recent symposia (Dirzo & Sarukhán 1984; Jackson, Buss & Cook 1985;

White 1985). The use of the approaches and techniques of demography to address questions pertaining to the field of plant community ecology is hardly novel (Harper 1977). The study of plant populations and the study of plant communities have, however, remained relatively separate undertakings, in contrast to the situation in zoological studies. It is my contention that both areas could profit from an increased interchange of ideas, and that studies at the interface between population and community ecology can yield useful insights in both areas.

The effects of the community, that is, of other plant species, upon an individual plant and upon plant populations, are nearly limitless. Likewise, the number of ways in which population approaches can be used in the study of plant communities is also very large. Rather than attempt a survey, this chapter will discuss, as an example, the general problem of the regulation of plant community structure and the closely related issue of the regulation of plant population size, and, in particular, the central role of spatial and temporal heterogeneity in the regulation of plant communities and populations. Three types of effects of environmental heterogeneity on populations and communities will be discussed: (1) environmental heterogeneity obscures the action of regulating factors from our detection, (2) it can promote the coexistence of species, and (3) it can be a source of random noise in the system.

## THE REGULATION OF PLANT POPULATIONS AND COMMUNITIES

It is necessary to begin this discussion with a clarification of terminology. The phrase population regulation, in its strictest sense, implies that there is an equilibrium population size and that the regulating factor(s) tend to produce an increase in numbers of individuals when the population drops below this size and a decrease when it rises above it. In this strict sense, only negative density-dependent factors such as competition and certain kinds of predation can be regulators of population size.

In the same way, regulation of community composition, in its strictest sense, implies that there is an equilibrium population size for each of the species comprising the community and hence that the relative abundances (frequencies) of species have equilibrium values. Thus a community-regulating factor will tend to make a species, which is more abundant than its equilibrium frequency, less common and a species which is rarer than its equilibrium frequency more common (the invasibility criterion of community stability (Chesson & Case 1986)). In other words, a community-regulating factor will behave in a negative frequency-dependent fashion. If two plant species do not affect each other, so that their populations are regulated independently, this population regulation will, by itself, also regulate their relative abundances. This situation will occur, for example, when two species have completely non-overlapping niches. To the extent that two species do affect each other's populations in a negative fashion (e.g. compete with each other), regulation of community structure also implies that negative frequency-dependent factors operate to regulate their relative abundances. In this situation, negative density-dependent factors (i.e. factors that cause a decrease in survival or reproduction as the sum of the densities of both species increases) are still necessary but they are not sufficient. For example, density-dependent regulation of two competing species without regulation of their relative frequencies would imply that the two species were ecological equivalents and that their relative abundances were subject only to random drift.

The regulation of populations and the regulation of community structure are therefore closely related. Note that these definitions of population and community regulation say nothing about how frequently a population or community is at or even near equilibrium. Note also that, for the sake of simplicity, stable limit cycles and similar complexities are ignored here. Finally, although relative frequency is an essential variable in the definition of community regulation, this does not imply that it is recommended as a variable to be studied directly; in general, the *absolute* densities of each species are more appropriate in practice.

The phrases population regulation and community regulation are sometimes used in a broader sense, to include the density-independent factors that set the equilibrium sizes of populations and the equilibrium values of the relative abundances of species. These phrases may even be used to describe any factor that affects population size or community composition. Such density-independent and frequency-independent factors include climatic factors, density-independent predation, and so on, and may be regular (e.g. seasonal), random (stochastic), or both. However, these phrases are used here in the strict sense described above.

There was a prolonged debate in the 1950s about the commonness and importance of population regulation in the strict sense (Andrewartha & Birch 1954; Nicholson 1957). It was partially resolved by the recognition that both sorts of factors may be simultaneously important, density-dependent factors tending to move the population towards an equilibrium size, and density-independent factors setting the equilibrium size. The study of the regulation of populations, in particular the role of density-dependent processes, is for some reason relatively out-of-fashion at the moment. It is nevertheless a subject of considerable interest, and one about which little is known regarding plant populations, as the review by Antonovics & Levin (1980) amply demonstrates.

In contrast, the closely related subject of the regulation of community composition is currently receiving a great deal of attention. Much of this attention has focussed on the relative importance of competition, partly in response to an earlier tendency to consider competition as the primary, or even the only, factor determining community composition (Connell 1983; Schoener 1983; Strong 1983; Strong *et al.* 1984; Diamond & Case 1986). This tendency seems not to have been as pronounced among plant ecologists as among some others, perhaps because so many plant ecologists work in grazed or mown vegetation. Other issues besides the relative importance of competition have been receiving attention as well. These include the role of spatial and temporal heterogeneity in the environment in the regulation of community composition and the degree to which community composition is regulated at all, in the strict sense (e.g. Diamond & Case 1986).

The first of these two issues, the effects of environmental heterogeneity upon the regulation of community structure and composition, together with the closely related topic of the effects of environmental heterogeneity on population regulation, are the subjects of the remainder of this chapter. Some comments will be made on the second of these issues, the degree of regulation, since it is partly determined by environmental heterogeneity.

## THE ROLES OF ENVIRONMENTAL HETEROGENEITY

One of the strongest impressions that comes from detailed study of any plant community is the very high degree of environmental variation, both spatial and temporal, confronted by the plants of a single population. Parallel to this environmental variability and perhaps both causing it and caused by it, is an equally high level of variability in plant density, community composition, and demographic parameters, the latter including probabilities of survival, growth-rates and fecundities.

The statement that very high levels of spatial and temporal variation in the environment and in population and community characteristics are the rule, not the exception, does not rest solely upon subjective impressions. Evidence can be found in almost any published field study in which the results are reported in sufficient detail, for example, where the results from separate quadrats are reported separately (e.g. Bishop, Davy & Jefferies 1978; Newell, Solbrig & Kincaid 1981; Antonovics & Primack 1982; Fowler 1984; Kalisz 1986), in addition to the evidence from all of the studies in which the effects of small-scale variation in some environmental factor were explicitly studied. Likewise, many studies contain reports of marked differences between years (e.g. Raynal & Bazzaz 1975; Klemow & Raynal 1983; Mack & Pyke 1983; Chapin & Shaver 1985; Kalisz 1986). In many cases the authors have considered this heterogeneity to be a statistical problem, a source of unwanted variation that obscures whatever phenomenon they are studying. However, any set of factors that can account for as much as 60% of the variation (Fowler 1981) is worth studying in its own right. Spatial and temporal heterogeneity must have major impacts upon all aspects of the dynamics and structure of plant populations and communities.

This section considers three potential impacts of environmental, spatial and temporal heterogeneity upon populations and communities, (1) as factors obscuring the action of competition and other regulating factors; (2) as sources of niche separation leading to coexistence; and (3) as sources of random 'noise' affecting populations and communities in a variety of ways.

## Heterogeneity as an obscuring factor

The potential of temporal fluctuations to mislead investigators has been recognized at least since the pioneering work of Weaver and his students, who described the fluctuations in the plant communities of the semi-arid grasslands of the central United States (west of the tallgrass-dominated true prairie) during and after the great drought of the 1930s (Albertson & Weaver 1942; Weaver & Albertson 1956). They found that as the drought progressed the more drought-tolerant, shortgrass species (*Buchloe dactyloides* and *Bouteloua gracilis*) expanded their local distributions downwards from the more xeric 'level uplands' to replace the less drought-tolerant, taller grasses (e.g. *Schizachyrium scoparium*) on slopes and in ravines (Fig. 13.1; nomenclature follows Gould 1975.). Weaver continued his studies through the 1940s and early 1950s and was therefore able to recognize correctly that these fluctuations did not represent a succession, but one-half of a recurring cycle.

While Weaver described the effects of cyclic droughts with an apparent periodicity of decades, temporal fluctuations need not be cyclic nor of such duration to obscure underlying phenomena or falsely to suggest successional change, as an example from my own work shows. In a study of a mown grassland in North Carolina, Fowler & Antonovics (1981) recorded marked fluctuations in the abundances of species between seasons and between years. These fluctuations were due both to changes in numbers of individuals and to changes in the sizes of individuals, as was also true in Weaver's studies. Although these fluctuations initially could have been misinterpreted as successional, they were not; the field in fact changed very little in subsequent years (Antonovics, personal communication).

The degree of temporal fluctuations in composition will vary between

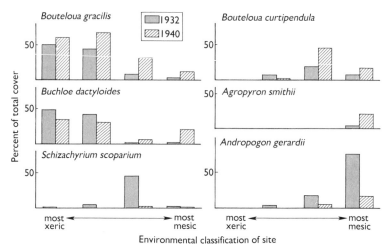

FIG. 13.1.   Changes in the relative basal cover of the most abundant grass species in western Kansas grasslands before (1932) and after (1940) a major drought. Most species declined in absolute cover during this period. Only relative basal cover is shown here (i.e. if all species were shown, the columns would sum to 100). (Redrawn from Weaver and Albertson 1956, Chapter 7, Table 2.)

communities, of course, but in general it is larger than the novice researcher expects. North Carolina, for example, is not generally considered to have a particularly variable climate, yet its climate is sufficiently variable between years to cause major shifts in population sizes from year to year. The appropriate responses, caution in interpretation of short-term results, and a call for long-term ($> 5$ year) studies, have been urged by many authors and can only be repeated here.

The obscuring effects of spatial heterogeneity in the environment and in population parameters on the detection of underlying events are less appreciated. In properly designed and analysed manipulative experiments, the obscuring effects of spatial heterogeneity will be limited to rendering the effects of treatments statistically non-significant. The potential for misleading conclusions in descriptive studies is greater. For example, a larger number of descriptive studies have looked for a negative relationship between local density and some measure of plant performance as evidence of the regulation of population size by density-dependent factors, often using the distance between plants as a surrogate for density (e.g. Klemow & Raynal 1981; Nobel 1981; Robberecht, Mahall & Nobel 1983; Connell, Tracey & Webb 1984; Hubbell & Foster 1986; references in Fowler 1986b). Some of these descriptive studies have involved more than one species, so they have also permitted comparisons between intraspecific and interspecific density-dependent effects to see if these differ in ways that can account for the coexistence of the species involved (e.g. Yeaton & Cody 1976; Yeaton, Travis & Gilinsky 1977).

All of these descriptive approaches have the same limitation: while the existence of a significant negative relationship between density and plant performance can reasonably be interpreted as evidence for negative density-dependent population regulation, the absence of such a relationship cannot be used as evidence for the absence of such regulation, due to the obscuring effects of spatial heterogeneity in the environment. As argued elsewhere (Fowler 1984), patchiness in the favourability of the environment can cancel out apparent effects of competition. Imagine, for example, that the environment resembles a chequer-board (Fig. 13.2). Resources are twice as dense on white squares as on black (or the rate of resource input is twice as high). If this condition has persisted for long enough, the density of plants on white squares will be twice that on black squares, all else being equal (scenario one). At this point, resources per plant will be constant over the whole chequer-board, and there will be no apparent relationship between the density of plants in a square and plant performance. If the relative resource levels have not been constant for long enough for the density of plants on the favourable white squares to rise to carrying capacity, but have persisted long enough for the increase in density to have begun, then there will be a *positive* relationship between density and plant performance (scenario two).

Another source of a positive relationship is a mode of dispersal that puts seeds into favourable patches at higher densities than into unfavourable patches; this could produce densities in favourable patches that are high enough for competition to cancel out completely the increased favourability of the environment in those patches (scenario three), or somewhat lower densities at which competition does not completely cancel out environmental favourability, thus producing positive correlations (scenario four).

The high levels of spatial and temporal heterogeneity experienced by plant populations make these scenarios likely, suggesting that we should not accept the absence of a negative relationship between density and plant performance as evidence of the absence of competition. Demonstrating that any one of these scenarios is in fact occurring, however, is difficult. It requires a separate, experimental test for the occurrence of competition, generally involving the experimental manipulation of the density of one or more species so as to make density independent of the existing spatial heterogeneity. Such experiments are uncommon, so it is not clear how commonly the scenarios just described may occur.

As a demonstration of the plausibility of these scenarios, the behaviour of seedlings of two grass species in a grassland in central Texas will serve,

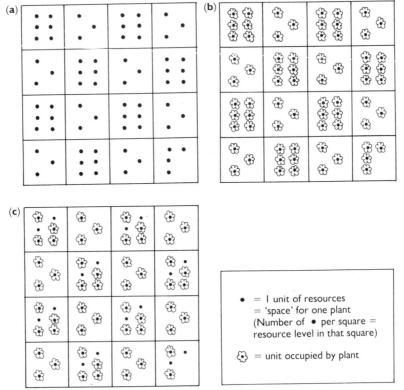

FIG. 13.2.   Schematic representation of possible relationships between resources available to
the individual and plant density; resource availability can be extrapolated to predict any
measure of individual performance. (a) Resource distribution. (b) Densities are in proportion to
resource levels, resulting in equal availability of resources to each individual; no apparent
relationship exists between density (plants/square) and individual performance (assumed to be
proportional to resource availability). (c) Densities have not equilibrated with resource levels;
there is a spurious positive correlation between density and individual performance.

although the experiments were not designed for this purpose. The author
demonstrated the occurrence of negative density-dependent interactions
among seedlings of *Bouteloua rigidiseta* and of similar effects from seedlings
of this species upon seedlings of *Aristida longiseta*, by adding seeds of
*B. rigidiseta* to an otherwise undisturbed community and observing a decrease
in the size of seedlings of both species in response to the increased density of
*B. rigidiseta* seedlings (Fowler 1986a). Plant size is closely correlated with
survival in these species, as in most others. This experiment did not measure
the intensity of negative density-dependent effects among seedlings at the
lower densities found in the unmanipulated community. It did demonstrate,

however, that negative density-dependent effects will occur where seedlings are close together, if we assume that the observed density-dependent effects occurred only between neighbouring seedlings, a reasonable assumption in the observed absence of grazing. There is no reason to expect this to be untrue in other years, although the frequency of seedlings with neighbours will vary among years as population size fluctuates.

However, the relationship between the number of neighbouring seedlings of either species within 2 cm and the performance (survival and growth) of seedlings of these species in the same site 1 and 2 years later was significantly positive (Fig. 13.3) (Fowler, in press). So was the relationship between seedling survival and the presence of a neighbouring juvenile plant. The most likely scenario to account for the latter finding is that the presence of a juvenile plant indicates that a patch is favourable, but that densities have not risen to a level that would cancel out the effects of a favourable physical environment (scenario two). Since the seedlings all germinated within a few weeks of each other, this cannot be the explanation of the positive relationship between survival and the presence of neighbouring seedlings. Instead, the fourth scenario outlined above seems plausible: that seed dispersal tends to move seeds to favourable patches, and simultaneously to aggregate them, but that, again, densities are not so high as to cancel out the effects of a favourable physical environment.

Clearly, all of these scenarios remain merely plausible hypotheses, in this and other communities. Experiments specifically designed to test these hypotheses will be required before we can judge to what extent spatial heterogeneity obscures the action of density-dependent effects from detection by purely descriptive (as opposed to manipulative) methods. The inherent plausibility of these scenarios, however, would appear to warrant greater caution in interpreting the failure of descriptive methods to detect density-dependent effects.

None of these comments should be construed as denigrating the value of descriptive methodologies. Logistic considerations often make these approaches the only feasible ones, especially if a survey of many sites or communities is desired, or if the community responds slowly to experimental manipulations (Diamond 1986). In addition, descriptive studies are often invaluable sources of hypotheses and of preliminary data.

### Heterogeneity as a factor promoting coexistence

Most of the known ways in which plant species can reduce niche (*sensu lato*) overlap and hence coexist are based upon the existence of temporal and spatial heterogeneity in the environment (Appendix 13.1). Note that this is a

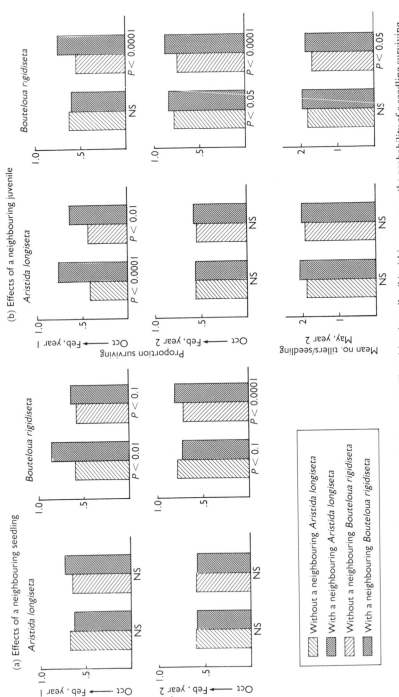

Fig. 13.3. The effects of the presence of one or more neighbouring seedlings (a) or juveniles (b) within 2 cm upon the probability of a seedling surviving, and upon its size. Most seedlings germinated in October or November each year. Juveniles were defined as plants already present before germination began in October of the given year, but less than five tillers in size the following May. Thus the surviving seedlings of year 1 became the juvenile neighbours of the seedlings germinating in year 2 and the seedlings of year 2 were not present in year 1. Calculated from the data described in Fowler (in press).

list of factors that promote coexistence, that is, move a community towards an equilibrium composition; there is no implication that any particular community is actually held near an equilibrium at any time by these factors. The importance of environmental heterogeneity in allowing the coexistence of plant species is perhaps not surprising in view of the limited number of resources for which plants compete. Of the factors outlined in Appendix 13.1, only those involving herbivory, allelopathy, and differential nutrient use can occur without reference to spatial or temporal variation in the environment. Several of these latter are merely speculative, that is, the theoretical possibility exists that they are important but no examples have yet been found.

The remaining factors outlined in Appendix 13.1 are dependent upon spatial and/or temporal variation in the environment, some of it created by other plants, which acts in concert with appropriate morphological or physiological differences between species in their responses to this variation. This variation may provide separate regulation of the populations of different species within a community, or, more likely, provide the negative frequency-dependent regulation of their abundances that allows them to coexist despite competing for the same resources. For example, to return to the grasslands studied by Weaver (Fig. 13.1), the taller *Schizachyrium scoparium* is evidently competitively dominant to the shorter *Buchloe dactyloides* in more mesic patches. Should *S. scoparium* become rare (relative to its equilibrium frequency) and *B. dactyloides* abundant in these patches, competition for light would tend to restore the equilibrium community composition, favouring the rarer species. The situation is evidently reversed in drier patches, where competition for water is more important and some patches may be too dry for *S. scoparium* even without competing species. To revert to the chequer-board metaphor, white (*S. scoparium*) wins on white (wetter) squares, black (*B. dactyloides*) on black (drier) squares, so both are maintained in the community. The negative frequency-dependent regulation arises from the chequer-board nature of the environment combined with the morphological and physiological differences between the species. However, a fluctuating environment, seed dispersal (the 'mass effect' of Shmida & Ellner 1984) and other stochastic factors prevent the formation of a community entirely devoid of stands (squares) containing both species. Indeed, in some plant communities, such factors may entirely conceal the chequer-board nature of the environment.

Note that both spatial and temporal heterogeneity promote coexistence by creating multiple equilibria. Different equilibrium population sizes and community compositions exist in different patches and at different times. In this way hypotheses involving them differ from classical models of population

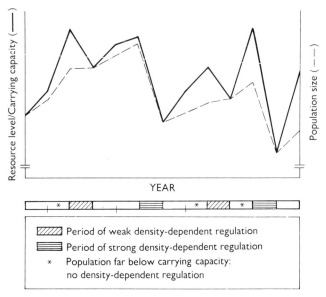

FIG. 13.5.   If resource levels fluctuate rapidly, the carrying capacity may sometimes increase more rapidly than population size, resulting in periods in which the population is so far below the carrying capacity of the site that density-dependent population regulation is weak or absent. These periods are indicated by asterisks.

temporal variation in the environment will reduce the frequency with which communities are regulated and the amount of time they spend near their long-term equilibrium composition.

In view of the climatic variability typical in many parts of the world, it seems probable that few plant populations or communities spend much time near an equilibrium condition. This may be especially true in arid regions, which are usually characterized by highly variable rainfall. Similar suggestions have been made by Wiens (1977) and Strong (1986 and references therein) with regard to animal populations. Grubb (1986) has suggested that density-dependent population regulation of sparse species may be particularly weak.

The author has obtained some evidence of reduced interspecific competition as a result of fluctuating resource levels in a grassland site in central Texas (Fowler 1986a). In this site, although the occurrence of some negative density-dependent effects was demonstrated, at least among seedlings (as described above), it was extremely weak (Fig. 13.6). Both intraspecific competition among individuals of the dominant grass *Bouteloua rigidiseta* and interspecific competition by *Bouteloua rigidiseta* against the very similar, abundant grass *Aristida longiseta*, were either very weak or non-

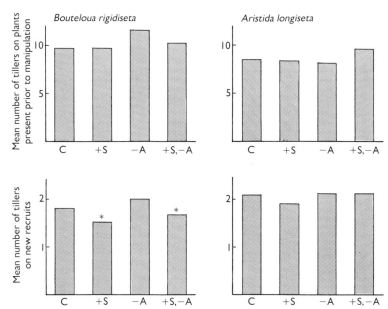

FIG. 13.6.   The effects of the addition of seeds and the removal of adults upon mean plant size
(as number of tillers/plant) of two grasses 1 year after treatment. Treatments were C = control;
+ S = addition of seeds of *Bouteloua rigidiseta*, which increased the number of seedlings of this
species by 2.7 times; − A = removal of 44% of existing adult *B. rigidiseta* individuals;
+ S, − A = both manipulations applied. Only the smaller size of seedlings of *B. rigidiseta* under
the + S and + S, − A treatments is significantly different from the control value. The larger size
of *B. rigidiseta* adults under the − A treatment is close to significance ($P < 0.052$). Survival and
fecundity are closely correlated with size in both species. (Redrawn with permission from
Fowler 1986a.)

existent, at all investigated stages of the life-cycle. Since it is difficult to
imagine how interspecific competition could be strong if intraspecific
competition between individuals of an abundant species is weak or absent,
these results can be interpreted as evidence that all interspecific competition
involving *B. rigidiseta*, not just the effects of competition from this species on
*A. longiseta*, were weak or absent. This weak competition can be ascribed to
a reduction in the size of the *B. rigidiseta* population due to a period of low
rainfall, followed by an increase in resources due to a period of relatively
high rainfall, water being a major limiting factor in this site. Furthermore, it
seems probable that this situation is common in this community, in view of
the temporal variability in rainfall in central Texas (Fig. 13.7).

As a source of stochastic variation ('noise'), environmental variation
should tend to reduce the rate of competitive exclusion of a competitively
subordinate species by a competitively dominant species. At the same time,

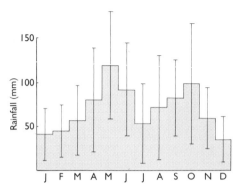

FIG. 13.7. Mean monthly rainfall in central Texas. The vertical bars represent one standard deviation around each monthly mean. (Calculated from US Weather Service records 1971–83, Fowler 1986a.)

it should also tend to reduce the effectiveness of factors that promote coexistence (Appendix 13.1). If both competitive exclusion and coexistence are sufficiently weakened, the composition of a community should not be regulated at all in the strict sense, not even in the sense of possessing multiple equilibria. Instead, it should drift in a fashion analogous to the genetic drift of neutral alleles; in this situation species would effectively be ecological equivalents. While it is difficult to imagine that this is the case for an entire plant community, it is more plausible to imagine that this model may describe the relationships among particular subsets of the species of a community. Similar models have been proposed by Hubbell (1979), Hubbell & Foster (1986), Huston (1979), and Shmida & Ellner (1984), among others. Grubb (1986) suggested that this situation may apply to the relationships between sparse species in chalk grassland and more generally to many speciose communities.

It is not at all clear yet to what extent the model just described is correct for natural plant communities or for subsets of species within natural communities. It will be intrinsically difficult to prove, since it postulates the absence of frequency-dependent effects, effectively demanding the proof of a null hypothesis. Further modelling, as well as experimentation, is clearly needed. In the meantime it seems important to distinguish between the absence of community regulation by frequency-dependent and density-dependent factors, and infrequent regulation by these factors. We must not cease to search for factors that tend to move populations and communities towards equilibria just because we realize that many communities are rarely at or near an equilibrium condition.

## ACKNOWLEDGMENTS

I thank S. Hubbell, P. Grubb, A. Watkinson and an anonymous reviewer for their comments. Support for writing this chapter and collecting the data described in it was provided by National Science Foundation grants BSR-8118968 and BSR-8600068.

APPENDIX 13.1.   Factors that can promote the coexistence of plant species in a community. Disturbances, as used here, include a wide variety of intermittent, small-scale events, but not grazing (included as herbivory) nor fire (which would be classified as a form of temporal variation in this scheme).

---

TEMPORAL VARIATION IN THE ENVIRONMENT
*Within a single season*
   Example: different use of small rainfall events by two grasses in a semi-arid North American grassland (Sala & Lauenroth 1982; Sala, Lauenroth & Reid 1982).

*Seasonal*
   Example: different growing seasons of cool-season (C3) and warm-season (C4) grasses in the same grassland (Dickinson & Dodd 1976; Monson & Williams 1982; Monson, Littlejohn & Williams 1983).

*Between years*
   Example: different grasses favoured at different points in the drought cycle in a similar grassland (Weaver & Albertson 1956).

SPATIAL VARIATION IN THE ENVIRONMENT: VERTICAL
*Different rooting zones*
   Example: differences in rooting depth among grass species in these grasslands and also in the more mesic true prairie further east; a few very deeply rooting forb species in both types of grassland; very shallowly rooted *Opuntia* cactus in the semi-arid grasslands (Weaver 1954; Weaver & Albertson 1956).

*Different canopy heights*
   Example: canopy trees *vs.* understorey herbs in any forest.

SPATIAL VARIATION IN THE ENVIRONMENT: HORIZONTAL
*Variation in soil water content*
   Example 1: different species living on badger mounds of different soil water content in the true prairie (Platt 1975; Platt & Weis 1977).
   Example 2: different species of *Solidago* along a slope gradient in the true prairie (Werner & Platt 1976).

*Variation in soil chemistry*
   Example: distribution of *Trifolium repens* in Welsh pastures (Snaydon 1962).

*Variation in light levels*
   Example: shrub-dependent *vs.* shrub-independent annuals in the Sonoran desert (Halvorsan & Patten 1975).

*Variation in temperature*
    Example: distribution of *Carnegiea gigantea* limited to microsites where rocks and shrubs
    protect young plants from freezing (Turner, Alcorn & Olin 1969; Steenbergh & Lowe 1977;
    Nobel 1980).

## ENVIRONMENTAL VARIATION IN BOTH SPACE AND TIME
*Disturbances*
    Example: a species restricted to badger mounds *vs.* dominant prairie grasses (Platt 1975;
    Platt & Weis 1977).

*Vegetational cycles not dependent on disturbance* (*sensu* Watt 1947)
    Example: different species favoured at different stages of an erosional cycle (formation and
    loss of soil in patches) in the Chihuahuan Desert (Muller 1940).

## HERBIVORY AND PATHOGENS
*Selective herbivore or pathogen* reduces the population of a competitive dominant below the level
    at which it excludes the competitive subordinate, but not to local extinction.

*Herbivore switches* to the more common plant species.

*Two species are regulated by two different herbivores* below the level at which competitive
    exclusion occurs.

## ALLELOPATHY
*Allelochemicals* affect competing species more than self.

## DIFFERENCES IN NUTRIENT USE
*Species are limited by different nutrients* (including fixed *vs.* gaseous nitrogen).

*Species are limited by different levels* of the same nutrients (Tilman 1982).

# REFERENCES

**Albertson, F. W. & Weaver, J. E. (1942).** History of the native vegetation of western Kansas
    during seven years of continuous drought. *Ecological Monographs,* **12**, 23–51.
**Andrewartha, H. G. & Birch, L. C. (1954).** *The Distribution and Abundance of Animals.* University
    of Chicago Press, Chicago.
**Antonovics, J. & Levin, D. A. (1980).** The ecological and genetic consequences of density-
    dependent regulation in plants. *Annual Review of Ecology and Systematics,* **11**, 411–52.
**Antonovics, J. & Primack, R. B. (1982).** Experimental ecological genetics in *Plantago.* VI. The
    demography of seedling transplants of *P. lanceolata. Journal of Ecology,* **70**, 55–75.
**Bishop, G. F., Davy, A. J. & Jefferies, R. L. (1978).** Demography of *Hieracium pilosella* in a Breck
    grassland. *Journal of Ecology,* **66**, 615–29.
**Chapin, F. S. & Shaver, G. R. (1985).** Individualistic growth responses of tundra plant species to
    environmental manipulations in the field. *Ecology,* **66**, 564–76.
**Chesson, P. L. (1986).** Environmental variation and the coexistence of species. *Community
    Ecology* (Ed. by J. Diamond & T. J. Case), pp. 240–56. Harper and Row, New York.
**Chesson, P. L. & Case, T. J. (1986).** Overview: nonequilibrium community theories: chance,
    variability, history, and coexistence. *Community Ecology* (Ed. by J. Diamond & T. J. Case),
    pp. 229–39. Harper and Row, New York.
**Connell, J. H. (1983).** On the prevalence and relative importance of interspecific competition:
    evidence from field experiments. *American Naturalist,* **122**, 661–96.

Connell, J. H., Tracey, J. G. & Webb, L. J. (1984). Compensatory recruitment, growth, and mortality as factors maintaining rain forest tree diversity. *Ecological Monographs,* 54, 141–64.

Diamond, J. (1986). Overview: laboratory experiments, field experiments, and natural experiments. *Community Ecology* (Ed. by J. Diamond & T. J. Case), pp. 3–22. Harper and Row, New York.

Diamond, J. & Case, T. J. (1986). *Community Ecology.* Harper and Row, New York.

Dickinson, C. E. & Dodd, J. L. (1976). Phenological pattern in the short grass prairie. *American Midland Naturalist,* 96, 367–78.

Dirzo, R. & Sarukhán, J. (1984). *Perspectives on Plant Population Ecology.* Sinauer, Sunderland, Massachusetts.

Fitter, A. H. (1987). Spatial and temporal separation of activity in plant communities: prerequisite or consequence of coexistence? *Organisation of Communities: Past and Present.* (Ed. by P. S. Giller & J. Gee), pp. 119–39. Blackwell Scientific Publications, Oxford.

Fowler, N. L. (1981). Competition and coexistence in a North Carolina grassland. II. The effects of the experimental removal of species. *Journal of Ecology,* 69, 843–54.

Fowler, N. L. (1984). Patchiness in patterns of growth and survival of two grasses. *Oecologia,* 62, 424–28.

Fowler, N. L. (1986a). Density-dependent population regulation in a Texas grassland. *Ecology,* 67, 545–54.

Fowler, N. L. (1986b). The role of competition in plant communities in arid and semiarid regions. *Annual Review of Ecology and Systematics,* 17, 89–110.

Fowler, N. L. (1988). What is a safe site?: Neighbor, litter, germination date, and patch effects. *Ecology,* 69, 947–61.

Fowler, N. L. & Antonovics, J. (1981). Competition and coexistence in a North Carolina grassland. I. Patterns in undisturbed vegetation. *Journal of Ecology,* 69, 825–41.

Gould, F. W. (1975). *The Grasses of Texas.* Texas A & M University Press, College Station, Texas.

Grubb, P. J. (1977). The maintenance of species-richness in plant communities: the importance of the regeneration niche. *Biological Review,* 52, 107–45.

Grubb, P. J. (1986). Problems posed by sparse and patchily distributed species in species-rich plant communities. *Community Ecology* (Ed. by J. Diamond & T. J. Case), pp. 207–25. Harper and Row, New York.

Halvorsan, W. L. & Patten, D. T. (1975). Productivity and flowering of winter ephemerals in relation to Sonoran desert shrubs. *American Midland Naturalist,* 93, 311–19.

Harper, J. L. (1977). *Population Biology of Plants.* Academic Press, London.

Hubbell, S. P. (1979). Tree dispersion, abundance and diversity in a tropical dry forest. *Science,* 203, 1299–309.

Hubbell, S. P. & Foster, R. B. (1986). Biology, chance and history and the structure of tropical rain forest tree communities. *Community Ecology* (Ed. by J. Diamond & T. J. Case), pp. 314–29. Harper and Row, New York.

Huston, M. (1979). A general hypothesis of species diversity. *American Naturalist,* 113, 81–101.

Jackson, J. B. C., Buss, L. W. & Cook, R. E. (1985). *Population Biology and Evolution of Clonal Organisms.* Yale University Press, New Haven.

Kalisz, S. (1986). Variable selection on the timing of germination in *Collinsia verna* (Scrophulariaceae). *Evolution,* 40, 479–91.

Klemow, K. M. & Raynal, D. J. (1981). Population ecology of *Melilotus alba* in a limestone quarry. *Journal of Ecology,* 69, 33–44.

Klemow, K. M. & Raynal, D. J. (1983). Population biology of an annual plant in a temporally variable habitat. *Journal of Ecology,* 71, 691–703.

**Mack, R. N. & Pyke, D. A. (1983).** The demography of *Bromus tectorum*: variation in time and space. *Journal of Ecology*, **71**, 69–93.

**Monson, R. K., Littlejohn, R. O. & Williams, G. J. (1983).** Photosynthetic adaptation to temperature in four species from the Colorado shortgrass steppe: a physiological model for coexistence. *Oecologia*, **58**, 43–51.

**Monson, R. K. & Williams, G. J. (1982).** A correlation between photosynthetic temperature adaptation and seasonal phenology patterns in the shortgrass prairie. *Oecologia*, **54**, 58–62.

**Muller, C. H. (1940).** Plant succession in the *Larrea-Flourensia* climax. *Ecology*, **21**, 206–12.

**Newell, S. J., Solbrig, O. T. & Kincaid, D. T. (1981).** Studies on the population biology of the genus *Viola*. III. The demography of *Viola blanda* and *Viola pallens*. *Journal of Ecology*, **69**, 997–1016.

**Nicholson, A. J. (1957).** The self-adjustment of population to change. *Cold Spring Harbor Symposia on Quantitative Biology*, **22**, 153–73.

**Nobel, P. S. (1980).** Morphology, nurse plants, and minimum apical temperatures for young *Carnegiea gigantea*. *Botanical Gazette*, **141**, 188–91.

**Nobel, P. S. (1981).** Spacing and transpiration of various sized clumps of a desert grass, *Hilaria rigida*. *Journal of Ecology*, **69**, 735–42.

**Platt, W. J. (1975).** The colonization and formation of equilibrium plant species associations on badger disturbances in a tall grass prairie. *Ecological Monographs*, **45**, 285–305.

**Platt, W. J. & Weis, I. M. (1977).** Resource partitioning and competition within a guild of fugitive prairie plants. *American Naturalist*, **111**, 479–513.

**Raynal, D. J. & Bazzaz, F. A. (1975).** The contrasting life-cycle strategies of three summer annuals found in abandoned fields in Illinois. *Journal of Ecology*, **63**, 587–96.

**Robberecht, R., Mahall, B. E. & Nobel, P. S. (1983).** Experimental removal of intraspecific competitors—effects on water relations and productivity of a desert bunchgrass, *Hilaria rigida*. *Oecologia*, **60**, 21–4.

**Sala, O. E. & Lauenroth, W. K. (1982).** Small rainfall events: an ecological role in semiarid regions. *Oecologia*, **53**, 301–4.

**Sala, O. E., Lauenroth, W. K. & Reid, C. P. P. (1982).** Water relations: a new dimension for niche separation between *Bouteloua gracilis* and *Agropyron smithii* in North American semi-arid grasslands. *Journal of Applied Ecology*, **19**, 647–57.

**Schoener, T. W. (1983).** Field experiments on interspecific competition. *American Naturalist*, **122**, 240–85.

**Shmida, S. & Ellner, S. P. (1984).** Coexistence of plant species with similar niches. *Vegetatio*, **58**, 29–55.

**Snaydon, R. W. (1962).** Microdistribution of *Trifolium repens* and its relation to soil factors. *Journal of Ecology*, **50**, 133–43.

**Steenbergh, W. F. & Lowe, C. H. (1977).** Ecology of the saguaro: II. Reproduction, germination, establishment, growth, and survival of the young plant. *National Park Service Scientific Monograph Series, Number 17*. US Gov. Printing Office, Washington, D.C.

**Strong, D. R. (1983).** Natural variability and the manifold mechanisms of ecological communities. *American Naturalist*, **122**, 636–60.

**Strong, D. R. (1986).** Density-vagueness: abiding the variance in the demography of real populations. *Community Ecology* (Ed. by J. Diamond & T. J. Case), pp. 257–268. Harper and Row, New York.

**Strong, D. R., Simberloff, D., Abele, L. G. & Thistle, A. B. (1984).** *Ecological Communities: Conceptual Issues and the Evidence*. Princeton University Press, Princeton, New Jersey.

**Tilman, D. (1982).** *Resource Competition and Community Structure*. Princeton University Press, Princeton, New Jersey.

**Turner, R. M., Alcorn, S. M. & Olin, G. (1969).** Mortality of transplanted saguaro seedlings. *Ecology*, **50**, 835–44.

Watt, A. S. (1947). Pattern and process in the plant community. *Journal of Ecology*, 35, 1–22.

Weaver, J. E. (1954). *North American Prairie*. Johnsen Publishing Company, Lincoln, Nebraska.

Weaver, J. E. & Albertson, F. W. (1956). *Grasslands of the Great Plains*. Johnsen Publishing Company, Lincoln, Nebraska.

Werner, P. A. & Platt, W. J. (1976). Ecological relationships of co-occurring goldenrods (*Solidago*: Compositae). *American Naturalist*, 110, 959–71.

White, J. (1985). *Studies on Plant Demography. A Festschrift for John L. Harper*. Academic Press, London.

Wiens, J. A. (1977). On competition and variable environments. *American Scientist*, 65, 590–97.

Yeaton, R. I. & Cody, M. L. (1976). Competition and spacing in plant communities: the northern Mojave Desert. *Journal of Ecology*, 64, 689–96.

Yeaton, R. I., Travis, J. & Gilinsky, E. (1977). Competition and spacing in plant communities: the Arizona upland association. *Journal of Ecology*, 65, 587–95.

# 14. THE INFLUENCE OF TREE POPULATION DYNAMICS ON FOREST SPECIES COMPOSITION

## T. C. WHITMORE

*Department of Plant Sciences, University of Oxford, South Parks Road, Oxford OX1 3RB, UK*

## SUMMARY

**1** In all forests there is a growth-cycle of gap, building and mature phases of duration 10–1000 years. Gaps vary in size from those created by one tree falling to those created by a cataclysm such as a cyclone or wild fire.

**2** Tree species fall into two ecological groups. Pioneer species grow up in big gaps and climax species in small ones.

**3** Seeds of the pioneer species commonly occur in the soil as a seed bank which is variously heterogeneous in space and time. Seeds germinate after gap creation and groups of pioneer trees are even-aged. Species differ in seed dispersal, seed longevity, seedling establishment, growth requirements and in life-span.

**4** Climax species develop seedling banks below closed forest canopy. Species differ notably in the size of gap required to stimulate seedling growth. Recruitment is continual and populations are multi-aged.

**5** Population processes, mainly of seed and seedling ecology at the gap-phase of the growth-cycle, are an important factor controlling floristic composition of a forest. Floristic stability and the importance of disturbances cannot be specified without considering spatial scale.

**6** The three regions of the world's rain forests show different responses to human disturbance. Forests of the East recover best because they have numerous gap-loving species which are abundant. The neotropical forests recover worst because such species are uncommon. African forests appear to be intermediate in ability to recover.

## INTRODUCTION

This chapter concerns population processes of tree species in forests and their role in the determination of forest community composition. There has been much interest in forest population and community dynamics during the last 20 years and numerous syntheses of the subjects (Whitmore 1978, 1982;

White 1979; Ewel 1980; Pickett 1980; Waring 1980; Oliver 1981; Reichle 1981; West, Shugart & Botkin 1981; Yamamoto 1981; Shugart 1984; Gomez-Pompa & del Amo 1985; Ogden 1985; Pickett & White 1985). The forests of North America have attracted most attention followed by tropical rain forests. The focus here is on the latter. Comparisons and contrasts are drawn with the behaviour of forests in other biomes in order to stimulate further research.

Trees regenerate in canopy gaps. First, therefore it is necessary to discuss the dynamic aspects of forest structure in order to define the role of gaps. This leads to a discussion of the responses of tree species to gaps, and especially of the ways in which they differ in seed and seedling ecology. Based on these population processes various generalizations about species composition in forest communities will be explored.


## FOREST DYNAMICS

### *The forest growth-cycle*

In all forests (and many other communities) there is a growth-cycle (Watt 1947; Cousens 1974; Whitmore 1975, 1978, 1982, 1984). Trees grow up in gaps and it is possible to recognize gap, building and mature phases of the growth-cycle. What grows up in a gap determines the floristic composition of the whole cycle which commonly lasts for a century or more, although in some secondary tropical rain forests it may last less than a decade (Saulei 1984, 1986). At the other extreme, where they are not cut short by fire, growth-cycles of over a millenium occur in some Pacific coast conifer forests of North America, for example in western red cedar (*Thuja plicata*) and coastal redwood (*Sequoia sempervirens*) forests of California (Franklin & Waring 1980; Franklin & Hemstrom 1981).The term 'gap-phase replacement' has come into common usage in recent discussions of forest dynamics. It usefully focusses on the most important phase of the growth-cycle.


### *Forest succession*

In all forests there is an interplay between seedlings of species with different light requirements for survival and growth, and gaps of different size from those caused by a single tree dying 'on its feet', to single treefalls, to huge open swathes created by some cataclysm.

The classic view of forest succession is of a stand of so-called pioneer species with light-demanding seedlings colonizing a large canopy gap and developing into building and then mature forest. Beneath them grow other, so-called, climax species with shade-tolerant seedlings. When the pioneers die the mature-phase canopy breaks up and forms small gaps through which the climax species grow. Thus the change from big gap to small gap is associated with a floristic change, which is succession. So long as no further catastrophe creates big gaps, subsequent growth-cycles are based on small gaps and climax species. In all but the most floristically impoverished forests successive cycles at a particular spot may be of different climax species, but at a large scale species composition is steady: climax is reached. The dynamic process is described as cyclic succession, or more accurately as cyclic replacement (Miles 1979). Fresh investigations repeatedly uncover the operation of this basic model or variations of it.

### Catastrophic disturbance

Many forests in different parts of the world have been shown to have a species composition which arose after a cataclysm, e.g. cyclones (typhoons or hurricanes) or mass earth movements associated with volcanoes or earthquakes (Whitmore 1974; Garwood, Janos & Brokaw 1979), and that will not be maintained unless another cataclysm occurs. For example, this is the case in some *Nothofagus* forests of south-central Chile and New Zealand (Veblen 1985) and in many forests in North America (Oliver 1981). Cataclysm-derived tropical rain forests occur on all continents. Within the last few years another source of catastrophic destruction of tropical rain forests has been discovered with the realization that even they may suffer rare droughts, occurring once a century or less. Vast fires occurred after a 7 month drought in western Malesia in 1982 (Beaman *et al.* 1985; Malingreau, Stephens & Fellows 1985). Signs of past extensive natural fires have recently been discovered as charcoal buried in the soil in northern South America (Sanford *et al.* 1985). In fact, charcoal has often been found in rain forest soil profiles but its significance not realized.

Another remarkable and extensive kind of cataclysm has also recently been reported. Just east of the Andes in Peru the swiftly flowing upper tributaries of the Amazon flood annually and move their courses laterally by up to 180 m (though mostly much less). Every year new alluvium is deposited on which primary forest succession occurs, but the forest is very likely to be swept away before attaining climax (Salo *et al.* 1986). The same phenomenon occurs on a lesser scale south of the central New Guinea cordillera (Johns 1986). These upper Amazonian and Papuan rain forests are thus a mosaic of *primary* successional forests of different age maintained by continual catastrophic destruction.

### Growth from sprouts

New trees may also arise from sprout shoots developed from stems or roots. This is a common method of regeneration in some temperate forests. The coppice system of silviculture, which was important for many centuries until recently in western European deciduous forest, depends on the ability of all the component species (without exception) to produce coppice shoots.

Nowhere in tropical rain forests do foresters base silviculture on coppice regrowth, although it is important in savanna woodlands, and until recently such vegetative regrowth was thought to be rare. To date, any study which shows a natural lowland rain forest in which vegetative reproduction plays a substantial role is lacking, though several recent studies have shown that sprout growth can be locally important (e.g. Putz *et al.* 1983) and that the fire which is often set by humans in forests felled for agriculture reduces the quantity and species diversity of sprouts (e.g. Stocker 1981; Uhl 1982). Heath forest in Malesia lacks a pioneer tree flora and Riswan (1982) showed by experiment that in a felled but unburned Kalimantan heath forest, vegetative shoots were a major source of the regrowth forest. The only similarities found so far to the English forests in which clonally reproducing *Ulmus* is invading (Rackham 1975) are three species of understorey tree in West Africa which, like *Rubus fruticosus*, spread by the layering of drooping branches and form dense patches several tens of metres in diameter (Jenik 1969), and the mangrove *Sonneratia alba* which spreads seawards by similar layering (Holmbrook & Putz 1982). Krummholz in the high mountains of New Guinea does show extensive layering (Gillison 1970), just as it does in temperate zones.

Regrowth of mature-phase forest by sprouts or by layering perpetuates the same species composition and thus succession does not take place. Sometimes, as will be described below, regrowth is a mixture of seedlings and sprouts.

### Simultaneous colonization

The classic mode of succession described above is of a suite of pioneer species being replaced by a suite of climax species. This was termed relay floristics (Egler 1954). Egler pointed out that as an alternative, forests may develop after the simultaneous colonization of all or most of their final species complement and examples have now been found in both temperate and tropical forests.

In a recently deglaciated valley in Alaska many species arrived as seed rain almost simultaneously, and a few trees soon grew to dominate the stand (Oliver, Adams & Zasoski 1985). In the Great Smoky Mountains the small canopy gaps ($15$–$150$ m$^2$) were all invaded by three tree species of different shade tolerance (White, MacKenzie & Busing 1985). In the Laurentian Highlands of Canada, species of a range of tolerances colonized more or less simultaneously after fire (Cogbill 1985). In all these non-tropical forests, site conditions determined which species came to dominate.

In a study on Barro Colorado Island, seedlings of climax forest species were already present at the time of gap creation and became only slightly augmented in numbers during the 5 years of observation. By contrast, the pioneers all appeared soon after gap creation. The two groups of species then grew up together (Brokaw 1985, 1987). At Atewa, Ghana, starting from bare land, 90% of pioneer species and 60% of climax species appeared in the first year (Swaine & Hall 1983). The two groups grew up together, progressively enriched by the remaining 40% of climax species. Thus, in both these tropical examples, colonization was almost but not completely simultaneous.

Forest succession after human disturbance may also start with simultaneous growth of the whole complement of tree species. Where the canopy is opened by timber extraction climax seedlings commonly survive, sprouts may be common and seeds of pioneer species soon germinate. Out of this mixture the developing forest is usually dominated by the pioneers, until eventually the slower growing, long-lived climax species grow through. Such simultaneous colonization occurs in logged forest in the northern hardwood forests of the Appalachians, whereas in the same region, species invade old abandoned pastures in sequence (Marquis 1967; Bormann & Likens 1979).

In many tropical forests seedlings of climax species which survive logging are found to be replaced in big gaps by pioneers, a fact foresters continually struggle against. What causes death of the climax species has not yet been studied although it is of great practical significance for silviculture. Also in many tropical forests photophilic climbers become common, either as seedlings or by vegetative growth from remnants of the former forest, and these can delay canopy regrowth.

## ECOLOGICAL GROUPS OF TREE SPECIES

The basic processes of forest succession depend upon the fact that the seedlings of tree species differ in their tolerance of canopy shade. As mentioned, there is a group of species which can only germinate and establish in the open and a group with shade-tolerant seedlings which can germinate and persist below a canopy (Whitmore, 1988; Swaine & Whitmore, 1988). No single nomenclature is applied to these two groups. Here they are referred to as pioneer and climax species respectively. Many other characteristics, such as growth rate, crown growth pattern and wood density differ within these two groups (van Steenis 1958; Whitmore 1975; Bazzaz & Pickett 1980; Boojh & Ramakrishnan 1982; Vazquez Yanes & Guevara Sada 1985; Shukla & Ramakrishnan 1986). In North America, ecophysiological research has confirmed that the degree of shade tolerance of seedlings is the most important single feature of the tolerance classes (Baker 1950).

Pioneers are normally replaced by shade-tolerant climax species but several temperate forests have been described in which pioneers replace themselves *in situ*. Near the treeline in south-central Chile, *Nothofagus* initially colonize bare land surfaces and pure stands develop. As these age, small canopy gaps form and the same species colonize them. At lower altitudes where other tree species coexist with the *Nothofagus*, the ecological niche of the latter is restricted to big gaps (Veblen *et al.* 1981). Similarly, at the northern forest limit in Europe, pure *Betula* forests develop and perpetuate themselves, whereas further south the same species are seral. In the Pacific north-west of America *Pseudotsuga menziesii* is a pioneer invading after fire and on certain sites it too replaces itself. On other sites shade-tolerant *Picea sitchensis* or *Tsuga heterophylla* invade after fire, acting as pioneers (Franklin & Hemstrom 1981). There seem to be no parallels to these cases yet discovered in tropical rain forests. One might conjecture that the humid tropics are everywhere so rich in tree species that such niche-transgression does not occur. Nevertheless, examples are now known of pioneers establishing in canopy gaps so small that one would, *a priori*, expect to find climax species in them. The scattered observations previously assembled (Whitmore 1982) which showed that conditions for pioneer establishment occur mainly in gaps over 1000 m$^2$ now have several exceptions, Brokaw 1985; Whitmore & Sidiyasa 1986).

The basic division of tree species into these two groups depends on a qualitative and readily observable difference with important consequences for population and community dynamics. It is discussed more fully by Swaine & Whitmore (1988) and appears to be a natural dichotomy without exceptions, but close scrutiny of seed and seedling ecology is needed, especially of niche-transgressing species, because published research sometimes omits key details.

*Population processes of pioneer species*

*Seeds*

Numerous studies made since the early 1970s (reviewed by Whitmore 1983; Vazquez-Yanes & Orozco Segovia 1984) have discovered a soil seed bank mainly of pioneer tree species below both primary and secondary rain forests. Seed rain has also been repeatedly demonstrated. Little is known about the longevity of seeds in the soil bank, or about spatial variation. It still remains to be discovered which features are common to all forests and which variations are on a universal theme. Concerning longevity, more seed was found under a forest in Thailand than the annual production within dispersal range, implying storage for over a year (Cheke, Nanakorn & Yankoses 1979). In a forest near Turrialba, Costa Rica, variations were found between species, from low soil storage and a high input rate to high storage and low input rate of seed (Young, Ewel & Brown 1987). Hopkins & Graham (1984) found losses of pioneer species from the seed bank under the canopy and in small gaps after soil disturbance, implying that there had been germination of these species without successful seedling establishment.

Concerning spatial pattern, the seed bank below a forest may be historical or it may be continually augmented by seed rain. Pioneer species characteristically have small seeds dispersed by either wind or animals, though in both cases most fall near the parent tree as a seed 'footprint'* (e.g. *Trema guineensis* in the Ivory Coast (Alexandre 1978)). One possibility is that the seed bank in a forest is made up of a series of footprints, each slightly larger than the crown(s) of a former single tree or group of parent trees, becoming fainter as the seeds die. Seed rain seems important in Costa Rican and Malaysian forest studies by Young, Ewel & Brown (1987) and Putz & Appanah (1987) because the seed bank diminishes with increasing distance into primary forest away from seed sources. Seed rain may be patchy. For example in pasture at Los Tuxtlas, Mexico, bird-dispersed seed is much commoner below relict trees which act as perches and as foci for forest restoration (Guevara, Purata & van der Maarel 1986). Whether seed is concentrated in a similar way below continuous rain forest has not been ascertained. In Penang, floristic dissimilarity in seedling composition in small gaps in primary forest increases with their distance apart (Raich 1987), which rules out homogeneous seed rain though it does not discriminate between other variables.

Where the amount of seed stored in the forest soil is large due to proximity to a seed source or to site history, recruitment into a gap from the seed bank may overwhelm that from newly arrived seeds or sprouts (Young, Ewel & Brown, 1987; Putz & Appanah, 1987). All these confusing variables at the seed stage of the tree's life-cycle have clear demographic implications at gap-phase replacement. There are yet more variables after the seeds have germinated.

* Whose depth (i.e. seed density) can vary from one place to another; often called a 'seed-shadow', which is a misleading term in that a shadow implies a negative, something lacking.

*Seedlings*

Several studies have now been conducted which demonstrate that species differ in their microsite requirements for seedling establishment. For example, in a large artificial clearing in north Venezuela six different types of microsites were created and differences demonstrated in the success of species planted on them as seed (Uhl *et al.* 1981). In French Guiana, *Cecropia obtusa* preferentially establishes on treefall mounds (Riera 1985). On Barro Colorado Island, Brokaw (1987) studied thirty natural gaps and found differences in the gap-sizes in which survivors of three different species occurred after 8–9 years, as well as differences in the timing of germination after gap formation. The different parts of a gap also provide different microsites and at La Selva, Costa Rica, they have been shown to be colonized by different species (Brandani, Hartshorn & Orians 1988). Species have been shown to have different susceptibilities to death from uprooting or snapping (Putz *et al.* 1983). Thus different species create different microsites in the gaps created by their death and therefore influence the species composition of the next forest growth-cycle.

Another important aspect of seedling ecology is that the degree of alteration after clearance of a tropical forest site influences the species composition of the pioneer forest which grows up. Thus site conditions select a subset of the locally available pioneer flora. This is shown in *Cecropia* and *Macaranga*, archetypal pioneers of the American and Eastern rain forests respectively, which are confined to more eutrophic sites (C. Jordan and T. C. Whitmore, unpublished data). Another example is that in the Osa Peninsula forests of Costa Rica, different pioneer tree and shrub species invade abandoned farms and forest gaps (Herwitz 1981), an observation repeatable in the rain forests of the East.

*Longevity*

A few rain forest pioneer tree species live for only about 10 years. They never grow large and the growth cycle is very short. At the other extreme are pioneer species which may reach 100 years old. These become big trees, and because of their fast growth and pale soft wood many are valuable timber species. New Guinea has examples of both extremes, e.g. *Pipturus argenteus* and *Eucalyptus deglupta* respectively, as well as many species of intermediate longevity and stature (e.g. *Macaranga* spp., *Trichospermum* spp.).

*Population structure*

Pioneer species occur as groups of even-aged trees. Within a group the trees will be more or less the same height because any individual which grows more slowly and becomes overtopped soon dies. Thus, stands of pioneers characteristically have an even canopy top. As the trees grow older girth becomes increasingly variable.

## Conclusion

Beyond the general rule that pioneers germinate from seed in big gaps there are seen to be numerous alternative possibilities as to which species colonize a gap and form the next tree population. It seems likely that future studies will unravel forest-specific combinations of these alternatives rather than uncover further universal general rules.

### Population processes of climax species

## Seeds and seedlings

Unlike pioneers, climax species usually germinate under a closed canopy (Swaine & Whitmore 1988). Many rain forest climax species have seeds with no capacity for dormancy which germinate immediately—so-called 'recalcitrant' seeds (Roberts 1973). Others, for example many Leguminosae, do have seeds with a dormant period. These are found in the soil seed bank, from which they may germinate either simultaneously or staggered over a period of time.

The seedlings establish under closed canopy and await 'release' by the development over them of a gap. Species differ in the size of gap needed to stimulate growth. There are as yet few detailed studies about this from rain forests, though foresters have accumulated much anecdotal information. At one extreme are species which never attain the top of the canopy and investigations of rain forest floristics have discovered whole genera and families confined to this synusia. In Asia, the Annonaceae, Ebenaceae, Euphorbiaceae, Myristicaceae and Rubiaceae are examples. Presumably, given the ability to compete below ground for nutrients and water, lower and middle canopy trees grow up to their mature size wherever there is room for crown development.

Among canopy-top climax species there is a spectrum from species with seedlings able to grow up in very small gaps, which are typically slow growers with dense, dark timber, to others which are only released in larger gaps, are relatively fast growers and have lighter, paler timber. There are several mortality factors to be avoided before a seedling is established and then released: seeds must arrive and germinate, the microsite must be suitable for seedling establishment, the seedling must survive predation, fungal pathogen attack, damage from trampling or falling branches and then, for most climax species, a gap must occur over it of a size within which it can succeed. Seedling populations of climax species may be evanescent in time. They are seldom uniform in space. The initial pattern arising after dispersal is commonly, perhaps always, a footprint around the parent tree, with density diminishing with distance. Density-dependent mortality from predation or disease has now been demonstrated several times (Howe & Smallwood 1982; Augspurger 1984; Clark & Clark 1984; Becker & Wong 1985; Howe, Schupp & Westley 1985; Sterner, Ribic & Schatz 1986) although it is not universal (e.g. several dipterocarps studied by Burgess (1969, 1975) and Chan (1980)). So far allelopathic effects have not been detected in rain forest by contrast *Eucalyptus regnans* in South Australia, (Ashton & Willis 1982). In a careful study over 17 years in a Queensland rain forest, seedling and sapling mortality was found to be independent of the proximity of a conspecific adult, i.e. future pattern will not repeat the present one (Connell, Tracey & Webb 1984).

The species composition of a forest patch is determined at the gap-phase, making this the most important phase of the growth-cycle, and chance operates in many different ways to control which of numerous species succeeds. This was described for the Far Eastern rain forests by Whitmore (1975) and was set in its broadest context for plant communities in general by Grubb (1977) who coined the useful term 'regeneration niche'. Now examples have been assembled from many other forests.

*Longevity*

The longevity of tropical rain forest trees cannot be measured directly because they do not have growth rings (Bormann & Berlyn 1981). It is notoriously unreliable to make age estimates for such species from girth measurements. It is now generally believed that some individuals become suppressed, grow slowly and never reach full size while others grow faster and become bigger, probably because they occupy more favourable microsites (e.g. Primack *et al.* 1985). This explains the common observation in high forest that average growth rate increases with tree size (Alder 1983).

Swaine, Lieberman & Putz (1987) studied eighteen long-term rain forest study plots. Annual mortality of trees was mainly in the range of 1–2%. This represents a two-fold difference in the number of trees dying. Projected half-life values ranged from 35–69 years. Gap creation at about 1% of total area per year was reported for the North American spruce fir forests over a broad range of latitudes (White, MacKenzie & Busing 1985) and also in the temperate deciduous forest there (Runkle 1982), though it was less in a

Kentucky forest (Romme & Martin 1982). It can be seen that in many forests, for a climax species to perpetuate itself recruitment of a new tree need only take place at most a few times per century.

*Population structure*

Typically a climax species has a population with more small than big girth individuals. Age may be roughly equated to size with the provisos given above. The decline of the population with age is sometimes exponential, implying equal probability of mortality for all sizes, as in the tropical forests reported by Swaine, Lieberman & Putz (1987). However, in two different temperate forests Hett & Loucks (1976) and Veblen, Ashton & Schlegel (1979) found a power function gave a closer fit to survivorship data with the parameters set to show a decreasing rate of mortality in successive size classes.

A declining curve of tree numbers against girth is the classic expression of a multi-aged stand, so long as recruitment rate is steady (Veblen 1988), and a normal curve represents an even-aged stand, resulting from one single pulse of recruitment. However, differential mortality may lead to size distribution totally obscuring age structure, as Ogden (1985) has clearly described for New Zealand Kauri (*Agathis australis*). Also, Ford (1975) and Lorimer (1985) have shown how size–class bimodality may develop as some plants come to dominate by faster growth.

There have as yet been few studies with a sufficiently long time-base to enable us to analyse recruitment and mortality rates for individual rain forest species. At Kade, Ghana, *Celtis mildbraedii*, studied over 14 years, maintained itself by low death- and recruitment-rates in comparison with the other species in the same forest which had more rapid turnover (Swaine, Hall & Alexander 1987). Thus, community species composition remained stable at this site despite differences in population turnover rates between species.

Two recent studies in temperate forests have also shown species composition being maintained with differences in turnover rates. In subalpine conifer forests in Colorado, *Abies lasiocarpa* has big subcanopy populations and high recruitment into the canopy, where it is 37% of the trees but also 76% of fallen trees. By contrast, *Picea engelmanii* has consistently lower treefall rates, fewer juveniles and a lower recruitment rate. By these different life-histories the species are maintaining their balance (Veblen 1986). In the Great Smoky Mountains *Abies fraseri* has a high density of small individuals and high canopy turnover rate in contrast to *Picea rubens* (White, MacKenzie & Busing 1985).

A different situation was observed in two tropical rain forests where population structure suggests that species composition was not stable. Firstly, at Sungai Menyala, Malaysia, the biggest tree species are not well represented in the pole-sized middle part of the canopy (Wyatt-Smith 1966). The deduction made is that this forest is changing in composition (Whitmore 1975). Secondly at Okomu, Nigeria, in a forest with commercially valuable Meliaceae the commonest family of emergents, 76% of the emergent species had smaller size classes missing. One-third of these species were very strong light-demanders (Jones 1950, 1955–56). There were many dead emergents. Species in the lower and middle strata were, by contrast, sufficiently numerous to maintain themselves. The implication is that this forest will change in composition unless a major disturbance occurs followed by fresh recruitment of the emergent light-demanding species.

Differences in turnover may also occur at seedling stage. On Kolombangara in the Solomon Islands, seedling populations of the main canopy-top trees were monitored for 6.6 years. *Calophyllum kajewskii* and *C. vitiense* had big evanescent populations in contrast to *Dillenia salomonensis* and *Parinari salomonensis* which both had smaller but long-lived populations (Whitmore 1974). The composition of the next forest growth-cycle on Kolombangara clearly depends on the timing and positioning of canopy gaps.

## THE NATURE OF THE FOREST COMMUNITY

The floristic variation in forests can be roughly arranged in a hierarchy of progressively more local factors (Prentice 1986; for tropical forests, Whitmore 1975, Swaine & Hall 1986; and for all communities, Sousa 1984). Biogeography, reflecting evolutionary and Earth-history (including past changes in climate) stands first, followed by variation at the forest formation level between grossly different habitats, for example tropical heath forest on podzolized siliceous sands (spodosols) versus mesic rain forest on ultisols and oxisols. This level of variation is usually sharply discontinuous. Third level variation is that within a forest formation and we may initially distinguish gross topography, ridges, hillsides and valleys (see for example the Barro Colorado Island study of Hubbell & Foster (1983)). This section focusses on the residual variation once all these others have been allowed for, namely that variation which is concerned with gap-phase replacement. It is at this lowest level that population processes operate. Both determinism and many kinds of chance play their role.

### Floristic patches

Two recent analyses of floristic pattern in rain forest have revealed small clumps of climax tree species in mature-phase forest which grew up together in gaps, a direct demonstration of the operation of the forest growth-cycle. One study was of a Sarawak heath forest where there was also a coarser floristic pattern determined by soil (Newbery, Renshaw & Brunig 1986). The other was at Tai, Ivory Coast, where there was a coarser pattern related to topography (Vooren 1985). At La Selva, Costa Rica, one of three undergrowth palm species investigated had clump sizes corresponding to treefall canopy gaps (Richards & Williamson 1975).

## Stability

The question has arisen from the numerous recent studies of forest dynamics whether community floristic composition is stable or continually changing, i.e. whether it is in equilibrium or not (Drury & Nisbet 1973; Hubbell 1979; Huston 1979; Pickett 1980; Finegan 1984; Ogden 1985; Veblen 1985). Forests in which external factors such as freak windstorms, earthquakes, fire or disease epidemics (Augspurger 1988), continually create big gaps, may never attain an equilibrium condition, although over a whole landscape the area of secondary forest may remain roughly constant (Sousa 1984). It is now usually accepted that areas of successional forest should be regarded as part of the total forest community, whose species list therefore includes pioneer species. The term 'shifting mosaic steady state' aptly describes the dynamic nature of the whole community (Bormann & Likens 1979). Within both big and small gaps, replacement will not commonly be by the same species as formed the previous canopy. Again, over a large area, species composition may remain stable. The question regarding floristic stability is too loose: it cannot be answered without defining spatial scale.

## Intermediate disturbance and species richness

Another question is how so many species coexist in some forests. The discussions above demonstrate many factors which contribute to species richness. Any discussion of this topic also has to specify the scale, because the different levels of variation in the community have different associated species. The 'intermediate disturbance hypothesis' (Margalef 1963; Connell 1978) postulated that the richest forests are those in a state of succession following disturbance because they contain both pioneer and climax species. Reflection will show that this hypothesis is bound to be true at some particular spatial scale, but it is not necessarily true at all scales (see examples in Whitmore 1982, p. 54).

## Rare species

How climax tree species manage to coexist in rain forests in parts of
Amazonia and Malesia at the rate of over 100 and sometimes over 200
species $\geqslant$ 10 cm diameter per hectare (see Whitmore 1984, Fig. 1.5) is one of
the questions which exercises a perennial fascination for tropical botanists.
It is unlikely to depend solely on the disturbance regime, which is restricted
in these forests to small gaps. There are commonly 400–600 tree stems per
hectare, so there are on average only three or four stems per species per
hectare, but because some species are invariably relatively common, many
always occur at a density of less than one tree per hectare.

## Differences between the three great rain forest regions

The different autecological properties of different tree species (viz. their
'silvics') are of great practical as well as scientific interest. Sustained timber
production depends on the regrowth of mature-phase forest, and its species
composition will depend on what grows up in the gaps created by timber
extraction. The greater the disturbance the more the light-demanding species
are favoured. Silviculture depends on autecological knowledge and upon
manipulation of the forest before, during or after the logging operation. But
humans can only work with what Nature has provided and there appear to
be real differences in the relative proportions of species with different
regeneration patterns between the three great regions of the world's tropical
rain forests, though generalizations based on present knowledge are of
necessity tentative.

   The Far Eastern rain forests have a long history of silviculture and
ecological study and are the main contemporary source of tropical hardwood
timber. In the dipterocarp rain forests of Malaysia, Brunei, the Philippines
and western Indonesia the floor is carpeted with dipterocarp seedlings for
much of the time. The family flowers and fruits gregariously once or twice a
decade and these events are followed by an influx of millions of seedlings per
hectare. The cohorts die away, different species at different rates, but they
are usually augmented by another seedling influx before they totally disappear
(Fox 1972; Whitmore 1975). A high percentage of the dipterocarps are of
relatively light-demanding species which respond to substantial canopy
openings by vigorous upgrowth. Dipterocarp rain forests are the tropical
silviculturist's dream because, so long as care is taken not to destroy the
seedling carpet during logging, they are so easy to regenerate.

   Silviculture consists of controlling the amount of canopy opening and damage, then freeing released seedlings from competition from climbers and non economic 'weed' tree species. These dipterocarp forests are unique in their high density of seedlings, so many of which are of light-demanding species. Further east, beyond Wallace's Line, dipterocarps are rare, except in a few small areas of New Guinea, but the forests contain numerous other climax species with relatively light-demanding seedlings which respond in a similar way, as well as numerous big pioneer species. New Guinea forests, especially those of Papua New Guinea, are continually wracked by cataclysmic destruction by earthquake, landslide, volcano, flooding rivers, fire and human destruction (Johns 1986). This is reflected in canopy composition, for these conditions ensure that big pioneers and light-demanding climax species are common amongst the canopy-top trees (e.g. *Anthocephalus chinensis, Campnosperma* spp., *Eucalyptus deglupta, Octomeles sumatrana, Pometia* spp.). The Solomons further south and east are at the northern edge of the south-west Pacific cyclone belt and the commercially valuable and formerly extensive *Campnosperma brevipetiolatum* forests as well as *Endospermum medullosum, Gmelina moluccana* and *Terminalia calamansanai* all grow up after cyclone destruction (Whitmore 1974). These species, especially the *Campnosperma* and *Terminalia*, have been planted after logging, for which their autecology well suits them.

South American rain forests comprise about half of the total global area of the biome. The timber industry and silviculture are still very rudimentary. Work in Brazil (Pitt 1961a, b; Jankauskis 1983; Silva, personal communication) and Suriname (Schulz 1960; de Graaf 1986) all shows a major difference from the Far Eastern forests. Seedlings of climax species are present but never at the density found after a dipterocarp gregarious fruiting. Moreover, most species are very shade-tolerant and respond to canopy-opening only by slow growth. The present mature-phase forest consists of the strongly to extremely shade-tolerant species which dominate the seedling bank. Members of the Leguminosae, subfamily Caesalpinoideae are prominent. There are few species or individuals of large pioneers. The practical consequence is that the massive canopy opening of a modern commercial logging operation is followed by a replacement forest of very different composition, comprising tree species uncommon in the undisturbed forest where they are largely confined to river banks (e.g. the big pioneers *Goupia glabra, Laetia procera*) or of commercially useless small pioneers (e.g. *Cecropia*), or by climber tangles. Silviculture is much more difficult than in the Far Eastern forests and, if and when large-scale timber exploitation commences, it may be very difficult to grow commercially useful replacement forest to make utilization sustainable. The extremely valuable *Cedrela* spp., and *Swietenia* spp., abundant in some parts of the American forests, are both big pioneers the seeds of which are not stored in soil. Their silviculture depends on ensuring that suitable seed trees survive exploitation.

In West African rain forests the most valuable commercial species are the big top-of-canopy Meliaceae. These are climax species and amongst them *Entandophragma* spp. and *Khaya* spp. have light-demanding seedlings, enabling them to grow well after modest timber extraction and canopy opening. There are also areas where very shade-tolerant caesalpinoid legume species form consociations in the top of the canopy and so are able to perpetuate *in situ* in a regime of small canopy gaps (Jones 1955–56). The rain forest outliers of Uganda are similar (Eggeling 1947). These African forests appear intermediate between the Eastern and American ones.


## ACKNOWLEDGMENTS

I find it very difficult to keep abreast of research in forest dynamics even just in the tropics and I am extremely grateful to the numerous scientists throughout the world who have sent me reprints of their work or answered my written enquiries. I apologize for not here mentioning every separate study. The Brazilian Academy of Sciences (CNPq) kindly made possible a short visit to Brazil. That visit and one to Ghana was also supported by the Royal Society. Both proved most valuable. I thank F. E. Putz, M. D. Swaine, T. T. Veblen, K. Young and an anonymous referee for comments on a draft, and C. Brotherton and C. Budden for much secretarial assistance.


## REFERENCES

Alder, D. (1983). *Growth and yield of the mixed forests of the humid tropics.* Consultancy report for FAO, Oxford.

Alexandre, D. Y. (1978). Observations sur l'ecologie de *Trema guineensis*. *Office de la Recherche Scientifique et Technique Outre-Mer en basse Côte d'Ivoire. Cahiers series Biologie*, **13**, 261–66.

Ashton, D. H. & Willis, E. J. (1982). Antagonisms in the regeneration of *Eucalyptus regnans* in the mature forest. *The Plant Community as a Working Mechanism* (Ed. by E. I. Newman), pp. 113–28. Blackwell Scientific Publications, Oxford.

Augspurger, C. A. (1984). Light requirements of neotropical tree seedlings: a comparative study of growth and survival. *Journal of Ecology*, **72**, 777–95.

Augspurger, C. A. (1988). The impact of pathogens on natural plant populations. *Plant Population Ecology* (Ed. by A. J. Davy, M. J. Hutchings & A. R. Watkinson), pp. 413–33. Blackwell Scientific Publications, Oxford.

Baker, F. S. (1950). *Principles of Silviculture.* McGraw Hill, New York.

Bazzaz, F. A. & Pickett, S. T. A. (1980). Physiological ecology of tropical succession: a comparative review. *Annual Review of Ecology and Systematics*, **11**, 287–310.

Beaman, R. S., Beaman, J. H., Marsh, C. W. & Woods, P. V. (1985). Drought and forest fires in Sabah in 1983. *Sabah Society Journal*, **8**, 10–30.

Becker, P. & Wong, M. (1985). Seed dispersal, seedling ecology, seed predation and juvenile mortality of *Aglaia* sp. (Meliaceae) in lowland dipterocarp rain forest. *Biotropica*, 17, 230–37.

Boojh, R. & Ramakrishnan, P. S. (1982). Growth strategy of trees related to successional status. I. Architecture and extension growth. II. Leaf dynamics. *Forest Ecology and Management*, 4, 359–74, 375–86.

Bormann, F. H. & Berlyn, G. (1981). *Age and growth rate of tropical trees: new directions for research*. Bulletin 94, Yale University School of Forestry and Environmental Studies, New Haven.

Bormann, F. H. & Likens, G. E. (1979). *Pattern and Process in a Forested Ecosystem*. Springer, New York.

Brandani, A., Hartshorn, G. S. & Orians, G. H. (1988). Internal heterogeneity of gaps and tropical tree species richness. *Journal of Tropical Ecology*, 4, 99–119.

Brokaw, N. V. L. (1985). Gap-phase regeneration in a tropical forest. *Ecology*, 66, 682–87.

Brokaw, N. V. L. (1987). Gap-phase regeneration of three pioneer tree species in a tropical forest. *Journal of Ecology*, 75, 9–19.

Burgess, P. F. (1969). Preliminary observations on the autecology of *Shorea curtisii* Dyer ex King in the Malay Peninsula. *Malayan Forester*, 32, 438.

Burgess, P. F. (1975). Silviculture in the hill forests of the Malay Peninsula. *Malaysian Forest Department Research Pamphlet* 66.

Chan, H. T. (1980). Reproductive biology of some Malaysian dipterocarps. *Tropical Ecology and Development* (Ed. by J. I. Furrado), pp. 169–75. International Society of Tropical Ecology, Kuala Lumpur.

Cheke, A. S., Nanakorn, W. & Yankoses, C. (1979). Dormancy and dispersal of seeds of secondary forest species under the canopy of a primary tropical rain forest in northern Thailand. *Biotropica*, 11, 88–95.

Clark, D. A. & Clark, D. B. (1984). Spacing dynamics of a tropical rain forest tree: evaluation of the Janzen:Connell model. *American Naturalist*, 124, 769–88.

Cogbill, C. V. (1985). Dynamics of the boreal forests of the Laurentian Highlands, Canada. *Canadian Journal of Forest Research*, 15, 252–61.

Connell, J. H. (1978). Diversity in tropical rain forests and coral reefs. *Science*, 199, 1302–10.

Connell, J. H., Tracey, J. G. & Webb, L. J. (1984). Compensatory recruitment, growth and mortality as factors maintaining rain forest tree diversity. *Ecological Monographs*, 54, 141–64.

Cousens, J. (1974). *An Introduction to Woodland Ecology*. Oliver and Boyd, Edinburgh.

de Graaf, N. R. (1986). *A Silvicultural System for Natural Regeneration of Tropical Rain Forest in Suriname*. Agriculture University, Wageningen.

Drury, W. H. & Nisbet, I. C. T. (1973). Succession. *Journal of the Arnold Arboretum*, 54, 331–68.

Eggeling, W. J. (1947). Observations on the ecology of Budongo rain forest, Uganda. *Journal of Ecology*, 34, 20–87.

Egler, F. E. (1954). Vegetation science concepts. 1. Initial floristic composition, a factor in old field vegetation development. *Vegetatio*, 4, 412–17.

Ewel, J. (1980). Tropical succession. *Biotropica*, 12, supplement.

Finegan, B. (1984). Forest succession. *Nature*, 312, 109–14.

Ford, E. D. (1975). Competition and stand structure in some even-aged plant monocultures. *Journal of Ecology*, 63, 311–34.

Fox, J. E. D. (1972). *The Natural Vegetation of Sabah and Natural Regeneration of the Dipterocarp Forests*. Ph.D. thesis, University of Wales.

Franklin, J. F. & Hemstrom, M. A. (1981). Aspects of succession in the coniferous forests of the Pacific northwest. *Forest Succession* (Ed. by D. C. West, H. H. Shugart & D. B. Botkin), pp. 212–29. Springer, New York.

**Franklin, J. F. & Waring, R. H. (1980).** Distinctive features of the northwestern coniferous forests. *Forests: Fresh Perspectives from Ecosystem Analysis* (Ed. R. H. Waring), pp. 59–86. Oregon State University Press, Corvallis.

**Garwood, N. C., Janos, D. P. & Brokaw, N. (1979).** Earthquake-caused landslides: a major disturbance to tropical forests. *Science*, **205**, 997–99.

**Gillison, A. N. (1970).** Structure and floristics of a montane grassland/forest transition, Doma Peaks region, Papua. *Blumea*, **18**, 71–79.

**Gomez-Pompa, A. & del Amo, S. R. (1985).** *Investigaciones Sobre la Regeneracion de Selvas Altas en Veracruz, Mexico* 2. INIREB, Xalapa.

**Grubb, P. J. (1977).** The maintenance of species richness in plant communities: the importance of the regeneration niche. *Biological Reviews*, **52**, 107–45.

**Guevara, S., Purata, S. E. & van der Maarel, E. (1986).** The role of remnant forest trees in tropical secondary succession. *Vegetatio*, **67**, 77–84.

**Herwitz, S. R. (1981).** Regeneration of selected tropical tree species in Corcovado National Park, Costa Rica. *University of California Publications in Geography*, 24.

**Hett, J. M. & Loucks, O. E. (1976).** Age structure models of balsam fir and eastern hemlock. *Journal of Ecology*, **64**, 965–74.

**Holmbrook, N. M. & Putz, F. E. (1982).** Vegetative seaward expansion of *Sonneratia alba* trees in a Malaysian mangrove forest. *Malaysian Forester*, **45**, 278–81.

**Hopkins, M. S. & Graham, A. W. (1984).** The role of soil seed banks in regeneration in canopy gaps in Australian tropical lowland rain forest–preliminary field experiments. *Malaysian Forester*, **47**, 146–58.

**Howe, H. F., Schupp, E. W. & Westley, L. C. (1985).** Early consequences of seed dispersal for a neotropical tree (*Virola surinamensis*). *Ecology*, **66**, 781–91.

**Howe, H. E. & Smallwood, J. (1982).** Ecology of seed dispersal. *Annual Review of Ecology & Systematics*, **13**, 201–28.

**Hubbell, S. P. (1979).** Tree dispersion, abundance and diversity in a tropical dry forest. *Science*, **215**, 1299–309.

**Hubbell, S. P. & Foster, R. B. (1983).** Diversity of canopy trees in a neotropical forest and implications for conservation. *Tropical Rain Forest Ecology and Management* (Ed. by S. L. Sutton, T. C. Whitmore & A. C. Chadwick), pp. 25–42. Blackwell Scientific Publications, Oxford.

**Huston, M. (1979).** A general hypothesis of species diversity. *American Naturalist*, **113**, 81–101.

**Jankauskis, J. (1983).** *Recuperacao de Florestas Tropicais Mecanicamente Explorada*. SUDAM, Belem.

**Jenik, J. (1969).** The life of *Scaphopetalum amoeum* A. Chev. *Preslia*, **41**, 109–12.

**Johns, R. J. (1986).** The instability of the tropical ecosystem in New Guinea. *Blumea*, **31**, 341–71.

**Jones, E. W. (1950).** Some aspects of natural regeneration in the Benin rain forest. *Empire Forestry Review*, **29**, 108–24.

**Jones, E. W. (1955–56).** Ecological studies on the rain forest of southern Nigeria. IV. The plateau forest of the Okomu forest reserve. IV (Continued). *Journal of Ecology*, **43**, 564–94; **44**, 83–117.

**Lorimer, C. G. (1985).** Methodological considerations in the analysis of forest disturbance history. *Canadian Journal of Forest Research*, **15**, 200–13.

**Malingreau, J. P., Stevens, G. & Fellows, L. (1985).** Remote sensing of forest fires: Kalimantan and north Borneo 1982–3. *Ambio*, **14**, 314–21.

**Margalef, R. (1963).** On certain unifying principles in ecology. *American Naturalist*, **97**, 73–88.

**Marquis, D. (1967).** Clear cutting in northern hardwoods. *United States Department of Agriculture Forest Science Research Paper*. NE85.

**Miles, J. (1979).** *Vegetation Dynamics*. Chapman & Hall, London.

Newbery, D. McC., Renshaw, E. & Brunig, E. F. (1986). Spatial pattern of trees in kerangas forest, Sarawak. *Vegetatio*, **65**, 77–89.

Ogden, J. (1985). An introduction to plant demography with special reference to New Zealand trees. *New Zealand Journal of Botany*, **23**, 751–72.

Oliver, C. D. (1981). Forest development in North America following major disturbance. *Forest Ecology and Management*, **3**, 153–68.

Oliver, C. D., Adams, A. B. & Zasoski, R. J. (1985). Disturbance patterns and forest development in a recently deglaciated valley in the northwestern Cascade Range of Washington USA. *Canadian Journal of Forest Research*, **15**, 221–32.

Pickett, S. T. A. (1980). Non-equilibrium co-existence of plants. *Bulletin of the Torrey Botanical Club*, **107**, 238–48.

Pickett, S. T. A. & White, P. S. (Eds.) (1985). *The Ecology of Natural Disturbance and Patch Dynamics*. Academic Press, New York.

Pitt, J. (1961a). *Application of Silvicultural Methods to Some of the Forests of the Amazon*. Expanded Technical Assistance Program FAO 1337. FAO, Brazil.

Pitt, J. (1961b). Amazon forests. Possible methods of regeneration and improvement. *Unasylva*, **15**, 63–9.

Prentice, C. (1986). Vegetation responses to past climatic variation. *Vegetatio*, **67**, 131–41.

Primack, R. B., Ashton, P. S., Chai, P. & Lee, H. S. (1985). Growth rate and population structure of Moraceae trees in Sarawak, East Malaysia. *Ecology*, **66**, 577–88.

Putz, F. E. & Appanah, S. (1987) Buried seeds, newly dispersed seeds, and the dynamics of a lowland forest in Malaysia. *Biotropica*, **19**, 326–33.

Putz, F. E., Coley, P. D., Lu, K., Montalvo, A. & Aiello, A. (1983). Uprooting and snapping of trees: structural determinants and ecological consequences. *Canadian Journal of Forest Research*, **13**, 1011–20.

Rackham, O. (1975). *Hayley Wood: Its History and Ecology*. Cambridgeshire & Isle of Ely Naturalists Trust Ltd, Cambridge.

Raich, J. W. (1987). *Canopy Openings, Seed Germination and Tree Regeneration in Malaysian Coastal Hill Dipterocarp Forest*. Ph. D. thesis, Duke University.

Reichle, D. E. (1981). *Dynamic Properties of Forest Ecosystems*. University Press, Cambridge.

Richards, P. & Williamson, G. B. (1975). Treefalls and patterns of understorey in a wet lowland tropical forest. *Ecology*, **56**, 1226–29.

Riera, B. (1985). Importance des buttes de deracinement dans la regeneration forestiere en Guyane Francais. *Revue Ecologie (Terre Vie)*, **40**, 321–29.

Riswan, S. (1982). *Ecological Studies on Primary, Secondary and Experimentally Cleared Mixed Dipterocarp Forest and Kerangas Forest in East Kalimantan, Indonesia*. Ph.D. thesis, Aberdeen University.

Roberts, E. H. (1973). Predicting the storage life of seeds. *Seed Science and Technology*, **1**, 499–574.

Romme, W. H. & Martin, W. H. (1982). Natural disturbance by treefalls in old growth mixed mesophytic forest: Lilley Cornett Woods, Kentucky. *Proceedings 4th Central Hardwoods Forest Conference, University of Kentucky, Lexington*, pp. 367–83.

Runkle, J. R. (1982). Patterns of disturbance in some old-growth mesic forests of eastern North America. *Ecology*, **63**, 1533–46.

Salo, J., Kalliola, R., Hakkinen, I., Makinen, Y., Niemala, P., Puhakka, M. & Coley, P. D. (1986). River dynamics and the diversity of Amazon lowland forest. *Nature*, **322**, 254–8.

Sanford, R. L., Saldarriaga, J., Clark, K. E., Uhl, C. & Herrera, R. (1985). Amazon rain-forest fires. *Science*, **227**, 53–5.

Saulei, S. M. (1984). Natural regeneration following clear-fell logging operations in the Gogol Valley, Papua New Guinea. *Ambio*, **13**, 351–54.

**Saulei, S. M. (1986).** *The Recovery of Tropical Lowland Rain Forest After Clear Fell Logging at the Gogol Valley, Papua New Guinea.* Ph.D. thesis University of Aberdeen.

**Schulz, J. P. (1960).** Ecological studies on the rain forest of northern Surinam. *The Vegetation of Suriname,* Vol. 2. North Holland, Amsterdam.

**Shugart, H. H. (1984).** *A Theory of Forest Dynamics.* Springer, New York.

**Shukla, R. P. & Ramakrishnan, P. S. (1986).** Architecture and growth strategies of tropical trees in relation to successional status. *Journal of Ecology,* **74,** 33–46.

**Sousa, W. P. (1984).** The role of disturbance in natural communities. *Annual Review of Ecology and Systematics,* **15,** 353–91.

**Steenis, C. G. G. J. van (1958).** Rejuvenation as a factor for judging the status of vegetation types: the biological nomad theory. *Study of Tropical Vegetation. Proceedings of the Kandy Symposium (1956) Humid Tropics Research,* pp. 212–15. UNESCO, Paris.

**Sterner, R. W., Ribic, C. A. & Schatz, G. E. (1986).** Testing for life historical changes in spatial patterns of four tropical tree species. *Journal of Ecology,* **74,** 621–33.

**Stocker, G. C. (1981).** Regeneration of a north Queensland rain forest following felling and burning. *Biotropica,* **13,** 86–92.

**Swaine, M. D. & Hall, J. B. (1983).** Early succession on cleared forest land in Ghana. *Journal of Ecology,* **71,** 601–28.

**Swaine, M. D. & Hall, J. B. (1986).** Forest structure and dynamics. *Plant Ecology in West Africa* (Ed. by G. W. Lawton), pp. 47–93. Wiley, Chichester.

**Swaine, M. D., Hall, J. B. & Alexander, J. J. 1987.** Tree population dynamics at Kade, Ghana (1968–1982). *Journal of Tropical Ecology,* **3,** 331–46.

**Swaine, M. D., Lieberman, D. & Putz, F. E. (1987).** The dynamics of tree populations in tropical forests: a review. *Journal of Tropical Ecology,* **3,** 359–66.

**Swaine, M. D. & Whitmore, T. C. (1988).** On the definition of ecological species groups in tropical rain forests. *Vegetatio (in press).*

**Uhl, C. (1982).** Recovery following disturbance of different intensities in the Amazon rain forest of Venezuela. *Interciencia,* **7,** 19–24.

**Uhl, C., Clark, K., Clark, H. & Murphy, P. (1981).** Early plant succession after cutting and burning in the upper Rio Negro region of the Amazon basin. *Journal of Ecology,* **69,** 631–49.

**Vazquez Yanes, C. & Guevara Sada, S. (1985).** Characterizacion de los grupos ecologicos de arboles de la selva humeda. *Investigaciones Sobre la Regeneracion de Selvas en Veracruz, Mexico* 2 (Ed. by A. Gomez-Pompa & S. R. del Amo), pp. 67–78. INIREB, Xalapa, Mexico.

**Vazquez Yanes, C. & Orozco-Segovia, A. (1984).** Ecophysiology of seed germination. *Physiological Ecology of Plants of the Wet Tropics* (Ed. by E. Medina, H. A. Mooney & C. Vazquez Yanes), pp. 37–50. Junk, The Hague.

**Veblen, T. T. (1985).** Stand dynamics in Chilean *Nothofagus* forests. *The Ecology of Natural Disturbance and Patch Dynamics* (Ed. by S. T. A. Pickett & P. S. White), pp. 35–51. Academic, New York.

**Veblen, T. T. (1986).** Treefalls and the co-existence of conifers in subalpine forests of the central Rockies. *Ecology,* **67,** 644–9.

**Veblen, T. T. (1988).** Regeneration dynamics. *Plant Succession* (Ed. by D. C. Glenn-Lewin), in press. Chapman & Hall, London.

**Veblen, T. T., Ashton, D. H. & Schlegel, F. M. (1979).** Tree regeneration strategies in a lowland *Nothofagus*-dominated forest in south-central Chile. *Journal of Biogeography,* **6,** 329–40.

**Veblen, T. T., Donoso, C. Z., Schlegel, F. M. & Escobar, B. R. (1981).** Forest dynamics in south-central Chile. *Journal of Biogeography,* **8,** 211–47.

**Vooren, A. P. (1985).** Patterns in tree and branch-fall in a West African rain forest. *Department of Silviculture Wageningen Agricultural University Report,* D 85–05.

**Waring, R. H. (1980).** *Forests: Fresh Perspectives from Ecosystem Analysis.* Oregon State University Press, Corvallis.

**Watt, A. S. (1947).** Pattern and process in the plant community. *Journal of Ecology*, **35**, 1–22.

**West, D. C., Shugart, H. N. & Botkin, D. B. (1981).** *Forest Succession, Concepts and Application.* Springer, New York.

**White, P. S. (1979).** Pattern, process and natural disturbance in vegetation. *Botanical Review*, **45**, 229–99.

**White, P. S., Mackenzie, M. D. & Busing, R. T. (1985).** Natural disturbance and gap phase dynamics in southern Appalachian spruce-fir forests. *Canadian Journal of Forest Research*, **15**, 233–40.

**Whitmore, T. C. (1974).** *Change with Time and the Role of Cyclones in Tropical Rain Forest on Kolombangara, Solomon Islands.* Commonwealth Forestry Institute, Paper 46.

**Whitmore, T. C. (1975).** *Tropical Rain Forests of the Far East.* Clarendon Press, Oxford.

**Whitmore, T. C. (1978).** Gaps in the forest canopy. *Tropical Trees as Living Systems* (Ed by P. B. Tomlinson & M. H. Zimmermann), pp. 639–55. University Press, Cambridge.

**Whitmore, T. C. (1982).** On pattern and process in forests. *The Plant Community as a Working Mechanism* (Ed. by E. I. Newman), pp. 45–59. Blackwell Scientific Publications, Oxford.

**Whitmore, T. C. (1983).** Secondary succession from seed in tropical rain forests. *Forestry Abstracts*, **44**, 767–69.

**Whitmore, T. C. (1984).** *Tropical Rain Forests of the Far East*, 2nd edn. Clarendon Press, Oxford.

**Whitmore, T. C. (1988).** Canopy gaps as the major determinants of forest dynamics and the two major groups of forest tree species. *Ecology*, (in press).

**Whitmore, T. C. & Sidiyasa, K. (1986).** Composition and structure of a lowland rain forest at Torant, northern Sulawesi. *Kew Bulletin*, **41**, 747–56.

**Wyatt-Smith, J. (1966).** Ecological studies on Malayan forests 1. *Malayan Forestry Department Research Pamphlet*, 52.

**Yamamoto, S. (1981).** Gap-phase dynamics in climax forests. A review. *Biological Science*, **33**, 8–16.

**Young, K. R., Ewel, J. J. & Brown, B. J. (1987).** Seed dynamics during forest succession in Costa Rica. *Vegetatio*, **71**, 157–74.

Finite population growth rates of *A. mexicanum* vary from under-equilibrium values in gaps to above-equilibrium in mature forest sites. Since the largest proportion of the forest is in the mature phase, the palm populations in the Los Tuxtlas forest as a whole are increasing. Data suggest that a density-dependent regulation of numbers operates in populations of the palm, through survivorship in the young categories.

## INTRODUCTION

The population dynamics of tropical rain forest trees takes place in a highly heterogeneous environment. Under the apparently stable conditions of the undisturbed forest, biotic and abiotic forces affecting survivorship, growth and reproduction of trees change from year to year, and as they grow from one life-cycle stage to the next. In the course of their lives, the performance of these individuals may also be affected by the formation of gaps created as neighbouring trees die and fall.

When a canopy gap is formed, abrupt alterations take place in the environment of the trees (Brokaw 1985a; Martínez-Ramos 1985). Physical and chemical conditions are modified, resource levels increase, the interference with neighbouring trees is relaxed and the diversity and strength of biotic interactions with animals and pathogens may change (Augspurger 1984; Dirzo 1984; Brokaw 1985a; Martínez-Ramos & Alvarez-Buylla 1986). As the gap ages, and as a result of the growth in vegetation induced by the canopy opening, a vegetation structure and microenvironment similar to predisturbance conditions is gradually restored (Brokaw 1985a). Further disturbances may cause individuals to experience additional series of gradual and abrupt environmental changes until their death.

The above considerations suggest that the contribution of an individual tree to population growth depends on how much it can withstand the environmental variability generated by the cycle of gap dynamics. In the ecological short-term, gaps may have profound consequences for the population dynamics and regulation of a tree species, and in the evolutionary long-term, the dynamics of gaps may have acted as a complex of selective pressures which have directed the evolution of the life-history of forest trees.

Evaluation of the consequences of gaps on tree populations necessitates study of the relationships between canopy disturbance regimes and the demographic traits of populations (e.g. germination, longevity, age at first reproduction, survivorship, growth and fecundity rates). The degree to which gaps can have a potential selective role on these traits depends on: (1) the extent of the disturbance, which affects the proportion of the population affected by a gap, (2) the rate at which gaps are created (i.e. the number of gaps opening per unit area, per unit time), (3) the probability that a given size of gap will affect individuals in one or more life-cycle stages, and (4) the degree of genetic variability which is available for selection to operate on a given life-history trait.

This chapter discusses some problems related to the demography of trees and treefall gap dynamics in tropical rain forests. Most of the data used comes from a long-term study (Sarukhán 1978; Martínez-Ramos 1985) being carried out at Los Tuxtlas region, south-eastern México, with the northern-most area of rain forest in the neotropics. Some aspects of the gap formation process which are relevant to the population dynamics of trees are first described and these are used to model the equilibrium status of gap dynamics at Los Tuxtlas. Then the demographic effects of canopy gaps on tree populations is covered and the way in which these effects may play a role in maintaining the high levels of species diversity found in tropical rain forests. Studies on the palm *Astrocaryum mexicanum* Liebm. are used to illustrate the demographic effects of gaps on the life-cycle of a long-lived, understorey tree. Finally, the relationship between the population dynamics of *A. mexicanum* and forest gap dynamics are discussed in an integrative approach.

## GAP DYNAMICS AS A PROCESS

Since Watt's (1947) seminal paper on 'pattern and process in the plant community', there has been an increasing consensus in considering forest communities, both of temperate and tropical forest, as mosaics of regenerating cyclical phases represented by patches of various sizes and ages (e.g. Hallé, Oldeman & Tomlinson 1978; Brokaw 1985a; Martínez-Ramos 1985). For tropical rain forests, Whitmore (1975, 1988) proposed a simple conceptual model of the regenerative process which he called the 'forest growth-cycle'. The model considers three phases: (1) a gap phase, occurring when treefall disturbances create canopy openings, (2) a building phase, which appears when gaps are colonized (both by advance regeneration or by seedlings germinated from seeds recently falling to the gap) and trees reach their maximum growth rate. After the opening of large gaps, an active successional process is developed in this phase. (3) A mature phase, established after the death of the short-lived, light-demanding species. As the gap gets older, the growth rates of most trees decrease and recruitment of most species declines.

A large proportion of the forest maintains a mosaic structure in terms of the percentage of the area covered by each of these phases. These percentages differ between forests and are related to differences in the severity with which the forest canopy is destroyed by natural forces (Whitmore 1975; Hallé,

Oldeman & Tomlinson 1978; Brokaw 1985a; Martínez-Ramos 1985; Runkle 1985).

From a population point of view, the study of forest mosaics is relevant, since: (1) the structure and dynamics of populations of a given species may differ between the three forest regeneration phases, (2) as a result of this, each phase may impose different environmental constraints upon population growth. For example, we can expect that resource availability declines from young gaps to the mature phases, as is the case for light (e.g. Chazdon & Fetcher 1984), and that plant interactions are stronger in the building phase than in the gap and mature phases (Brokaw 1985b).

At Los Tuxtlas, the natural regeneration of the forest mosaic is being studied in a permanent 5 hectare plot. This study site was established in 1982, as a $100 \times 500$ m transect running from the forest edge into the forest (Martínez-Ramos 1985). All individuals (defined as 1 m in height or more) of the seventy most abundant tree species were identified and mapped. Gaps (*sensu* Brokaw 1982) larger than 5 $m^2$ which developed during a 4 year period (1982–85) were recorded, measured and located on maps. The stems of *Astrocaryum mexicanum* palms were used to date treefalls in all parts of the forest at a scale of 25 $m^2$ (Martínez-Ramos, Alvarez-Buylla, Sarukhán & Piñero 1988). The palm stems may record gap events up to 100 years old.

On average, the gap phase (i.e. canopy openings less than 1 year old) covered about 2% of the forest area, the building phase (openings with ages ranging from 1 to 35 years) 34%, and the mature phase 64%. These percentages varied strongly in space, since the steepest area sites and those nearer the forest edges showed between two and six times more area in gap phase than flat areas located far from the forest edges of which only 0.69% was in gap phase. Thus, there was a noticeable edge effect on the disturbance rates occurring in this forest (Martínez-Ramos 1985; Sarukhán, Piñero & Martínez-Ramos 1985).

The annual proportion of forest area opened to gaps also showed a high variability between years. Gap formation over a period of 70 years prior to 1982 has been analysed with the aid of the *A. mexicanum* stems (Martínez-Ramos *et al.* 1988). It was found that the proportion of forest affected by gaps annually has a mean value of 2.3% with values ranging widely between 0.1% and 9.2% per year. The variation was found to be related to the total annual rainfall. However, no regular, predictable pattern of variation was found in the annual rates of gap creation.

Another source of variation can be discerned in the frequency with which canopy openings of different sizes occur in space and time. Again using the *A. mexicanum* stems as sensors of gap formation it was estimated that gaps smaller than 100 $m^2$ were produced at rates about three orders of magnitude higher than the largest gaps recorded, which were in the region of 1500 m$^2$ (Martínez-Ramos & Alvarez-Buylla, unpublished data). At least one gap smaller than 100 m$^2$ was produced every year in a given hectare of the forest; gaps between 100 and 500 m$^2$ occurred at intervals of 2 to 6 years in 5 hectare plots. Larger gaps ($>$ 500 m$^2$) appeared at a rate of one every 15 or more years in a 5 hectare tract of forest. Mapping an area of 75 hectares or more would be required to ensure the recording of one such gap every year. The forest turnover rate, i.e. the time elapsing between two consecutive small gaps appearing on the same spot of the forest (*sensu* Hartshorn 1978) is about 50 years, while for gaps larger than 300 m$^2$ the rate is about four times longer.

The above forest turnover rates may be overestimations of the real rates since the area actually affected by a canopy opening is larger than the vertical projection of the opening onto the ground (Runkle 1985). In effect, it has been found for Los Tuxtlas, using the distributions of seedlings and saplings of pioneer species as indicators of the extent of the 'gap environment', that the effective gap size is about 3.5 times larger than the corresponding canopy opening (Popma *et al.* 1988).

The forest growth-cycle is shown in Fig. 15.1. Yearly transition probabilities between phases indicate the existence of three regenerative routes: (1) an 'integrated route', which applies only to large gaps ($>$ 50 m$^2$) and includes building and mature phases, (2) a 'secondary route', through which forest patches in a building phase return to the gap phase, and (3) a 'mature route', in which small gaps ($<$ 50 m$^2$) return to the mature phase without undergoing a building phase. Nearly 2.2% of a mature forest is opened every year to gaps. About half of the canopy openings in a year

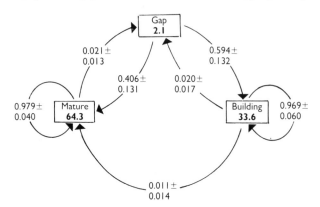

FIG. 15.1. The 'forest growth-cycle' at Los Tuxtlas rain forest, south-eastern Mexico. Bold figures are percentages of forest area in each regenerative phase. Arrows indicate transitions between phases; figures are means and one S.D. of the annual transition probabilities. Based on analysis of an area of 5 hectares.

produce gaps of under 50 m$^2$ in size. These small gaps follow the 'mature regeneration route' while larger gaps follow the 'integral' and 'secondary' routes.

A forest patch following a given regeneration route may alter into a different regeneration route if its probability of experiencing the formation of a new large or small gap changes with time. Causes of such a change include: (1) its species composition (trees of some species are more susceptible to damage by wind and rain than others), (2) the age of trees present (a large gap may occur if senile individuals are present), and (3) the degree to which the activity of natural enemies affects the vigour of large trees in the patch (Brokaw 1985a).

In the long term and over a large-scale, a forest may reach a state of equilibrium in which the proportions of the forest following the different regenerative routes is in balance. Using the data of Fig. 15.1, a Markovian model of the forest growth-cycle at Los Tuxtlas was developed (Martínez-Ramos 1985). Assuming that transition probabilities between regeneration phases remain constant through time, the model predicts a forest mosaic in which, at equilibrium, the area of the building phase is significantly larger than that actually observed in the forest mosaic (Table 15.1). This difference is partly explained by the fact that forest patches located towards the edge of the forest are losing mature-phase forest at relatively higher rates than the forest interior because of very frequent treefalls (Martínez-Ramos 1985). Under the assumptions of the model (Horn 1975), this result suggests that the forest mosaic at Los Tuxtlas is in a non-equilibrium state, as is the case also for the forest at Barro Colorado Island, Panama (Hubbell & Foster 1986a).

## THE DEMOGRAPHIC RESPONSE TO GAP DYNAMICS: THE CASE OF *ASTROCARYUM MEXICANUM*

The population structure and dynamics of many rain forest tree species are closely dependent on gap-phase regeneration which produces pulses of recruitment (Sarukhán, Piñero & Martínez-Ramos 1985); such pulses can be identified both in time and space. They vary in their frequency and distribution for each species (Hubbell & Foster 1986a, 1987), probably reflecting the way in which different species utilize gaps of different sizes. At one extreme are the species that depend for recruitment on larger gaps, which occur less predictably and at large time and space intervals. The populations of these species are patchily distributed, with age structures within each patch closely reflecting the disturbance history of a given gap. This is the case of the pioneer, short-lived, strongly light-demanding species, such as

TABLE 15.1. The observed forest mosaic structure and the expected structure at equilibrium obtained from a Markovian model of the forest growth-cycle. Observed and expected structures are significantly different ($\chi^2 = 125.5$; $P < 0.001$)). See text for further details.

| Forest stage | Age (years) | % of the forest area | | |
|---|---|---|---|---|
| | | Observed (o) | Expected (e) | (o–e) |
| Gap | <1 | 2.06 | 2.06 | 0 |
| Building | 1–35 | 33.59 | 39.01 | −6.58 |
| Mature | >35 | 64.35 | 58.93 | 6.58 |

*Cecropia obtusifolia* (Alvarez-Buylla 1986). It could be said that these species perceive their environment in a coarse-grained manner. At the other extreme are the shade-tolerant, long-lived species which can regenerate within small gaps (Brokaw 1985b). These occur with a high spatial and temporal frequency, and therefore have a more even spatial distribution. These so-called persistent species perceive their environment as fine-grained.

Pioneer and persistent species are the extremes of a wide range of gap-phase regeneration behaviours found in species-rich tropical rain forest communities. Much controversy has arisen about the mechanisms by which the high diversity of these forests is maintained (Hubbell & Foster 1986b); also, much debated is the question of whether there are a few or many types of regeneration behaviours. Two points of view have emerged. The first, 'the equilibrium hypothesis', proposes that many species can coexist because populations are regulated by intraspecific competition; each species has a narrow and specific regeneration niche (*sensu* Grubb 1977), so that competitive exclusion between species plays a minor role. Forest gaps are considered an important component of the regeneration niche of species (e.g. Ricklefs 1977; Strong 1977; Pickett 1983), and it has been suggested that each species has particular gap requirements (i.e. different sizes or ages of the gaps or even different areas within the gaps) for regeneration (Denslow 1980; Orians 1982).

The second point of view, the 'non-equilibrium hypothesis', considers that populations of most species are controlled by density-independent mechanisms. Individuals of a given species will have, at different times in the same spot of the forest, different mixtures of neighbouring species. The identity of these neighbours will depend on the chance events that determine the presence and growth of seedlings and saplings. It is proposed (Hubbell & Foster 1986b) that species diversity is generated at geographical scales and that the species richness of a particular forest is maintained by local

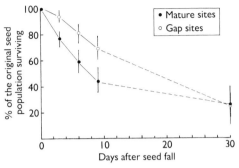

FIG. 15.2.   Postdispersal loss rates of *Astrocaryum mexicanum* seeds in the soil of gap and mature sites. Seed population size at the beginning of observations was 0.21 seeds m$^{-2}$ in the gap and 0.08 seeds m$^{-2}$ in mature sites.

the end of the first month of observations (Fig. 15.2). For the stages from seedlings to the oldest palms, there is greater mean annual survival for plants growing in the mature forest compared with those in gaps. Mortality consistently decreases with the age of palms (Fig. 15.3a). Survivorship of

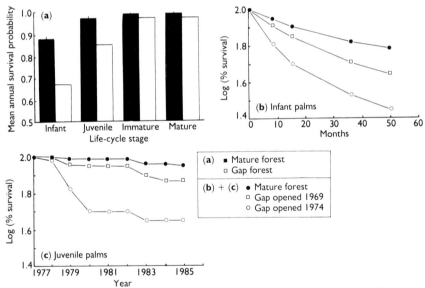

FIG. 15.3.   Survival patterns of *Astrocaryum mexicanum* palms at the forest mosaic of Los Tuxtlas, Mexico. (a) the annual survival probability for palms of different life-cycle stages, growing in mature forest and in a gap; vertical lines represent two standard errors; infants = palms between 1 and 8 years old; juveniles = palms between 9 and 17 years old; immatures = palms between 18 and 22 years old; matures = palms between 23 and 100 years old. (b) Survivorship curves for infant palms found in disturbed sites and in mature patches of the forest. (c) Survivorship curves for juvenile palms.

infant and juvenile palms is greater for all ages in older gaps and in mature forest (Fig. 15.3b and c).

Additionally, the event of gap formation is itself an important cause of mortality. More than one-third of mature palms which died in a 12 year period were killed by falling branches or trees (see also Piñero, Martínez-Ramos & Sarukhán 1984). Leaf-area reductions produced by treefall blows may also increase the risk of dying in young palms (Mendoza, Piñero & Sarukhán 1987).

### Growth patterns

Palms in the immature and mature stages grow significantly faster in gap conditions than in the mature forest (Fig. 15.4a); the lack of statistically significant growth differences between infant palms exposed to both conditions is possibly due to the fact that light conditions at or near ground level in the gaps become similar to those in the mature forest only about 2 years after the gap is opened. Humidity and air temperature measurements

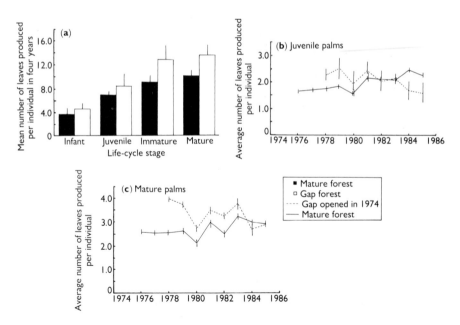

FIG. 15.4. Growth patterns of *Astrocaryum mexicanum* in the forest mosaic of Los Tuxtlas, Mexico. (a) The annual mean number of leaves produced per individual as a function of the life-cycle stage. (b) The yearly mean rate of leaf production for juveniles growing in a gap opened in 1974 (– – –) and in mature forest (——); vertical lines indicate two standard errors. (c) The yearly mean rate of leaf production for mature palms.

at seedling height in gaps show that conditions may approach the values in mature sites within 2 years (Fetcher *et al.* 1987). A similar situation applies for juvenile palms, although differences in growth can be observed during the first 6 years after gap formation (Fig. 15.4b). Differences in growth rates between both immature and mature palms growing in undisturbed forest or gaps, last for about 12 years after the formation of the gap (Fig. 15.4c).

These results imply that light is an important limiting resource for the growth of *A. mexicanum* palms. Observations on the rate of leaf production between individuals of mature forest and gaps as a function of their crown size (number of functional leaves) show that there is a statistically significant relationship ($P < 0.01$) between the number of leaves of palms in mature forest and their rate of leaf production, while this relationship is absent for palms growing in gaps ($P > 0.15$). These results suggest that having more leaves is a clear advantage for light interception, under the conditions of limited light in the mature forest whereas this is irrelevant under the high level of incident light typical in a gap.

### Patterns in reproduction

Growing in mature forest sites or gaps does not alter the time of year in which flowers are formed, or the period for maturation of inflorescences and infructescences. However, other reproductive traits, such as the probability of reproducing (Sarukhán, Martínez-Ramos & Piñero 1984) and the number of fruits produced, differ considerably between the two habitats.

The mean number of infructescences produced annually is consistently greater for all sizes of palms growing in gaps than for those in the mature forest (Fig. 15.5a). That reproduction seems to be closely linked to the amount of energy fixed, is strongly suggested by the fact that the differences in reproduction due to a greater light availability in the gaps can be measured for 12 years after gap formation (Fig. 15.5b), as is the case for leaf-production rates (Fig. 15.4c). Moreover, reproductive rate is a function of growth measured as rate of leaf production (Fig. 15.6); since the mean number of leaves per crown remains fairly constant ($14.73 \pm 3.13$ S.D.) throughout life after palms reach 25 years of age, the turnover of leaves is greater under conditions of high light availability.

The process of gap maturation is reflected in changes in the reproductive behaviour of the palm population. Fig. 15.7 shows that the palm population present in a gap exhibits a change in the frequency distribution of number of infructescences per palm per year. The early stages after gap formation have a wide distribution of values of infructescence production, with a mode of

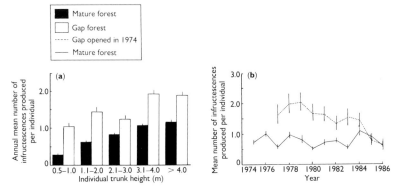

FIG. 15.5. Reproductive patterns of *Astrocaryum mexicanum* palms at the forest mosaic of Los Tuxtlas, Mexico. (a) Annual mean fecundity for mature palms of different trunk height classes. (b) Yearly mean fecundity for palms growing in mature forest and in a gap opened in 1974. Vertical lines indicates two standard errors.

between two and three infructescences. The later stages of the gap have a narrower distribution with a mode of one. This is typical of *A. mexicanum* in populations from mature forest sites.

## CORRELATED CHANGES IN GAP DYNAMICS AND POPULATION DYNAMICS

A comparison of numerical fluxes of populations of *A. mexicanum* is demonstrated in Fig. 15.8 from mature sites and two gap sites of different age. Gap 1 is a site which was opened in 1974, and gap 2 is a site which was opened in 1969. Data for the mature sites are averages of six permanent

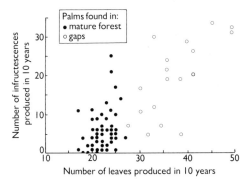

FIG. 15.6. The reproductive performance of *Astrocaryum mexicanum* palms as a function of their leaf production rate. The line of best fit through the data has the equation $y = -14.6 + 0.9x$; $n = 80$; $r = 0.80$; $P < 0.01$.

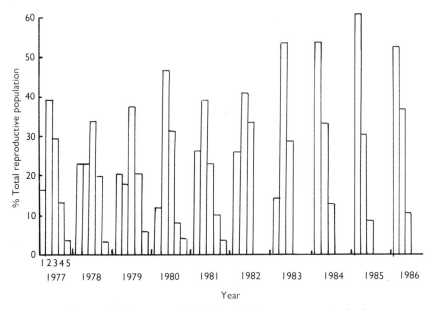

Fig. 15.7.   Changes in the frequency distributions of infructescence production in an *Astrocaryum mexicanum* population during the aging process of a forest gap. For each year, histograms show the percentage of the reproductive population producing 1, 2, 3, 4 or 5 infructescences. The gap was formed in 1974.

plots. All sites are 600 m$^2$ in size. The values in the diagrams represent yearly means obtained from 10 years of observations.

The values of the population flux for the mature sites resemble more closely those of gap 2 than that of the younger gap 1. In general, the main differences between mature and gap sites are in the reproductive output, which is more than three times higher in the younger gap than in the other sites; this is compensated by a lower probability of a seed becoming a seedling in gap 1 than in the other sites. Also, the probability of an immature palm in the gap sites becoming a mature individual is two to three times as high as in the mature site. Mature palms have a higher probability of surviving to the next year in the mature site than in the gap sites. This situation is repeated for all other life-stages.

The finite population growth rates ($\lambda$), obtained from projected Lefkovich matrices (using the data of Fig. 15.8), give values of 0.989 for gap 1, 1.003 for gap 2, and 1.012 for the mature sites averaged. The significant differences in survivorship, growth and reproduction found between gaps and the mature forest, indicate that the palm populations grow at different rates in the three phases of the forest mosaic, and that growth rate is positive in mature forest.

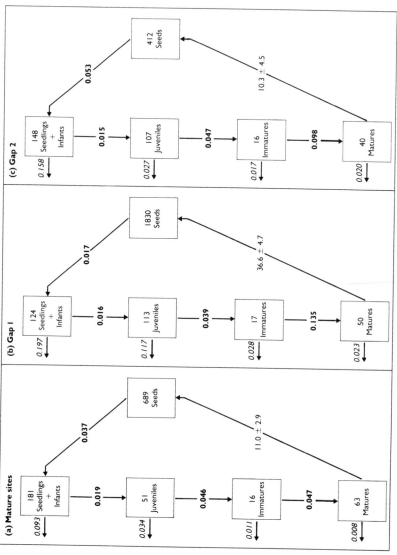

Fig. 15.8. Population flux diagrams of *Astrocaryum mexicanum* palms in mature sites and in gaps at the Los Tuxtlas rain forest, Mexico. Gap 1 was opened in 1974, gap 2 in 1969. The numbers of individuals (per 600 m²) at the different life-cycle stages are indicated in boxes. Bold figures are annual transition probabilities between life-cycle stages; mean annual mortality rates are in italic numbers. The mean annual number of fruits produced per mature palm is indicated with one S.D.

In order to integrate the population dynamics of the palm and the gap dynamics of the forest a new Lefkovich matrix was constructed which incorporated the Markovian process of the 'forest growth-cycle' (Fig. 15.9). The demographic data of gap 1 and that of the mature sites were included. In this case, the transition probabilities were multiplied by the probability that a given forest patch in gap or mature phases remains or changes into a different phase in 1 year. The demographic effects of a 300 m$^2$ canopy gap on individuals of the palm last only for some 12 years, so that in 1 year 0.087 of populations in forest patches having a disturbance age between 0 to 12 years return to a mature forest phase. Palm populations in mature forest patches change to the gap-phase state with an annual probability of 0.005, considering gap sizes larger than 300 m$^2$.

The overall population growth of the palm within this hypothetical forest composed of two different forest patches is similar ($\lambda = 1.0114$) to that obtained for the mature sites. However, the model predicts that the population growth may be dramatically affected if the frequency and size of gap change (Fig. 15.9). Considering a given gap size as a constant, population growth becomes negative at a certain rate of gap creation. This gap opening rate may be very low if gaps created are large (e.g. if the gap effects on the palm last 66 years) but very infrequent in time, and very fast if gaps are small and frequent. In addition, the model predicts that population growth may even be positive at gap formation rates as fast as 0.2 (turnover rates of only 5 years!) if the demographic effects of the gaps last only 1 year.

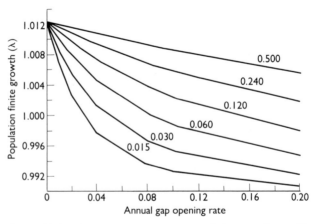

FIG. 15.9.   The sensitivity of *Astrocaryum mexicanum* population growth rates ($\lambda$) to changes in gap dynamics. Numbers alongside the curves indicate the proportion of the population found in gap phases which returns annually to the mature forest condition.

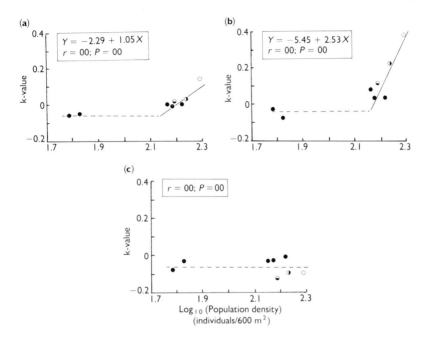

FIG. 15.10. The relationship between population density and population growth in *Astrocaryum mexicanum*. Population density is the total number of juvenile, immature and mature palms per 600 m$^2$. Population growth is expressed as $k$-values [$\text{Log}_{10}$ (density in 1975) − $\text{Log}_{10}$ (density in 1982)]. (○) Gap opened in 1974, (◓) gap opened in 1969, (◑) mature site disturbed in 1979, (●) mature sites. Parameters of linear regressions (excluding the low density plots) are shown. (a) $k$-values for all population, (b) $k$-values for juvenile individuals, (c) $k$-values for immature and mature palms.

This may represent the situation in a forest with a canopy which is very frequently opened by branches snapping off trees, opening small gaps which are quickly closed by the lateral growth of neighbouring trees.

All these results suggest that populations of *A. mexicanum* may be at equilibrium or able to increase under a wide range of forest gap regimes. The gap dynamics occurring at Los Tuxtlas is characterized by a high frequency of small gaps. Most *A. mexicanum* individuals during an average life-time of about 130 years are affected on average by eight small gaps (< 100 m$^2$); by contrast, they experience the influence of only one large (> 300 m$^2$) gap in the course of their life. If these conditions are maintained, we must expect the palm population to increase in the forest as a whole. However, the above models assume population growth without regulatory mechanisms. The authors have found evidence suggesting that gaps constitute an important element in the overall regulation of population size of *A. mexicanum* palms in the forest mosaic. This takes place through two main processes occurring in gaps, namely a greater mortality of younger palms and an increased seed production.

The reduced survivorship of *A. mexicanum* individuals in gaps could be interpreted as a physiological negative effect of the high incidence of light acting on a shade-tolerant species. A closer analysis, however, suggests that density-dependent regulation operates on the youngest stages in the life-cycle (Fig. 15.10b). The population growth rate of adult palms, on the other hand, does not seem to be regulated in a density-dependent fashion (Fig. 15.10c). However, by growing faster in gaps, juveniles move more rapidly into the adult category. This produces an increase in the density of adult palms in gaps of a few years of age. The fact that adults regulate the recruitment of juveniles in high density stands of palms (Sarukhán, Piñero & Martínez-Ramos 1985) suggests that the formation of height hierarchies is decisive in controlling the transition of individuals from one category to the next. Attempts are being made to understand how much these density-dependent relationships alter the predictions of the population growth in the forest mosaic obtained from the models without regulation.

*Astrocaryum mexicanum* represents an exceptional case among rain forest trees, because of its high population densities. Conspecific interactions in this species are more frequent than interactions with individuals of different species of the understorey. Consequently, intense intraspecific competition may be expected. This sharply contrasts with those canopy species in which most contacts between neighbours are interspecific (Hubbell & Foster 1986b). We do not yet have demographic information on the population dynamics of species that interact primarily with members of other species in a gap.

More demographic studies on these topics are needed. The large spatial scales and long time scales over which tropical tree populations develop (Hubbell & Foster 1983, 1987) impose serious field and logistic limitations which must be coped with in order to obtain a better understanding of the population and community ecology of tropical rain forest. Perhaps more than in other areas, plants in the tropics are waiting to be counted to help us to understand why they are where they are!

## ACKNOWLEDGMENTS

We thank Elena Alvarez-Buylla, A. López, J. Nuñez-Farfán, R. Dirzo, E. González-Soriano, Ana Mendoza, F. Cervantes-Reza, O. Chavez, R. Palafox and A. Watkinson for field assistance. E. Alvarez-Buylla made valuable comments on the manuscript. This work was carried out partially with the aid of a grant from the Consejo Nacional de Ciencia y Tecnología (CONACyT, México) and supported by the Estación de Biología Tropical Los Tuxtlas, Instituto de Biología, Universidad Nacional Autónoma de México.

## REFERENCES

Alvarez-Buylla, E. (1986). *Demografía y Dinámica Poblacional de* Cecropia obtusifolia *Bertol. en la Selva de Los Tuxtlas,* Ver. Tesis de Maestría. Facultad de Ciencias, UNAM, México, D.F.

Augspurger, C. K. (1984). Seedling survival among tropical tree species: interactions of dispersal distance, light gaps, and pathogens. *Ecology*, **65**, 1705–12.

Augspurger, C. K. (1988). Impact of pathogens on natural plant populations. *Plant Population Ecology* (Ed. by A. J. Davy, M. J. Hutchings & A. R. Watkinson), pp. 413–33. Blackwell Scientific Publications, Oxford.

Brokaw, N. V. L. (1982). The definition of treefall gap and its effect on measures of forest dynamics. *Biotropica*, **14**, 158–60.

Brokaw, N. V. L. (1985a). Treefalls, regrowth and community structure in tropical forests. *The Ecology of Natural Disturbance and Patch Dynamics* (Ed. by S. T. A. Pickett & P. S. White), pp. 53–69. Academic Press, New York.

Brokaw, N. V. L. (1985b). Gap-phase regeneration in a tropical forest. *Ecology*, **66**, 682–87.

Brokaw, N. V. L. (1987). Gap-phase regeneration of three pioneer tree species in a tropical forest. *Journal of Ecology*, **75**, 9–19.

Chazdon, R. & Fetcher, N. (1984). Photosynthetic light environments in a lowland tropical rain forest in Costa Rica. *Journal of Ecology*, **72**, 553–64.

Clark, D. A. & Clark, D. B. (1984). Spacing dynamics of a tropical rain forest tree: evaluation of the Janzen-Connell model. *American Naturalist*, **124**, 769–88.

Denslow, J. (1980). Gap partitioning among tropical forest trees. *Biotropica*, **12** (Supplement), 47–55.

Dirzo, R. (1984). Insect-plant interactions: some ecophysiological consequences of herbivory. *Physiological Ecology of Plants of the Wet Tropics* (Ed. by E. Medina, H. A. Mooney & C. Vázquez-Yanez), pp. 209–24. Junk Publishers, The Hague.

Fetcher, N., Oberbauer, S. F., Rojas, G., & Strain, B. R. (1987). Efecto del régimen lumínico sobre la fotosínteis y el crecimiento en plántulas de árboles de un bosque lluvioso de Costa Rica. *Revista de Biología Tropical*, **35** (Suplemento 1), 97–110.

Garwood, N. C. (1983). Seed germination in a seasonal tropical forest in Panama; a community study. *Ecological Monographs*, **53**, 159–81.

Grubb, P. J. (1977). The maintenance of species richness in plant communities: the importance of the regeneration niche. *Biological Reviews*, **52**, 107–45.

Hallé, F., Oldeman, R. A. A. & Tomlinson, P. B. (1978). *Tropical Trees and Forest: an Architectural Analysis*. Springer-Verlag, Berlin.

Hartshorn, G. (1978). Tree falls and tropical forest dynamics. *Tropical Trees as Living Systems* (Ed. by P. B. Tomlinson & M. H. Zimmerman), pp. 617–38. Cambridge University Press, London.

Horn, H. S. (1975). Markovian process of forest succession. *Ecology and Evolution of Communities* (Ed. by M. L. Cody & J. Diamond), pp. 196–213. Belknap, Cambridge, Massachusetts.

Hubbell, S. P. & Foster, R. (1983). Diversity of canopy trees in a neotropical forest and implications for conservation. *Tropical Rain Forest: Ecology and Management* (Ed. by S. C. Sutton, T. C. Whitmore & A. C. Chadwick), pp. 25–41. Blackwell Scientific Publications, Oxford.

**Hubbell, S. P. & Foster, R. (1986a).** Canopy gaps and the dynamics of a tropical forest. *Plant Ecology* (Ed. by M. Crawley), pp. 77–96. Blackwell Scientific Publications, Oxford.

**Hubbell, S. P. & Foster, R. (1986b).** Biology, chance, and history and the structure of tropical rain forest tree communities. *Community Ecology* (Ed. J. Diamond & T. J. Case), pp. 41–62. Harper & Row, New York.

**Hubbell, S. P. & Foster, R. (1987).** La estructura espacial en gran escala de un bosque tropical. *Revista de Ecología Tropical*, **35** (Suplemento 1), 7–22.

**Janzen, D. H. (1970).** Herbivores and the number of tree species in tropical forest. *American Naturalist*, **104**, 501–28.

**Martínez-Ramos, M. (1980).** *Aspectos sinecológicos del proceso de renovación natural de una selva alta perennifolia.* Tesis Profesional, Universidad Nacional Autónoma de México.

**Martínez-Ramos, M. (1985).** Claros, ciclos vitales de los arboles tropicales y la regeneración natural de las selvas altas perennifolias. *Investigaciones Sobre la Regeneración de Selvas Altas en Veracruz, Mexico. Vol. II* (Ed. by A. Gómez-Pompa & S. Del Amo), pp. 191–239. INIREB- Alhambra, México, D.F.

**Martínez-Ramos, M. & Alvarez-Buylla, E. (1986).** Seed dispersal, gap dynamics and tree recruitment: the case of *Cecropia obtusifolia* at Los Tuxtlas, México. *Frugivory and Seed Dispersal* (Ed. by A. Estrada & T. H. Fleming) pp. 169–86. Dr W. Junk, Dordrecht.

**Martínez-Ramos, M., Alvarez-Buylla, E., Sarukhán, J. & Piñero, D. (1988).** Treefall age determination and gap dynamics in a tropical forest. *Journal of Ecology*, in press.

**Mendoza, A., Piñero, D., & Sarukhán, J. (1987).** Effects of experimental defoliation on growth, reproduction and survival of *Astrocaryum mexicanum*. *Journal of Ecology*, **75**, 545–54.

**Nuñez-Farfán, J., & Dirzo, R. (1988).** Within-gap spatial heterogeneity and seedling performance in a Mexican tropical forest. *Oikos*, **51**, 274–84.

**Orians, G. H. (1982).** The influence of tree falls on tropical forest tree species richness. *Tropical Ecology*, **23**, 255–79.

**Pickett, S. T. A. (1983).** Differential adaptation of tropical species to canopy gaps and its role in community dynamics. *Tropical Ecology*, **24**, 69–84.

**Piñero, D., Martínez-Ramos, M. & Sarukhán, J. (1984).** A population model of *Astrocaryum mexicanum* and a sensitivity analysis of its finite rate of increase. *Journal of Ecology*, **72**, 977–91.

**Piñero, D., Sarukhán, J. & González, E. (1977).** Estudios demográficos en plantas: *Astrocaryum mexicanum* Liebm. I. Estructura de las Poblaciones. *Boletín de la Sociedad Botánica de México*, **37**, 60–118.

**Popma, J., Bongers, F., Martínez-Ramos, M. & Veneklaas, E. (1988).** Pioneer species distribution in treefall gaps in Neotropical rain forest; a gap definition and its consequences. *Journal of Tropical Ecology*, **4**, 77–88.

**Ricklefs, R. E. (1977).** Environmental heterogeneity and plant species diversity: a hypothesis. *American Naturalist*, **111**, 376–81.

**Runkle, J. R. (1985).** Disturbance regimes in temperate forests. *The Ecology of Natural Disturbance and Patch Dynamics* (Ed. by S. T. A. Pickett & P. S. White), pp. 17–33. Academic Press, New York.

**Sarukhán, J. (1978).** Studies on the demography of tropical trees. *Tropical Trees as Living Systems* (Ed. by P. B. Tomlinson & M. H. Zimmerman), pp. 163–84. Cambridge University Press, London.

**Sarukhán, J. (1980).** Demographic problems in tropical trees. *Demography and Evolution in Plant Populations* (Ed. by O. T. Solbrig), pp. 168–88. University of California Press, Berkeley.

**Sarukhán, J., Martínez-Ramos, M. & Piñero, D. (1984).** The analysis of demographic variability at the individual level and its population consequences. *Perspectives on Plant Population Ecology* (Ed. by R. Dirzo & J. Sarukhán), pp. 83–106. Sinauer Associates, Sunderland, Massachusetts.

**Sarukhán, J., Piñero, D. & Martínez-Ramos, M. (1985).** Plant demography: a community-level interpretation. *Studies on Plant Demography: A Festschrift for John L. Harper* (Ed. by J. White), pp. 17–31. Academic Press, London.

**Strong, D. R. (1977).** Epiphyte loads, treefalls, and perennial forest disruption: a mechanism for maintaining higher tree species richness without animals. *Journal of Biogeography*, **4**, 215–18.

**Vázquez-Yanes, C. & Orozco-Segovia, A. (1984).** Ecophysiology of seed germination in the tropical humid forests of the world: a review. *Physiological Ecology of Plants of the Wet Tropics* (Ed. by E. Medina, H. Mooney & C. Vazquez-Yanes), pp. 37–50. Dr. Junk Publishers, The Hague, Netherlands.

**Watt, A. S. (1947).** Pattern and process in the plant community. *Journal of Ecology*, **35**, 1–22.

**Whitmore, T. H. (1975).** *Tropical Rain Forest of the Far East*. Clarendon Press, Oxford.

**Whitmore, T. H. (1988).** The influence of tree population dynamics on forest species composition. *Plant Population Ecology* (Ed. by A. J. Davy, M. J. Hutchings & A. R. Watkinson), pp. 271–91. Blackwell Scientific Publications, Oxford.

# 16. SOME ECOLOGICAL PROPERTIES OF INTIMATE MUTUALISMS INVOLVING PLANTS

## RICHARD LAW

*Department of Biology, University of York, York YO1 5DD, UK*

## SUMMARY

**1** Mutualistic microorganisms (endophytes), living in intimate associations with plant roots, occur very widely in terrestrial communities, over a broad range of latitudes and habitats. Although most terrestrial plant species associate with at least one kind of endophyte, the number of endophyte taxa is generally very small.

**2** The endophytes usually have an obligate requirement for their hosts. They rapidly colonize root systems under experimental conditions, and their densities appear to be regulated by their hosts.

**3** Little is known about the effect of endophytes on host population dynamics. Under experimental conditions, plant growth will occur in the absence of endophytes, so the associations are not physiologically obligate for the hosts, but growth is usually enhanced in their presence. The effect of endophytes on plant growth in the field is less clear.

**4** A simple model of the population dynamics of host–endophyte associations suggests that endophytes may respond rapidly to changes in host density but not vice versa.

**5** A common feature of host–endophyte associates is a lack of specificity in mutual recognition. The connectance of these guilds is therefore high, and the endophytes may be 'keystone' species. This lack of specificity in mycorrhizas gives rise to hyphal connections between plants, allowing redistribution of resources within the host-plant guild.

**6** Despite their lack of specificity, endophyte strains, once established in plant roots, differ in their effectiveness, so far as plant growth is concerned. A simple model does not, however, support the contention that the mutualisms are particularly vulnerable to invasion by endophyte strains of low effectiveness ('cheaters').

## INTRODUCTION

The study of mutualistic interactions between species has been neglected by population biologists in general (May 1982), and by plant population biologists in particular (e.g. Harper 1977; Silvertown 1982; White 1985; Crawley 1986). This is in contrast with a longstanding interest of plant population biologists in competition, and a rapidly increasing interest in the interactions between plants and antagonists such as herbivores and pathogens.

The focus, here, is on a group of mutualisms which themselves represent a forgotten corner of this neglected subject in plant population biology. These are the associations of certain microorganisms with the roots of plants, namely, vesicular-arbuscular mycorrhizas (VAMs), ericoid mycorrhizas and ectomycorrhizas, together with nitrogen-fixing associations of *Rhizobium* with legumes and *Frankia* with certain woody plants. See Bond (1963), Marks & Kozlowski (1973), Torrey (1978), Akkermans & Roelofsen (1980), Allen & Allen (1981), Mosse, Stribley & LeTacon (1981), Harley & Smith (1983), Read (1983), and Powell & Bagyaraj (1984) for reviews of the biology of these associations. Lichens and hermatypic corals have some features in common with these mutualisms, and might be added as honorary members in view of their many plant-like features (Harper, Rosen & White 1986).

The choice of these mutualisms is not entirely wilful. Firstly, the mutualisms represent obvious, important lacunae in plant population biology, and deserve much more attention. Secondly, the associations are long lasting and highly integrated; they can, therefore, indicate biological features which emerge when mutualism plays a major, continuing role in the lives of organisms. One such feature—inconspicuousness—is probably a cause of their neglect by plant population biologists. In extreme cases, the distinction between partners can be so blurred that no more than a ghost of their separate identities remains—like the grin of Lewis Carroll's Cheshire Cat, to misuse a metaphor of Smith (1979). Thirdly, some of these intimate mutualisms (VAMs, ectomycorrhizas and legume nodules) are sufficiently important in agriculture and silviculture to have been studied in considerable detail; indeed, it is probably true that there is more information available on the interaction between *Rhizobium* and legumes than on any other mutualistic (or antagonistic) association involving plants. Although this literature is not, for the most part, directed towards the central issues of plant population biology, there is much that can be learnt from it.

This chapter reviews several recurring properties of these mutualisms between microorganisms and plant roots. These properties include: a low diversity of endophytes, wide geographical distribution, an absence of strong effects on plant population dynamics, relatively little specificity in recognition of partners, but rather more specificity in the effects of endophyte strains on the growth of host plants. General reviews of the ecology of mutualism are given by Boucher, James & Keeler (1982), and chapters in Boucher (1985).

## MUTUALISTIC ENDOPHYTES

Before embarking on an analysis of plant population dynamics in association with mutualistic microorganisms, some general biological background is needed. The microorganisms occur around and within the roots of host plants, and it is a measure of how intimate the associations are, that they usually involve extensive penetration of host cells (with the exception of ectomycorrhizal fungi). There is, therefore, an important asymmetry between the partners; the microorganisms live, to a greater or lesser degree, within the mutualistic environments of their hosts, whereas the plants live entirely in the antagonistic environment outside (Law & Lewis 1983; Law 1985). In recognition of this difference, the plants and microorganisms are referred to here as 'hosts' and 'mutualistic endophytes', respectively. Ectomycorrhizas, it should be noted, do not fit well into this classification, because the fungi form sheaths around the outside of their host roots and grow less within their hosts, intracellular penetration in particular being virtually absent (Marks & Kozlowski 1973). In spite of this, 'endophyte' is taken to encompass ectomycorrhizal fungi below, for want of a better collective term.

With the exception of ectomycorrhizal fungi, these mutualistic endophytes comprise a very small number of taxa (Table 16.1). Taxonomists have been able to distinguish no more than about ten genera, despite the long evolutionary histories of endophytes in association with plants, during which time the hosts have undergone extensive radiation and speciation (Law & Lewis 1983). It seems unlikely that this is a taxonomic artefact, and it has been suggested that special evolutionary constraints arising from the mutualistic environments of endophytes could be responsible for this low rate of evolution (Law & Lewis 1983; Law 1985). The much greater taxonomic diversity of ectomycorrhizal fungi, in conjunction with their sheathing growth habit, lends some support to this suggestion.

The taxonomic uniformity of most mutualistic endophytes does not, however, limit their geographical distributions or their importance in the

TABLE 16.1.   Approximate numbers of host and endophyte taxa involved in intimate mutualisms. Details of calculations given in Law & Lewis (1983).

| | Endophyte | | Host | |
| --- | --- | --- | --- | --- |
| | Genera | species | Genera | species |
| VAMs | 4 | 100 | 11 000 | 225 000 |
| Ericoid mycorrhizas | 2 | 2 | 25 | 2 000 |
| Ectomycorrhizas | 100 | 3 500 | 150 | 5 000 |
| *Rhizobium* and plants | 2 | 4 | 600 | 17 500 |
| *Frankia* and plants | 1 | ? | 17 | 150 |

biosphere. VAM fungi are present in almost all terrestrial communities (Mosse, Stribley & LeTacon 1981; Harley & Smith 1983, p. 400). Ericoid mycorrhizal fungi dominate the mycorrhizas of heathland communities at higher lattitudes (Read 1983). Ectomycorrhizal fungi are important components of forest communities at higher altitudes or latitudes (Singer & Morello 1960; Meyer 1973) and are also known to occur in association with certain lowland tropical rain forest trees (Janos 1983; Alexander & Hogberg 1986). Most species of the worldwide-distributed family Leguminosae that have been examined possess root nodules with *Rhizobium* or *Bradyrhizobium* (Allen & Allen 1981, p. xxviii).

This wide geographical distribution of mutualistic endophytes needs to be emphasized because population biologists have been influenced by the greater importance of certain conspicuous mutualisms in tropical, as opposed to temperate, communities (Farnworth & Golley 1974, p. 29 *et seq.*). May (1978, 1981, p. 98) attributed the latitudinal change to an apparent vulnerability of obligate mutualists to environmental fluctuations when modelled by simple systems of differential equations. Evidently, this does not apply to the mutualisms considered here, which occur at high latitudes and in pioneer and disturbed communities (Lewis 1973, 1982; Hutson, Law & Lewis 1985).

In passing, it is worth noting that our perception of factors controlling species-diversity would be somewhat different had community ecologists been more interested in mycorrhizas. The lowland, tropical rain forest, which epitomizes a diverse community, contains but a small number of VAM fungal taxa, capable of infecting a wide range of host species (Janos 1980; Malloch, Pirozynski & Raven 1980). It is as we move from the VAM forests of the tropics to the ectomycorrhizal forests of higher latitudes, that diversity of the mycorrhizal fungi increases (Malloch, Pirozynski & Raven 1980). While it would be unrealistic to envisage VAM and ectomycorrhizal fungi as members of a single group, there is little sign that their diversity decreases with increasing latitude, as is characteristic of most other groups. Species living in mutualistic environments appear to operate according to a set of rules different from those of species in antagonistic environments.

## DYNAMICS OF MUTUALISTIC HOST–ENDOPHYTE ASSOCIATIONS

### *Data*

#### *Growth of endophyte populations*

An endophyte population may consist of two parts, one free-living in the soil and rhizosphere and the other in direct contact with plant tissue. The importance of the free-living population and its significance in population dynamics varies from association to association. Free-living *Rhizobium* appears to be autonomous, being able to maintain populations in the rhizosphere of non-host plants, albeit at densities sometimes much lower than in the presence of hosts (Reyes & Schmidt 1979; Nutman & Hearne 1980; Alexander 1984). Both VAM and ectomycorrhizal fungi may produce spores after growth with host plants. Although these spores can intiate new infections following germination, they are believed to be less important than hyphae as a source of infection (e.g. Fries 1987). Spores of ericoid mycorrhizal fungi have not been found free-living, although hyphal fragments (oidia) present in many soils can give rise to new infections. Notice that hyphae of mycorrhizal fungi can grow from one plant root to another (Heap & Newman 1980a; Chiarello, Hickman & Mooney 1982; Read, Francis & Finlay 1985), providing an important path of infection that by-passes the free-living stage. Information on *Frankia* is scant, but the free-living stage is thought to be confined to spores. It appears from this that, with the putative exceptions of *Rhizobium* and ericoid mycorrhizal fungi, the endophytes have obligate requirements of their hosts, to complete their typical life-cycles.

The second component of an endophyte population—that in direct contact with plant tissue—starts to grow when infection of a suitable host takes place. In the case of *Rhizobium* and *Frankia*, there follows an organized series of steps during which the endophyte proliferates within cortical cells, culminating in the development of a root nodule. The life-time of a nodule is itself limited, and propagules released when it senesces replenish the free-living population. Synchronous senescence of many nodules, as in leguminous crops, may be accompanied by a noticeable surge in the free-living population (Reyes & Schmidt 1979; cf. Beringer *et al.* 1979). Nodules of legumes from temperate zones are believed to be predominantly annual, but those of tropical legumes and of the hosts of *Frankia* are more likely to be perennial (Allen & Allen 1981, p. xx).

Primary infection of a plant root by a VAM fungus occurs when a hypha from a germinated spore or from another root comes into contact with the root; this appears to be the result of random collision, there being little evidence of attraction towards the root until very close to it (Mosse & Hepper 1975; Powell 1976). Proliferation of the mycelium in the root cortex and externally in the soil then takes place. As the external mycelium grows, it may establish secondary infections on the same plant, as well as producing primary and secondary infections on further plants and spores free-living in the soil. Notice, however, that the number of entry points connecting the internal and external mycelium is believed to remain small, perhaps as low as one per primary or secondary infection, and certain authors have envisaged growth of the endophyte population in terms of these discrete infection units (Cox & Sanders 1974; Smith & Walker 1981). The whole process of colonizing root systems occurs rapidly under experimental conditions, most susceptible root tissue being colonized within about 50 days (Tinker 1975; Buwalda *et al.* 1982; Sanders & Sheikh 1983). Ectomycorrhizal fungi, on the other hand, sometimes colonize root systems more slowly, with the result that the roots may outgrow the mycelium (Bowen & Theodorou 1973, Read, Francis & Finlay 1985).

As one would expect from the intimate and often obligate associations of endophytes with their hosts, there is evidence that endophyte population densities are regulated by their hosts. The evidence for this is of two kinds. Firstly, experiments on growth of VAM fungi show that the fraction of host root-tissue infected tends towards an asymptote (Furlan & Fortin 1973; Tinker 1975; Sanders & Sheikh 1983). This asymptote is itself host-species dependent, some species being much more heavily infected than others (Buwalda *et al.* 1982). The extent of infection also depends on the nutrients available to the host, being lower on plants supplied with extra phosphorus (Sanders 1975). It has been proposed that the early stages of fungal growth depend on metabolites released from plant roots, and that this release is greater when root phosphorus concentration is low and membrane permeability is high (Graham, Leonard & Menge 1981). Secondly, observations on VAM populations in the field indicate that rather little change in density occurs through time, in spite of the potential for rapid turnover (Read, Koucheki & Hodgson 1976; Sparling & Tinker 1978). There is, however, evidence of greater changes in density of free-living *Rhizobium* in temperate zones (Wilson 1930); in certain cases this may be attributable to seasonal release of individuals from senescing nodules (Reyes & Schmidt 1979).

It should be understood that, notwithstanding the intimate association of endophytes with their hosts, the endophytes do have lives of their own (*sensu* May 1981, p. 98), in much the same way as do parasites, parasitoids and pathogens. They therefore have their own dynamic features which need to be distinguished from those of their hosts.

*Growth of host-plant populations*

The difficulties sometimes experienced in growing host populations without effective endophytes, suggest that the endophytes have important effects on host dynamics (Rayner 1938; Norris 1958; Lange & Parker 1960; Vozzo & Hacskaylo 1971). There is, however, very little information in the literature concerning the effect of mutualistic endophytes on the dynamics of their host-plant populations. The reason for this is that the physiological ecologists and agronomists who have been responsible for most of the work on these symbioses have been more interested in basic features of their biology and their potential for improving crop yield. For the most part, information on growth-rates of individual plants and harvested biomass is as near as the data come to that required for study of population dynamics. Some exceptions to this, in which effects of endophytes on host reproduction or mortality have been observed are: Vozzo & Hacskaylo (1971), Daft & Okusanya (1973), Khan (1975) and Janos (1980), Miller, Jarstfer & Pillai (1987).

The study of individual plants under sterile conditions provides overwhelming evidence that host growth in nutrient-deficient soils is enhanced in the presence of mutualistic endophytes (Table 16.2). In view of the positive relationship which normally exists between plant size and survival and production of offspring, it is reasonable to infer that endophytes enhance the per capita rate of increase of hosts under these conditions. Notice, though, that the endophyte is potentially a major carbon sink, and it may, under some circumstances, reduce the rate of growth of its host (Furlan & Fortin 1973; Trappe 1977; Smith 1980; Stribley, Tinker & Rayner 1980; Bethlenfalvay, Bayne & Pacovsky 1983). Notice, also, that some plant growth occurs in the absence of endophytes, so the association is not obligate so far as the host is concerned under these somewhat artificial conditions.

Under more natural conditions, however, the evidence for enhanced growth of hosts with endophytes is less strong. Major gains in growth of inoculated plants (relative to uninoculated ones) in agricultural or natural communities are reliably achieved only when the natural levels of effective inoculum are low (Rayner 1938; Jenkins, Vincent & Waters 1954; Bryan & Andrew 1955; Vozzo & Hacskaylo 1971; Kleinschmidt & Gerdemann 1972; Lambert & Cole 1980). In other communities the gains are often small or negligible. This is not altogether surprising, because uninoculated plants are often able to make use of natural sources of inoculum. What is more surprising is that, at least in VAMs, there is little discernible decrease in plant growth or nutrition when the density of endophytes is reduced by chemical treatment (Finlay 1985; Fitter 1986; McGonigle, unpublished data). The data seem to suggest that, above a minimum threshold of effective inoculum, mutualistic endophytes have little positive effect on plant growth in the field and, therefore, little effect on the dynamics of their hosts, insofar as this depends on plant size.

There are several reasons why VAM fungi might appear to have little effect on host population dynamics. Firstly, the data themselves may not yet be reliable guides to the processes taking place. Secondly, grazing of the mycelium by soil fungivores such as nematodes and collembolans may render the endophyte ineffective (Warnock, Fitter & Usher 1982; Finlay 1985;

TABLE 16.2. Examples of effects of endophytes on growth of host plants

| Sterile conditions | Non-sterile conditions (glasshouse and field) | Semi-natural communities |
|---|---|---|
| **VAMs** | | |
| + Daft & Nicolson 1966 | + Mosse, Hayman & Ide 1969 | +/o/− Powell 1977 |
| + Baylis 1967 | + Mosse & Hayman 1971 | + Powell 1979 |
| + Hayman & Mosse 1971 | + Kleinschmidt & Gerdemann 1972 | + Hayman & Mosse 1979 |
| + Daft & Okusanya 1973 | + Khan 1975 | + Lambert & Cole 1980 |
| +/− Furlan & Fortin 1973 | +/o Janos 1980 | +/o Rangeley, Daft & Newbould 1982 |
| +/− Bethlenfalvay, Bayne & Pacovsky 1983 | | o McGonigle unpublished data |
| **Ericoid mycorrhizas** | | |
| + Brook 1952 | | |
| + Morrison 1957 | | |
| +/o Stribley, Read & Hunt 1975 | | |
| +/o Stribley & Read 1976 | | |
| **Ectomycorrhizas** | | |
| + Lamb & Richards 1971 | + Kessell 1927 | |
| + Marx & Bryan 1971 | + McComb 1938 | |
| +/o Malajczuk, McComb & Loneragan 1975 | + Rayner 1938 | |
| + Alexander 1981 | + Vozzo & Hacskaylo 1971 | |
| **Rhizobium and plants** | | |
| + Corby 1967 | + Bryan & Andrew 1955 | + Jenkins, Vincent & Waters 1954 |
| | o Corby 1967 | |
| | o Ham, Cardwell & Johnson 1971 | |
| | + Hera 1976 | |
| | +/o Subba Rao 1976 | |

+, positive; −, negative; o, no effect on plant growth.

McGonigle & Fitter 1987); such grazing would be particularly serious close to the root surface, destroying the limited entry points between external and internal mycelia. Thirdly, the endophyte populations may contain a substantial proportion of 'cheaters' (Soberon & Martinez del Rio 1985), which are ineffective so far as their hosts are concerned (see below). Fourthly, as discussed in the next section, a lack of strong effects on host dynamics is compatible with a mutualistic interaction, if the dynamics of the partners operate on different time scales; the endophyte could 'bounce back' from perturbation fast enough for there to be little effect on the host.

## *Theory*

Below, an outline of a model for the dynamics of host–endophyte symbioses is given. It is intended primarily as an aid to thinking about general properties of the associations, and is deliberately kept as simple as possible. However, it also shows that the kind of information needed to fill in the biological detail would not be too difficult to obtain. Some steps have already been taken in this direction in the modelling of VAMs (Tinker 1975; Smith & Walker 1981; Buwalda *et al.* 1982; Sanders & Sheikh 1983), although such work has been aimed, for the most part, at the endophyte dynamics. The model follows a popular line of reasoning about mutualisms, in which the density of one species (the host) is ultimately limited by factors other than the second (the endophyte), whereas the equilibrium density of the second is directly proportional to the density of the first (May 1978, 1981, p. 96; Post, Travis & DeAngelis 1981; Soberon & Martinez del Rio 1981; Wells 1983).

The dynamics of host ($x_1$) and endophyte ($x_2$) population densities are given by a pair of differential equations of general form:

$$\frac{dx_1}{dt} = x_1\{-d_1 + R_1(x_1,x_2)\}$$
$$\frac{dx_2}{dt} = x_2\{-d_2 + R_2(x_1,x_2)\} \tag{1}$$

where $d_1$ and $d_2$ are instantaneous, per capita death-rates and $R_1$ and $R_2$ are instantaneous, per capita rates of reproduction. Separation of the per capita rate of increase into these birth and death components is intended simply to emphasize births and deaths as the processes controlling population dynamics. For simplicity, only the reproductive rates are assumed to be functions of host and endophyte density; this assumption can be readily modified to suit particular kinds of symbioses. The exact nature of state variables $x_1$ and $x_2$ is a matter of practical convenience. The number of infection units has been used in modelling VAM fungal dynamics, and the

host mutants unable to associate with endophytes; for example, non-nodulating legume mutants are well known (Young, Johnston & Brewin 1982; Devine 1984), and, *a priori*, there is no reason why non-mycorrhizal host mutants should not occur. Such mutants, introduced into natural communities, could shed much light on the effects of mutualistic endophytes on host population dynamics.

## NON-SPECIFIC RECOGNITION

Exotic plants often associate with the local indigenous guilds of mutualistic endophytes. There are many striking examples of this phenomenon. For example, *Rhododendron ponticum* readily invades heather moorlands in UK, making use of the local ericoid mycorrhizal fungi. Maize is infected by several VAM fungal species from the forest palm *Oenocarpus panamensis*, in Panama (Puga 1985). Bowen (1963) observed ectomycorrhizal development on the roots of *Pinus radiata*, when grown in twenty-eight soils from areas of *Eucalyptus* in South Australia. Lange & Parker (1961) found that *Lupinus digitatus*, originally from the Mediterranean and naturalized in South Western Australia, was extensively nodulated by local strains of *Rhizobium*.

Although the phenomenon is by no means an invariable rule (e.g. Rayner 1938; Norris 1958; Lange & Parker 1960; Mikola 1970; Vozzo & Hacskaylo 1971), observations of this kind are commonplace, and there is a danger of forgetting how remarkable they are. The associations presumably require coordinated action of both host and inhabitant genomes—genomes which may have been isolated for many millions of years in evolution. That such interactions are successful at all, would seem to imply an extraordinary degree of stabilizing selection. This catholicism of aliens for mutualistic endophytes makes an interesting contrast to their interaction with indigenous parasites and pathogens; these are often unable to attack exotic species, and may enable aliens to spread unchecked, sometimes with dire consequences.

There are good grounds, *a priori*, to expect mutualistic endophytes to be less host-specific than parasites and pathogens. In the coevolutionary interplay between endophytes and their various species of host under infertile conditions, host mutants that resist infection are unlikely to be at a selective advantage; consequently, there should be little selection for the different endophyte mutants which would overcome the specific resistance mechanisms of different host species (Johnson 1976; Vanderplank 1978; Harley & Smith 1983, p. 357 *et seq.*). Rather, one would expect selection in favour of host genotypes allowing infection by the endophytes currently predominant in the community. In effect, the presence of several host species reduces the likelihood of specialization which could occur in the presence of only one.

This positive, interspecific, frequency-dependent selection is potentially a powerful, cohesive evolutionary force in guilds of mutualistic species (Law 1985; Law & Koptur 1986).

The most detailed information at present available on the control of specificity in host recognition is to be found in the literature on *Rhizobium*, because of the interest this species has aroused among molecular biologists (reviewed by Kondorosi & Kondorosi 1986). For the most part, the suite of genes involved in the association with the legume host lie on a small number of large plasmids. The genes fall into two groups, *nod* genes controlling the early stages of interaction before appearance of a visible nodule, and *fix* genes controlling the later stages of nodule development and nitrogen fixation. There are two clusters of *nod* genes. A cluster of genes, *nod ABC* and *D*, have sequences which apear to be conserved across species (Rossen, Johnston & Downie 1984; Kondorosi, Banfalvi & Kondorosi 1984; Torok *et al.* 1984). It has been thought that the products of these genes have effects on root-hair curling which are common to different species of *Rhizobium*, but there is now evidence that the small differences in *nod D* are sufficient to influence host range (Spaink *et al.* 1987). Another cluster, *hsn ABC* and *D*, also controls the range of hosts inoculated, perhaps at the stage of infection-thread formation; these differ across *Rhizobium* species (Horvath *et al.* 1986). These data give us a fascinating glimpse of the mechanisms by which community connectance (*sensu* May 1974, p. 63) is controlled at a molecular level in *Rhizobium*-legume guilds. A model of host recognition, proposed by Kondorosi & Kondorosi (1986), indicates that a high degree of precision is possible; moreover, this is only one side of the coin, because the host plant also has genetic control over the range of acceptable endophytes (e.g. Hardarson & Jones 1979; Young, Johnston & Brewin 1982).

Evidently, there is the potential for highly specific recognition in the *Rhizobium*-legume association, and indeed some cases of specificity are well known, particularly involving legumes used in agriculture (e.g. Young, Johnston & Brewin 1982; Devine 1984). At the other end of the scale, it is also clear that some strains of *Rhizobium* and species of host are remarkably indiscriminate in their partners (Wilson 1944; Lange 1961; Allen & Allen 1981, p. xvii; Crow, Jarvis & Greenwood 1981), and it is probably most realistic to envisage a wide spectrum of specificities in this symbiosis. Since host range of *Rhizobium* can be radically altered by single gene mutations (Djordjevic, Schofield & Rolfe 1985; Horvath *et al.* 1986) and endophyte range can evolve rapidly in the host (Nangju 1980), the connectance of these guilds is presumably liable to rapid evolutionary change depending on local conditions.

Comparable information on recognition phenomena in other intimate mutualisms is lacking. Nonetheless, non-specificity is a recurring feature of these other associations (see Harley & Smith 1983, p 357 *et seq.* for a review of mycorrhizas). Some illustrative data are given in Table 16.3; note that the numerical values are, of necessity, lower limits, dependent upon the number of host taxa tested. A caveat to bear in mind here is that mutual recognition of the partners is simply the first in a series of developmental steps—the opening of the door to host tissues. Once access to the host is achieved, development of the association is contingent upon expression of genes in both partners, and some specificity in subsequent growth of both host and inhabitant is likely, depending on how well coordinated they are. As discussed in the next section, such specificity is well documented, and spans a continuum from associations which are highly effective so far as the host is concerned, to ones which are completely ineffective.

The non-specific recognition of many intimate mutualists is of considerable ecological interest. Firstly, it makes a striking contrast with the relatively high levels of specificity of biotrophic antagonistic species which live at a comparable level of intimacy with their hosts (Brian 1976). This suggests that these mutualistic interactions have, characteristically, greater community connectance (*sensu* May 1974, p. 63) than antagonistic ones. Notwithstanding the difficulty in observing effects on single host species, the endophytes may have important effects on interactions among host species; they are, potentially, 'keystone' species (*sensu* Paine 1969) in the structure of communities. Secondly, non-specificity highlights the importance of the biotic environment in the evolution of the structure of ecological communities. Interactions between species—the building blocks of community structure—are themselves dynamic variables in evolutionary time; mutualism evidently fosters links among species where antagonism breaks them down.

Non-specificity of mycorrhizal fungi is of special ecological interest, because the hyphal growth form of these species allows them to have simultaneous, physical connections with the roots of more than one individual of the same or different species. These interconnections give one plant access to another's resources (Heap & Newman 1980b; Chiarello, Hickman & Mooney 1982; Whittingham & Read 1982; Francis & Read 1984; Read, Francis & Finlay 1985), and the recipient may, as a result, achieve a greater size (Whittingham & Read 1982; Francis, Finlay & Read 1986). A potential consequence of this is that plants which would otherwise be suppressed by their neighbours may be able to grow more and survive for longer. Such an effect has been observed in an experimental community synthesized from calcareous grassland species (Grime *et al.* 1987); in the presence of VAM fungi, yield of the dominant species (*Festuca ovina*) declined, whereas that of some subordinate species increased (e.g. *Scabiosa columbaria, Hieracium*

TABLE 16.3. Number of host taxa successfully inoculated by endophyte strains.

| | Number of endophyte strains tested | Number of host species* successfully inoculated by a particular endophyte strain | | | | | | | Reference |
|---|---|---|---|---|---|---|---|---|---|
| VAMs | 1 | 20 | | | | | | | Mosse 1973 |
| | 12 | 21 | 8 | 7 | 5 | 5 | 4 | 3 | Gerdemann & Trappe 1974 |
| | 11 | 26(28)† | 14(19) | 14(16) | 13(17) | 7(9) | 7(9) | 4(4) | Johnson 1977 |
| | 3 | 3(5) | 2(2) | 1(9) | 0(9) | | | | |
| | 8 | 18(19) | 4(4)(6 isolates) | 1(4) | 1(4) | | | | Graw, Moawad & Rehm 1979 |
| Ericoid mycorrhizas | 6 | 6(6)(6 isolates) | | | | | | | Pearson & Read 1973 |
| Ectomycorrhizas | 1 | 8(8) | | | | | | | Laiho 1970 |
| | 3 | 19(21) | 14(21) | | | | | | Marx & Bryan 1970 |
| | 27 | 7(7)(11 isolates) | 6(7)(2 isolates) | 5(7) | 4(7)(3 isolates) | | | | Molina & Trappe 1982 |
| | | 3(7)(6 isolates) | 2(7)(3 isolates) | 0(7) | | | | | |
| *Frankia* and plants | | 5(7) | 4(8) | 4(4) | 3(6) | 3(6) | 3(3) | 1(8) | Akkermans & Roelofsen 1980* |
| | | 1(5) | 1(4) | | | | | | |

* Number of host genera for *Frankia*, based on a summary of cross-inoculation data from the literature.
† Terms in parenthesis are the numbers of host species (genera for hosts of *Frankia*) on which inoculation was attempted; where no such term is given, the authors did not state the number.

*pilosella, Plantago lanceolata*). It is as though a common mycorrhizal network exists in the soil into which the roots of different plants can be 'plumbed'. The sink created by suppressed plants may be replenished by carbon from the dominants, lessening the competitive imbalance between species, and enabling more species to coexist (see also Janos 1983).

## INVASION BY OTHER STRAINS OF ENDOPHYTE

Population biologists hold the view that mutualisms are particularly vulnerable to invasion by 'cheaters' which reap the benefits of association without any return to their partners (e.g. Axelrod & Hamilton 1981; Soberon & Martinez del Rio 1985). The emphasis of this rather than the reverse process—in which mutualists invade antagonistic associations—may provide more insights into the workings of the human mind than into the natural world. Nonetheless, mutualistic endophytes, with their characteristic time lag between entering host tissues and becoming effective, certainly do lend themselves to invasions by cheaters, as well as invasions by strains more beneficial to their hosts. ('Effective' is used here in its conventional sense in the endophyte literature, i.e. beneficial to the plant host.) For example, mutations which disrupt the chemical pathways of nitrogen fixation in the *Rhizobium*–legume symbiosis make the association ineffective even though the endophyte population may be well established within the cells of its host. *Glomus tenue*, a widely occurring, fine, VAM endophyte which has rarely been shown to enhance host growth, could well be a cheater (Fitter 1985).

There is abundant evidence for variation in effectiveness in associations involving mutualistic endophytes. Some of the data are summarized in Table 16.4. Clearly, effectiveness is best envisaged as a continuum, from the strains most beneficial to growth of a particular host to those most deleterious; this variation is of particular interest in agriculture and silviculture in view of the potential effects on plant yield (Date 1976; Trappe 1977; Powell 1982; Halliday 1984). Notice that strong interactions have sometimes been observed between host variety and endophyte strain, indicating some degree of specificity between the partners once the endophyte has entered host tissue (Nemec 1978; Menge *et al.* 1980; Mytton & Livesey 1983). Notice also that, in VAMs, effectiveness may depend on the amount of external mycelium associated with an infection unit and the rate of formation of new infection units, as well as on the transfer of nutrients per unit of mycelium.

The conditions under which mutant strains of endophyte invade these symbioses can be determined from an extension of the model in equations (1), (2) and (3), by incorporating a further equation to describe the dynamics

TABLE 16.4. Examples of the effect of endophyte strains on growth and/or nutrition of host plants.

| Host | Number of endophyte strains | Effect of endophytes on host | Reference |
|---|---|---|---|
| **VAMs** | | | |
| *Allium cepa* | 4 | Three endophytes gave similar increments in host growth; the fourth, with little external mycelium and a low rate of growth, was ineffective | Sanders *et al.* 1977 |
| *Trifolium subterraneum* | 3 | Host growth response varied from no increase to ×2.5 that of controls with natural inoculum; thought to be a consequence of the rate of formation of mycorrhizal roots | Abbott & Robson 1978 |
| *Citrus* (6 varieties) | 3 | Endophytes differed in their effects on host growth, with interactions between endophyte strain and *Citrus* variety | Nemec 1978 |
| *Glycine max* | 19 | Large differences between endophyte strains in effects on host growth, from highly effective to ineffective or mildly deleterious | Carling & Brown 1980 |
| *Persea americana* | 2 | Uptake of nutrients by the host differed between endophyte strains, apparently related to their rate of growth or ability to infect | Menge *et al.* 1980 |
| *Allium cepa* (3 varieties) | 6 | Endophyte strains had different effects on host growth, the effects depending on host variety | Powell, Clarke & Verberne 1982 |
| **Ectomycorrhizas** | | | |
| *Pinus elliottii* *Pinus radiata* | 8 | Stimulus to host growth was ×1.5 to ×11 that of uninoculated control, depending on endophyte species | Lamb & Richards 1971 |
| *Pinus ponderosa* | 4 | All endophyte species were ineffective, some apparently deleterious | Trappe 1977 |
| *Pseudotsuga menziesii* | 4 | All endophyte species were ineffective, some apparently deleterious | Trappe 1977 |
| *Tsuga heterophylla* | 4 | A small stimulus to host growth was observed, differing between endophyte species | Trappe 1977 |
| ***Rhizobium*—plants** | | | |
| *Trifolium incarnatum* | 5 | Host growth response varied from no increase to ×6 that of an uninoculated control, the increment being dependent on addition of fertilizer | Jenkins, Vincent & Waters 1954 |
| *Lupinus digitatus* | 39 | Nitrogen content of host varied from no increase to ×4 that of an uninoculated control | Lange & Parker 1961 |
| *Trifolium repens* (5 varieties) | 5 | Host growth weakly dependent on endophyte strain, but there was a strong interaction between host variety and endophyte strain | Mytton & Livesey 1983 |

To elucidate the short- and long-term (evolutionary) dynamics of plant–endophyte guilds is a major and interesting task for the future.

## ACKNOWLEDGMENTS

I am grateful to P. D. Crittenden, A. H. Fitter, J. P. Grime, J. L. Harper, V. Hutson, D. P. Janos, D. H. Lewis, T. McGonigle, D. J. Read, D. P. Stribley, C. T. Wheeler, M. H. Williamson and J. P. Young for their generosity in sharing many ideas relating to this work. I thank P. D. Crittenden, A. H. Fitter, K. Giller, D. P. Janos, D. H. Lewis, T. McGonigle and M. H. Williamson for critical comments on a draft of the manuscript.

## REFERENCES

Abbott, L. K. & Robson, A. D. (1978). Growth of subterranean clover in relation to the formation of endomycorrhizas by introduced and indigenous fungi in a field soil. *New Phytologist*, 81, 575–85.

Akkermans, A. D. L. & Roelofsen, W. (1980). Symbiotic nitrogen fixation by Actinomycetes in *Alnus*-type root nodules. *Nitrogen Fixation* (Ed. by W. D. P. Stewart & J. R. Gallon), pp. 279–99. Academic Press, London.

Alexander, I. J. (1981). The *Picea sitchensis + Lactarius rufus* mycorrhizal association and its effects on seedling growth and development. *Transactions of the British Mycological Society*, 76, 417–23.

Alexander, I. J. & Högberg, P. (1986). Ectomycorrhizas of tropical angiospermous trees. *New Phytologist*, 102, 541–49.

Alexander, M. (1984). Ecology of *Rhizobium*. *Biological Nitrogen Fixation: Ecology, Technology and Physiology* (Ed. by M. Alexander), pp. 39–50. Plenum Press, New York.

Allen, O. N. & Allen, E. K. (1981). *The Leguminosae. A Source Book of Characteristics, Uses and Nodulation*. University of Wisconsin Press, Madison.

Axelrod, R. & Hamilton, W. D. (1981). The evolution of cooperation. *Science*, 211, 1390–96.

Baylis, G. T. S. (1967). Experiments on the ecological significance of phycomycetous mycorrhizas. *New Phytologist*, 66, 231–43.

Beringer, J. E., Brewin, N., Johnston, A. W. B., Schulman, H. M. & Hopwood, D. A. (1979). The *Rhizobium*-legume symbiosis. *Proceedings of the Royal Society, London, Series B*, 204, 219–33.

Bethlenfalvay, G. J., Bayne, H. G. & Pacovsky, R. S. (1983). Parasitic and mutualistic associations between a mycorrhizal fungus and soybean: the effect of phosphorus on host plant-endophyte interactions. *Physiologia Plantarum*, 57, 543–48.

Bond, G. (1963). The root nodules of non-leguminous angiosperms. *Symposium of the Society for General Microbiology*, 13, 72–91.

Boucher, D. H. (1985). *The Biology of Mutualism: Ecology and Evolution*. Croom Helm, London.

Boucher, D. H., James, S. & Keeler, K. H. (1982). The ecology of mutualism. *Annual Review of Ecology and Systematics*, 13, 315–47.

Bowen, G. D. (1963). The natural occurrence of mycorrhizal fungi for *Pinus radiata* in South Australian soils. *C. S. I. R. O. Divisional Report*, 6/63.

Bowen, G. D. & Theodorou, C. (1973). Growth of ectomycorrhizal fungi around seeds and roots. *Ectomycorrhizae: their Ecology and Physiology*. (Ed. by G. C. Marks & T. T. Kozlowski), pp. 107–50. Academic Press, New York.

**Brian, P. W. (1976).** The phenomenon of specificity in plant disease. *Specificity in Plant Diseases* (Ed. by A. Graniti), pp. 15–22. Plenum Press, New York.

**Brockwell, J. & Dudman, W. F. (1968).** Ecological studies of root-nodule bacteria introduced into field environments. II. Initial competition between seed inocula in the nodulation of *Trifolium subterraneum* L. seedlings. *Australian Journal of Agricultural Research,* **19**, 749–57.

**Brook, P. J. (1952).** Mycorrhiza of *Pernettya macrostigma*. *New Phytologist,* **51**, 388–97.

**Bryan, W. W. & Andrew, C. S. (1955).** Pasture studies on the coastal lowlands of subtropical Queensland. II. The interrelation of legumes, *Rhizobium*, and calcium. *Australian Journal of Agricultural Research,* **6**, 291–98.

**Buwalda, J. G., Ross, G. J. S., Stribley, D. P. & Tinker, P. B. (1982).** The development of endomycorrhizal root systems III. The mathematical representation of the spread of vesicular-arbuscular mycorrhizal infection in root systems. *New Phytologist,* **91**, 669–82.

**Caldwell, B. E. (1969).** Initial competition of root-nodule bacteria on soybeans in a field environment. *Agronomy Journal,* **61**, 813–15.

**Carling, D. E. & Brown, M. F. (1980).** Relative effect of vesicular-arbuscular mycorrhizal fungi on the growth and yield of soybeans. *Soil Science Society of America Journal,* **44**, 528–32.

**Chiarello, N., Hickman, J. C. & Mooney, H. A. (1982).** Endomycorrhizal role for interspecific transfer of phosphorus in a community of annual plants. *Science,* **217**, 941–43.

**Corby, H. D. L. (1967).** Progress with the legume bacteria in Rhodesia. *Proceedings of the Grassland Society of South Africa,* **2**, 75–81.

**Cox, G. & Sanders, F. (1974).** Ultrastructure of the host-fungus interface in a vesicular-arbuscular mycorrhiza. *New Phytologist,* **73**, 901–12.

**Crawley, M. J. (1986).** *Plant Ecology.* Blackwell Scientific Publications, Oxford.

**Crow, V. L., Jarvis, B. D. W. & Greenwood, R. M. (1981).** Deoxyribonucleic acid homologies among acid-producing strains of *Rhizobium*. *International Journal of Systematic Bacteriology,* **31**, 152–72.

**Daft, M. J. & Nicolson, T. H. (1966).** Effect of *Endogone* mycorrhiza on plant growth. *New Phytologist,* **65**, 343–50.

**Daft, M. J. & Okusanya, B. O. (1973).** Effect of *Endogone* mycorrhiza on plant growth. VI. Influence of infection on the anatomy and reproductive development in four hosts. *New Phytologist,* **72**, 1333–39.

**Date, R. A. (1976).** Principles of *Rhizobium* strain selection. *Symbiotic Nitrogen Fixation in Plants* (Ed. by P. S. Nutman), pp. 137–150. Cambridge University Press, Cambridge.

**Dean, A. M. (1983).** A simple model of mutualism. *American Naturalist,* **121**, 409–17.

**Devine, T. E. (1984).** Genetics and breeding of nitrogen fixation. *Biological Nitrogen Fixation: Ecology, Technology and Physiology* (Ed. by M. Alexander), pp. 127–154. Plenum Press, New York.

**Djordjevic, M. A., Schofield, P. R. & Rolfe, B. G. (1985).** Tn5 mutagenesis of *Rhizobium trifolii* host-specific nodulation genes result in mutants with altered host-range ability. *Molecular and General Genetics,* **200**, 463–71.

**Farnworth, E. G. & Golley, F. B. (1974).** *Fragile Ecosystems. Evaluation of Research and Applications in the Neotropics.* Springer-Verlag, New York.

**Finlay, R. D. (1985).** Interactions between soil micro-arthropods and endomycorrhizal associations of higher plants. *Ecological Interactions in the Soil* (Ed. by A. H. Fitter), pp. 319–31. Blackwell Scientific Publications, Oxford.

**Fitter, A. H. (1985).** Functioning of vesicular-arbuscular mycorrhizas under field conditions. *New Phytologist,* **99**, 257–65.

**Fitter, A. H. (1986).** Effect of benomyl on leaf phosphorus concentration in alpine grasslands: a test of mycorrhizal benefit. *New Phytologist,* **103**, 767–76.

**Francis, R., Finlay, R. D. & Read, D. J. (1986).** Vesicular-arbuscular mycorrhiza in natural vegetation systems IV. Transfer of nutrients in inter- and intra-specific combinations of host plants. *New Phytologist*, **102**, 103–11.

**Francis, R. & Read, D. J. (1984).** Direct transfer of carbon between plants connected by vesicular-arbuscular mycorrhizal mycelium. *Nature, London*, **307**, 53–56.

**Fries, N. (1987).** Ecological and evolutionary aspects of spore germination in the higher Basidiomycetes. *Transactions of the British Mycological Society*, **88**, 1–7.

**Furlan, V. & Fortin, J. A. (1973).** Formation of endomycorrhizae by *Endogone calospora* on *Allium cepa* under three temperature regimes. *Naturaliste canadien* **100**, 467–77.

**Gerdemann, J. W. & Trappe, J. M. (1974).** The Endogonaceae in the Pacific Northwest. *Mycologia Memoir* 5. Published by the New York Botanical Garden.

**Graham, J. H., Leonard, R. T. & Menge, J. A. (1981).** Membrane-mediated decrease in root exudation responsible for phosphorus inhibition of vesicular-arbuscular mycorrhiza formation. *Plant Physiology*, **68**, 548–52.

**Graw, D., Moawad, M. & Rehm, S. (1979).** Untersuchungen zur Wirts- und Wirkungsspezifität der V A-Mykorrhiza. *Zeitschrift für Acker und Pflanzenbau*, **148**, 85–98.

**Grime, J. P., Mackey, J. M. L., Hillier, S. H. & Read, D. J. (1987).** Floristic diversity in a model system using experimental microcosms. *Nature London*, **328**, 420–22.

**Halliday, J. (1984).** Principles of *Rhizobium* strain selection. *Biological Nitrogen Fixation: Ecology, Technology and Physiology* (Ed. by M. Alexander), pp. 155–71. Plenum Press, New York.

**Ham, G. E., Cardwell, V. B. & Johnson, H. W. (1971).** Evaluation of *Rhizobium japonicum* inoculants in soils containing naturalized populations of rhizobia. *Agronomy Journal*, **63**, 301–03.

**Hardarson, G. & Gareth Jones, D. (1979).** The inheritance of preference for strains of *Rhizobium trifolii* by white clover. *Annals of Applied Biology*, **92**, 329–33.

**Harley, J. L. & Smith, S. E. (1983).** *Mycorrhizal Symbiosis*. Academic Press, London.

**Harper, J. L. (1977).** *Population Biology of Plants*. Academic Press, London.

**Harper, J. L., Rosen, B. R. & White, J. (1986).** Preface: the growth and form of modular organisms. *Philosophical Transactions of the Royal Society, London, Series B*, **313**, 3–5.

**Hayman, D. S. & Mosse, B. (1971).** Plant growth responses to vesicular-arbuscular mycorrhiza. I. Growth of *Endogone*- inoculated plants in phosphate-deficient soils. *New Phytologist*, **70**, 19–27.

**Hayman, D. S. & Mosse, B. (1979).** Improved growth of white clover in hill grasslands by mycorrhizal inoculation. *Annals of Applied Biology*, **93**, 141–48.

**Heap, A. J. & Newman, E. I. (1980a).** Links between roots by hyphae of vesicular-arbuscular mycorrhizas. *New Phytologist*, **85**, 169–71.

**Heap, A. J. & Newman, E. I. (1980b).** The influence of vesicular-arbuscular mycorrhizas on phosphorus transfer between plants. *New Phytologist*, **85**, 173–79.

**Hera, C. (1976).** Effect of inoculation and fertilizer application on the growth of soybeans in Rumania. *Symbiotic Nitrogen Fixation in Plants* (Ed. by P. S. Nutman), pp. 269–279. Cambridge University Press, Cambridge.

**Horvath, B., Kondorosi, E., John, M., Schmidt, J., Török, I., Györgypal, Z., Barabas, I., Wieneke, U., Schell, J. & Kondorosi, A. (1986).** Organization, structure and symbiotic function of *Rhizobium meliloti* nodulation genes determining host specificity for Alfalfa. *Cell*, **46**, 335–43.

**Hutson, V., Law, R. & Lewis, D. (1985).** Dynamics of obligate mutualisms—effects of spatial diffusion on resilience of the interacting species. *American Naturalist*, **126**, 445–49.

**Ireland, J. A. & Vincent, J. M. (1968).** A quantitative study of competition for nodule formation. *Transactions of the Ninth International Congress on Soil Science*, **2**, 85–93.

**Janos, D. P. (1980).** Vesicular-arbuscular mycorrhizae affect lowland tropical rain forest plant growth. *Ecology*, **61**, 151–62.

**Janos, D. P. (1983).** Tropical mycorrhizas, nutrient cycles and plant growth. *Tropical Rain Forest: Ecology and Management* (Ed. by S. L. Sutton, T. C. Whitmore & A. C. Chadwick), pp. 327–45. Blackwell Scientific Publications, Oxford.

**Jenkins, H. V., Vincent, J. M. & Waters, L. M. (1954).** The root-nodule bacteria as factors in clover establishment in the red basaltic soils of the Lismore District, New South Wales. III. Field inoculation trials. *Australian Journal of Agricultural Research,* **5,** 77–89.

**Johnson, H. W., Means, U. M. & Weber, C. R. (1965).** Competition for nodule sites between strains of *Rhizobium japonicum* applied as inoculum and strains in the soil. *Agronomy Journal,* **57,** 179–85.

**Johnson, P. N. (1977).** Mycorrhizal endogonaceae in a New Zealand forest. *New Phytologist,* **78,** 161–70.

**Johnson, R. (1976).** Genetics of host-parasite interactions. *Specificity in Plant Diseases* (Ed. by A. Graniti), pp. 45–61. Plenum Press, New York.

**Kessell, S. L. (1927).** Soil organisms. The dependence of certain pine species on a biological soil factor. *Empire Forestry Journal,* **6,** 70–74.

**Khan, A. G. (1975).** The effect of vesicular arbuscular mycorrhizal associations on growth of cereals. II Effects on wheat growth. *Annals of Applied Biology,* **80,** 27–36.

**Kleinschmidt, G. D. & Gerdemann, J. W. (1972).** Stunting of citrus seedlings in fumigated nursery soils related to the absence of endomycorrhizae. *Phytopathology,* **62,** 1447–53.

**Kondorosi, E., Banfalvi, Z. & Kondorosi, A. (1984).** Physical and genetic analysis of a symbiotic region of *Rhizobium meliloti*: identification of nodulation genes. *Molecular and General Genetics,* **193,** 445–52.

**Kondorosi, E. & Kondorosi, A. (1986).** Nodule induction on plant roots by *Rhizobium. Trends in Biochemical Sciences,* **11,** 296–99.

**Koske, R. E. (1981).** A preliminary study of interactions between species of vesicular-arbuscular fungi in a sand dune. *Transactions of the British Mycological Society,* **76,** 411–16.

**Laiho, O. (1970).** *Paxillus involutus* as a mycorrhizal symbiont of forest trees. *Acta Forestalia Fennica,* **106,** 1–72.

**Lamb, R. J. & Richards, B. N. (1971).** Effect of mycorrhizal fungi on the growth and nutrient status of slash and radiata pine seedlings. *Australian Forestry,* **35,** 1–7.

**Lambert, D. H. & Cole, H. (Jr.) (1980).** Effects of mycorrhizae on establishment and performance of forage species in mine spoil. *Agronomy Journal,* **72,** 257–60.

**Lange, R. T. (1961).** Nodule bacteria associated with the indigenous Leguminosae of South-Western Australia. *Journal of General Microbiology,* **26,** 351–59.

**Lange, R. T. & Parker, C. A. (1960).** The symbiotic performance of lupin bacteria under glasshouse and field conditions. *Plant and Soil,* **13,** 137–46.

**Lange, R. T. & Parker, C. A. (1961).** Effective nodulation of *Lupinus digitatus* by native rhizobia in South-Western Australia. *Plant and Soil,* **15,** 193–98.

**Law, R. (1985).** Evolution in a mutualistic environment. *The Biology of Mutualism: Ecology and Evolution* (Ed. by D. H. Boucher), pp. 145–70. Croom Helm, London.

**Law, R. & Koptur, S. (1986).** On the evolution of non-specific mutualism. *Biological Journal of the Linnean Society,* **27,** 251–67.

**Law, R. & Lewis, D. H. (1983).** Biotic environments and the maintenance of sex—some evidence from mutualistic symbioses. *Biological Journal of the Linnean Society,* **20,** 249–76.

**Lewis, D. H. (1973).** The relevance of symbiosis to taxonomy and ecology, with particular reference to mutualistic symbioses and the exploitation of marginal habitats. *Taxonomy and Ecology* (Ed. by V. H. Heywood), pp. 151–72. Academic Press, London.

**Lewis, D. H. (1982).** Mutualistic lives. *Nature, London,* **297,** 176.

**Malajczuk, N., McComb, A. J. & Loneragan, J. F. (1975).** Phosphorus uptake and growth of mycorrhizal and uninfected seedlings of *Eucalyptus calophylla* R. Br. *Australian Journal of Botany,* **23,** 231–38.

**Malloch, D. W., Pirozynski, K. A. & Raven, P. H. (1980).** Ecological and evolutionary significance of mycorrhizal symbioses in vascular plants (a review). *Proceedings of the National Academy of Science USA,* **77,** 2113–18.

**Marks, G. C. & Kozlowski, T. T. (1973).** *Ectomycorrhizae: their Ecology and Physiology.* Academic Press, New York.

**Marx, D. H. & Bryan, W. C. (1970).** Pure culture synthesis of ectomycorrhizae by *Thelephora terrestris* and *Pisolithus tinctorius* on different conifer hosts. *Canadian Journal of Botany,* **48,** 639–43.

**Marx, D. H. & Bryan, W. C. (1971).** Influence of ectomycorrhizae on survival and growth of aseptic seedlings on Loblolly pine at high temperature. *Forest Science,* **17,** 37–41.

**May, R. M. (1974).** *Stability and Complexity in Model Ecosystems* (2nd Edition). Princeton University Press, Princeton.

**May, R. M. (1978).** Mathematical aspects of the dynamics of animal populations. *Studies in Mathematical Biology Part II: Populations and Communities* (Ed. by S. A. Levin), pp. 317–66. Mathematical Association of America.

**May, R. M. (1981).** *Theoretical Ecology: Principles and Applications,* (2nd edn). Blackwell Scientific Publications, Oxford.

**May, R. M. (1982).** Mutualistic interactions among species. *Nature, London,* **296,** 803–4.

**McComb, A. L. (1938).** The relation between mycorrhizae and the development and nutrient absorbtion of pine seedlings in a prairie nursery. *Journal of Forestry,* **36,** 1148–54.

**McGonigle, T. P. & Fitter, A. H. (1987).** Evidence that Collembola suppress plant benefit from vesicular-arbuscular mycorrhizas (VAM) in the field. *Proceedings of the 7th North American Conference on Mycorrhizas, Gainesville Florida.*

**Menge, J. A., LaRue, J., Labanauskas, C. K. & Johnson, E. L. V. (1980).** The effect of two mycorrhizal fungi upon growth and nutrition of avocado seedlings grown with six fertilizer treatments. *Journal of the American Society of Horticultural Science,* **105,** 400–4.

**Meyer, F. H. (1973).** The distribution of ectomycorrhizae in native and man-made forests. *Ectomycorrhizae: their Ecology and Physiology* (Ed. by G. C. Marks & T. T. Kozlowski), pp. 79–105. Academic Press, New York.

**Mikola, P. (1970).** Mycorrhizal inoculation in afforestation. *International Review of Forestry Research,* **3,** 123–96.

**Miller, R. M., Jarstfer, A. G. & Pillai, J. K. (1987).** Biomass allocation in an *Agropyron smithii–Glomus* symbiosis. *American Journal of Botany,* **74,** 114–22.

**Molina, R. & Trappe, J. M. (1982).** Patterns of ectomycorrhizal host specificity and potential among Pacific Northwest conifers and fungi. *Forest Science,* **28,** 423–58.

**Morrison, T. M. (1957).** Host-endophyte relationships in mycorrhizas of *Pernettya macrostigma. New Phytologist,* **56,** 247–57.

**Mosse, B. (1973).** Advances in the study of vesicular-arbuscular mycorrhiza. *Annual Review of Phytopathology,* **11,** 171–96.

**Mosse, B. & Hayman, D. S. (1971).** Plant growth responses to vesicular-arbuscular mycorrhiza. II In unsterilized field soils. *New Phytologist,* **70,** 29–34.

**Mosse, B., Hayman, D. S. & Ide, G. J. (1969).** Growth response of plants in unsterilized soil to inoculation with vesicular-arbuscular mycorrhiza. *Nature, London,* **224,** 1031–32.

**Mosse, B. & Hepper, C. (1975).** Vesicular-arbuscular mycorrhizal infections in root organ cultures. *Physiological Plant Pathology,* **5,** 215–23.

**Mosse, B., Stribley, D. P. & LeTacon, F. (1981).** Ecology of mycorrhizae and mycorrhizal fungi. *Advances in Microbial Ecology,* **5,** 137–210.

**Mytton, L. R. & Livesey, C. J. (1983).** Specific and general effectiveness of *Rhizobium trifolii* populations from different agricultural locations. *Plant and Soil,* **73,** 299–305.

**Nangju, D. (1980).** Soybean response to indigenous *Rhizobia* as influenced by cultivar origin. *Agronomy Journal,* **72,** 403–6.

**Nemec, S. (1978).** Response of six citrus rootstocks to three species of *Glomus*, a mycorrhizal fungus. *Proceedings of the Florida State Horticultural Society*, **91**, 10–14.

**Norris, D. O. (1958).** A red strain of *Rhizobium* for *Lotononis bainesii* Baker. *Australian Journal of Agricultural Research*, **9**, 629–32.

**Nutman, P. S. & Hearne, R. (1980).** Persistence of nodule bacteria in soil under long-term cereal cultivation. *Rothamsted Experimental Station Report for 1979, Part 2*, 77–90.

**Paine, R. T. (1969).** A note on trophic complexity and community stability. *American Naturalist*, **103**, 91–93.

**Pearson, V. & Read, D. J. (1973).** The biology of mycorrhiza in the Ericaceae I. The isolation of the endophyte and synthesis of mycorrhizas in aseptic culture. *New Phytologist*, **72**, 371–79.

**Post, W. M., Travis, C. C. & De Angelis, D. L. (1981).** Evolution of mutualism between species. *Differential Equations and Applications in Ecology, Epidemics and Population Problems* (Ed. by S. N. Busenberg & K. L. Cooke), pp. 183–201. Academic Press, New York.

**Powell, C. Ll. (1976).** Development of mycorrhizal infections from *Endogone* spores and infected root segments. *Transactions of the British Mycological Society*, **66**, 439–45.

**Powell, C. Ll. (1977).** Mycorrhizas in hill country soils. III Effect of inoculation on clover growth in unsterile soils. *New Zealand Journal of Agricultural Research*, **20**, 343–48.

**Powell, C. Ll. (1979).** Inoculation of white clover and ryegrass seed with mycorrhizal fungi. *New Phytologist*, **83**, 81–85.

**Powell, C. Ll. (1982).** Selection of efficient VA mycorrhizal fungi. *Plant and Soil*, **68**, 3–9.

**Powell, C. Ll. & Bagyaraj, D. J. (1984).** *VA Mycorrhiza*. CRC Press Inc., Boca Raton, Florida.

**Powell, C. Ll., Clarke, G. E. & Verberne, N. J. (1982).** Growth response of four onion cultivars to several isolates of VA mycorrhizal fungi. *New Zealand Journal of Agricultural Research*, **25**, 465–70.

**Puga, C. (1985).** *Influence of vesicular-arbuscular mycorrhizae on competition between corn and weeds in Panama*. MS. Thesis, University of Miami, Coral Gables, Florida.

**Rangeley, A., Daft, M. J. & Newbould, P. (1982).** The inoculation of white clover with mycorrhizal fungi in unsterile hill soils. *New Phytologist*, **92**, 89–102.

**Rayner, M. C. (1938).** The use of soil or humus inocula in nurseries and plantations. *Empire Forestry Journal*, **17**, 236–43.

**Read, D. J. (1983).** The biology of mycorrhiza in the Ericales. *Canadian Journal of Botany*, **61**, 985–1004.

**Read, D. J., Francis, R., & Finlay, R. D. (1985).** Mycorrhizal mycelia and nutrient cycling in plant communities. *Ecological Interactions in Soil* (Ed. by A. H. Fitter), pp. 193–217. Blackwell Scientific Publications, Oxford.

**Read, D. J., Koucheki, H. K. & Hodgson, J. (1976).** Vesicular-arbuscular mycorrhiza in natural vegetation systems I. The occurrence of infection. *New Phytologist*, **77**, 641–53.

**Reyes, V. G. & Schmidt, E. L. (1979).** Population densities of *Rhizobium japonicum* Strain 123 estimated directly in soil and rhizospheres. *Applied and Environmental Microbiology* **37**, 854–58.

**Rossen, L., Johnston, A. W. B. & Downie, J. A. (1984).** DNA sequence of the *Rhizobium leguminosarum* nodulation genes *nodAB* and *C* required for root hair curling. *Nucleic Acids Research*, **12**, 9497–508.

**Sanders, F. E. (1975).** The effect of foliar-applied phosphate on the mycorrhizal infections of onion roots. *Endomycorrhizas* (Ed. by F. E. Sanders, B. Mosse & P. B. Tinker), pp. 261–76. Academic Press, London.

**Sanders, F. E. & Sheikh, N. A. (1983).** The development of vesicular-arbuscular mycorrhizal infection in plant root systems. *Plant and Soil*, **71**, 223–46.

**Sanders, F. E., Tinker, P. B., Black, R. L. B. & Palmerley, S. M. (1977).** The development of endomycorrhizal root systems: I. Spread of infection and growth-promoting effects with four species of vesicular-arbuscular endophyte. *New Phytologist*, **78**, 257–68.

Silvertown, J. W. (1982). *Introduction to Plant Population Ecology*. Longman, London.

Singer, R. & Morello, J. H. (1960). Ectotrophic tree mycorrhizae and forest communities. *Ecology*, 41, 549–51.

Smith, D. C. (1979). From extracellular to intracellular: the establishment of a symbiosis. *Proceedings of the Royal Society London, Series B*, 204, 115–30.

Smith, S. E. (1980). Mycorrhizas of autotrophic higher plants. *Biological Reviews*, 55, 475–510.

Smith, S. E. & Walker, N. A. (1981). A quantitative study of mycorrhizal infection in *Trifolium*: separate determination of the rates of infection and of mycelial growth. *New Phytologist*, 89, 225–40.

Soberon, J. M. & Martinez del Rio, C. (1981). The dynamics of a plant-pollinator interaction. *Journal of Theoretical Biology*, 91, 363–78.

Soberon, J. M. & Martinez del Rio, C. (1985). Cheating and taking advantage in mutualistic associations. *The Biology of Mutualism: Ecology and Evolution* (Ed. by D. H. Boucher), pp. 192–216. Croom Helm, London.

Spaink, H. P., Wijffelman, C. A., Pees, E., Okker, R. J. H. & Lugtenberg, B. J. J. (1987). *Rhizobium* nodulation gene *nodD* as a determinant of host specificity. *Nature*, 328, 337–40.

Sparling, G. P. & Tinker, P. B. (1978). Mycorrhizal infection in Pennine grassland. I. Levels of infection in the field. *Journal of Applied Ecology*, 15, 943–50.

Stribley, D. P. & Read, D. J. (1976). The biology of mycorrhiza in the Ericacae. VI. The effects of mycorrhizal infection and concentration of ammonium nitrogen on growth of cranberry (*Vaccinium macrocarpon* Ait.) in sand culture. *New Phytologist*, 77, 63–72.

Stribley, D. P., Read, D. J. & Hunt, R. (1975). The biology of mycorrhiza in the Ericaceae. V. The effects of mycorrhizal infection, soil type, and partial soil sterilization (by gamma-irradiation) on growth of cranberry (*Vaccinium macrocarpon*, Ait.). *New Phytologist*, 75, 119–30.

Stribley, D. P., Tinker, P. B. & Rayner, J. H. (1980). Relation of internal phosphorus concentration and plant weight in plants infected by vesicular-arbuscular mycorrhizas. *New Phytologist*, 86, 261–66.

Subba Rao, N. S. (1976). Field response of legumes in India to inoculations and fertilizer applications. *Symbiotic Nitrogen Fixation in Plants* (Ed. by P. S. Nutman), pp. 255–68. Cambridge University Press, Cambridge.

Tinker, P. B. H. (1975). Effects of vesicular-arbuscular mycorrhizas on higher plants. *Symposium of the Society for Experimental Biology*, 29, 325–49.

Török, I., Kondorosi, E., Stepkowski, T., Pósfai, J. & Kondorosi, A. (1984). Nucleotide sequence of *Rhizobium meliloti* nodulation genes. *Nucleic Acids Research*, 12, 9509–24.

Torrey, J. G. (1978). Nitrogen fixation by actinomycete-nodulated angiosperms. *Bioscience*, 28, 586–92.

Trappe, J. M. (1977). Selection of fungi for ectomycorrhizal inoculation in nurseries. *Annual Review of Phytopathology*, 15, 203–22.

Vanderplank, J. E. (1978). *Genetic and Molecular Basis of Plant Pathogenesis*. Springer-Verlag, Berlin.

Vozzo, J. A. & Hacskaylo, E. (1971). Inoculation of *Pinus caribaea* with ectomycorrhizal fungi in Puerto Rico. *Forest Science*, 17, 239–45.

Warnock, A. J., Fitter, A. H. & Usher, M. B. (1982). The influence of a springtail *Folsomia candida* (Insecta, Collembola) on the mycorrhizal association of leek *Allium porrum* and the vesicular-arbuscular mycorrhizal endophyte *Glomus fasciculatus*. *New Phytologist*, 90, 285–92.

Wells, H. (1983). Population equilibria and stability in plant-animal pollination systems. *Journal of Theoretical Biology*, 100, 685–99.

White, J. (1985). *Studies on Plant Demography. A Festschrift for John L. Harper*. Academic Press, London.

**Whittingham, J. & Read, D. J. (1982).** Vesicular-arbuscular mycorrhiza in natural vegetation systems. III. Nutrient transfer between plants with mycorrhizal interconnections. *New Phytologist,* **90,** 277–84.

**Wilson, J. K. (1930).** Seasonal variation in the number of two species of *Rhizobium* in soil. *Soil Science,* **30,** 289–96.

**Wilson, J. K. (1944).** Over five hundred reasons for abandoning the cross-inoculation groups of the legumes. *Soil Science,* **58,** 61–69.

**Young, J. P. W., Johnston, A. W. B. & Brewin, N. J. (1982).** A search for peas (*Pisum sativum* L.) showing strain specificity for symbiotic *Rhizobium leguminosarum. Heredity,* **48,** 197–201.

# 17. THE INFLUENCE OF MICROORGANISMS, PARTICULARLY *RHIZOBIUM*, ON PLANT COMPETITION IN GRASS–LEGUME COMMUNITIES

R. TURKINGTON[1], F. B. HOLL[2], C. P. CHANWAY[2] AND J. D. THOMPSON[3]

[1] *Department of Botany, University of British Columbia, Vancouver, British Columbia, V6T 2B1, Canada and*

[2] *Department of Plant Sciences, University of British Columbia, Vancouver, British Columbia, V6T 2A2, Canada and*

[3] *Department of Botany, University of Liverpool, Liverpool, L69 3BX, UK*

## SUMMARY

**1** In natural conditions plants continuously interact with a broad array of microorganisms both above and below ground. These microorganisms can influence plant performance by supplying nutrients, by producing substances with phytohormonal activity, or by immobilizing essential nutrients. This in turn influences the nature of any interactions with neighbouring plants.

**2** Actively growing plant roots exert a strong influence on soil microorganism populations. The nature and amount of root exudates varies with plant age and species and thus have a considerable effect on the host plant's growth and development.

**3** A review of the literature on the ecology of grass–legume communities is presented. Particular emphasis is placed on the ecological and evolutionary patterns described for the *Lolium perenne–Trifolium repens–Rhizobium trifolii* association. The thesis is developed that grasses indirectly influence the growth of their neighbouring *Trifolium repens* by their direct effect on soil microorganisms, particularly *Rhizobium trifolii*.

**4** A series of studies is described that were designed to investigate various components of the grass–clover–*Rhizobium* 'tri-symbiosis'. These involved sampling ramets of *Trifolium repens* from an old pasture from areas dominated by different grasses, or by different clones of the same grass. At each site tillers of the dominant grass were also sampled along with the root nodules from the *T. repens*. Factorial experiments were done in which ramets of *T.*

*repens*, tillers of grasses, and strains of *R. trifolii* were grown together in all possible combinations.

**5** The productivity of *T. repens* depends upon the specific association of *T. repens* genotype and its grass, and the *Rhizobium* strain. Inoculation with *R. trifolii* reduced the competitive ability of *Dactylis glomerata* and *Holcus lanatus* against *T. repens*.

**6** When *T. repens* was inoculated with *R. trifolii* isolated from its own nodules it tended to out-yield non-homologous combinations by only 10–15% ($P >$ 0.05). The *T. repens* in *Lolium–Trifolium–Rhizobium* homologous combinations out-yielded non-homologous ones by up to 35%, but similar yield advantages (up to 30%) could still be achieved by using most ramets of *T. repens* provided they were growing with a homologous *Lolium–Rhizobium* pair.

**7** We conclude that although *Rhizobium* is symbiotic with *T. repens*, in forage mixtures it may not necessarily be the significant factor influencing the growth of the clover. Rather, it is the specific association of *Rhizobium–Lolium* genotypes that have the greatest impact on *T. repens* yield—a conclusion consistent with our hypothesis that grasses indirectly influence the growth of their neighbouring *T. repens* by their indirect effect on *Rhizobium*.

## INTRODUCTION

'The primary conclusion from this research is that plant species [in British lowland grassland] can influence the abundance of root surface bacteria and fungi and of mycorrhizas on one another. This could provide an indirect means by which one plant influences another. If the root microorganisms of plant A are influencing its growth, then any alteration of their abundance caused by plant B will result in an indirect effect of B on A's growth.' (Newman *et al.* 1979).

Under natural conditions plants continuously interact with other organisms, including predators, neighbouring plants, and a broad array of microorganisms both above ground and below ground. In the later case, associations with soil microorganisms can influence both plant performance and the nature of any interaction with neighbouring plants, and thus such associations become a potentially important factor in plant population biology. Botanists have tended to concentrate on the ways in which plants are influenced by other plants and by microorganisms and soil microbiologists have focussed on the ways in which microorganisms are influenced by other microorganisms and by higher plants. Much less attention has been given to the question of how plant responses to each other may be mediated by soil microorganisms. This chapter will focus on this question by describing the ways in which previous studies have examined the generation of vegetation patterns in grasslands and pastures by processes involving microorganisms.

## GENERAL LITERATURE REVIEW

No attempt will be made here to summarize the extensive literature on the ways in which plants and microorganisms influence each other, but rather, broad fields of research are described and the reader is directed to selected references and reviews.

### *Microorganisms and the rhizosphere*

Microorganisms are known to have a critical influence on physiological processes in the ecosystem. They are invariably present within the zone of influence of the root, the rhizosphere, of growing plants. The microbial population of the soil is particularly concentrated in the rhizosphere; counts up to $10^{10}$ organisms $g^{-1}$ have been recorded (Barber 1978). For the top 30 cm of a mineral soil under grassland vegetation, Parkinson (1973) calculated the mass of bacteria to range from 32–76 g dry weight $m^{-2}$ and that of fungi from 84–117 g dry weight $m^{-2}$.

The numbers and types of organisms in the soil fluctuate depending on many factors including soil depth, soil moisture, nutrient status, organic matter content, light intensity and temperature, but the primary influence on bacterial numbers and composition is the age, nutritional status, species and the proximity of roots of associated plants (reviewed by Katznelson 1965; Rovira 1978). Roots strongly and selectively stimulate the multiplication of bacteria, fungi, actinomycetes and free-living nematodes. Indirectly this causes an increase in algae and protozoa because the increase in bacteria and fungi means that more food is available for them (Brown 1975).

### *Microorganisms and plant growth*

Rovira (1978) has reviewed the influence of rhizosphere organisms on plants. One of the most beneficial contributions of soil microorganisms to the ecosystem is the supply of nutrients necessary for plant growth (Nicholas 1965; Barber 1978; Dommergues & Krupa 1978). Especially important are those that fix nitrogen, such as *Rhizobium, Azotobacter, Clostridium,* and *Bacillus* (Mishustin 1970; Haynes 1980; Gaskins, Albrecht & Hubbell 1985) and those that enhance the phosphorus uptake by plants, such as mycorrhiza (Barea & Azcon-Aguilar 1983; Coleman, Reid & Cole 1983; Tinker 1984).

Root-associated nitrogen fixation by free-living diazotrophic bacteria has been reported for a variety of microbial species (Knowles 1977). While host plant. The research of Mytton and his colleagues has demonstrated that the performance of *T. repens* is strongly influenced by the identity of its rhizosphere microorganisms, specifically the strains of *Rhizobium trifolii*. In addition, the work of Harper, Turkington and co-workers has described many patterns of population differentiation among grasses and *T. repens*—that desperately require an explanation. It is the authors' thesis that grasses indirectly influence the growth of their neighbouring *T. repens* by their direct effect on soil microorganisms. To address this question the authors have conducted a series of studies to investigate various parts of the grass–clover–*Rhizobium* system.

Ramets of *T. repens* were collected in a 45 year old pasture from patches dominated by one of *D. glomerata, H. lanatus,* and *L. perenne.* At each collection site, tillers of the dominant grass were also collected (a homologous pair) along with root nodules from the *T. repens.* Each of the *T. repens* ramets had different strains of *Rhizobium*, identified by visual comparison of total protein patterns produced by SDS-PAGE (Thompson, Turkington & Holl, unpublished). Using sterilized (autoclaved) soil and equipment a factorial experiment was done in a glasshouse in which the three clover 'types' were grown in all possible combinations with their three *Rhizobium* strains, in monoculture and in mixture with the three grasses. There were significant differences in *T. repens* yield across treatments (Fig. 17.4), the most notable influence being the different grass species. However, there were also significant *T. repens* type × *Rhizobium* strain interaction effects, and interestingly, a grass species × *Rhizobium* interaction; the grass × *Rhizobium* interaction is central to our arguments pertaining to the next experiment. Thus, *T. repens* productivity is influenced by interactions within a three-way grass–clover–*Rhizobium* association. The highest *T. repens* yield was in association with *L. perenne*, regardless of *Rhizobium* strain, and in general, the *T. repens* collected from the *Holcus* patch was higher yielding than *T. repens* from other origins. But notably, homologous combinations of *T. repens* and its *Rhizobium* collected from the *Holcus* patch have a higher dry weight than non-homologous combinations involving the *Holcus*-inoculum, regardless of the competing grass species (Fig. 17.4c). In contrast, the homologous *T. repens-Rhizobium* combinations from both *Lolium* and *Dactylis* patches do not out-yield other combinations (Fig. 17.4b and d). These data reflect on the findings of Evans (1986), who showed that *T. repens* from *Holcus*-dominated areas of the same pasture is morphologically larger than *T. repens* from *Lolium*-dominated and *Dactylis*-dominated areas. Compared with the uninoculated controls, *Rhizobium* inoculation (by any strain) improved *T. repens* yield in mixture with *Dactylis* and *Holcus* relative to that with *Lolium*. Thus, it is apparent that inoculation by *Rhizobium* reduces the competitive edge of these

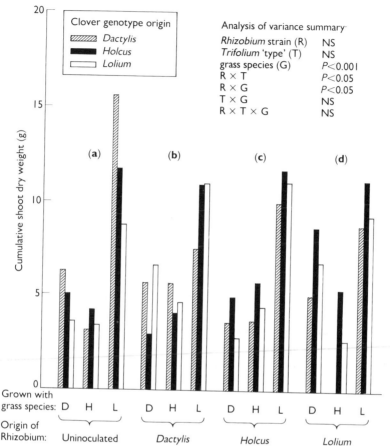

FIG. 17.4 Cumulative shoot dry weights of three genotypes of *Trifolium repens* when grown factorially with three species of grass; D, *Dactylis glomerata*; H, *Holcus lanatus*; L, *Lolium perenne*, three strains of *Rhizobium trifolii*, and an uninoculated control.

two species, a finding that concurs with results of de Wit, Tow & Ennik (1966), Harkess, Hurt & Frame (1970), and Turkington & Aarssen (1984).

Additional experiments were designed to focus specifically on one grass species, *Lolium perenne*, and to study the effect of root and rhizosphere microorganisms on above-ground relationships between *L. perenne* and *T. repens*. Pairs of plants (one *L. perenne* and one *T. repens*) growing side-by-side were collected from *L. perenne*-dominated areas of a 45 year old pasture; *R. trifolii* was isolated from root nodules of the *T. repens*. Using sterilized soil, a factorial experiment was done in a glasshouse in which the three factors were the origin of *L. perenne* tillers, *T. repens* ramets, and *R. trifolii* strains. Tillers, ramets and strains from the same origin were called a homologous group.

**Vallis, I. (1978).** Nitrogen relationships in grass/legume mixtures. *Plant Relations in Pastures* (Ed. by J. R. Wilson), pp. 190–201. CSIRO, East Melbourne, Australia.

**Veeger, C. & Newton, W. E. (Eds.) (1984).** *Advances in Nitrogen Fixation.* Proceeding of the 5th International Symposium on Nitrogen Fixation, Noordwijkerhout, Netherlands. Junk, The Hague.

**Verma, D. P. S. & Long, S. (1983).** The molecular biology of *Rhizobium*-Legume symbiosis. *International Review of Cytology Supplement 14* (Ed. by W. J. Kwang), pp. 211–45. Academic Press, New York.

**Vincent, J. M. (Ed.) (1982).** *Nitrogen Fixation in Legumes.* Academic Press, New York.

**Vose, P. B., & Ruschel, A. P. (Eds.) (1980).** *Associative Dinitrogen Fixation* Vol. 1 and Vol. 2. CRC Press, Boca Raton, Florida.

**Whittingham, J. & Read, D. J. (1982).** Vesicular-arbuscular mycorrhiza in natural vegetation systems. III. Nutrient transfer between plants. *New Phytologist,* **90,** 277–284.

**Wilson, J. R. (Ed.) (1978).** *Plant Relations in Pastures.* CSIRO, Melbourne, Australia.

**Wit, C. T. de, Tow, P. G. & Ennik, G. C. (1966).** Competition between legumes and grasses. *Verslagen van Landbouwkundige Onderzoekingen,* **687,** 3–30.

**Witty, J. F. (1979).** Over estimate of $N_2$-fixation in the rhizosphere by the acetylene reduction method. *The Soil-Root Interface.* (Ed. by J. L. Hardy & R. Scott Russel), pp. 137–44. Academic Press, London.

**Woldendorp, J. W. (1978).** The rhizosphere as part of the plant-soil system. *Structure and Functioning of Plant Populations* (Ed. by A. H. J. Freysen & J. W. Woldendorp), pp. 237–68. North Holland Publishing Company, New York.

**Young, N. R., Hughes, D. M. & Mytton, L. R. (1986).** The response of white clover to different strains of *Rhizobium trifolii* in hill land reseeding: a second trial. *Plant and Soil,* **94,** 277–84.

**Young, N. R. & Mytton, L. R. (1983).** The response of white clover to different strains of *Rhizobium trifolii* in hill land reseeding. *Grass and Forage Science,* **38,** 13–19.

# 18. HERBIVORES AND PLANT POPULATION DYNAMICS

## MICHAEL J. CRAWLEY

*Department of Pure and Applied Biology, Imperial College, Silwood Park, Ascot, Berkshire, SL5 7PY, UK*

## SUMMARY

**1** Herbivores can have pronounced effects on all aspects of plant performance, from seed production, seed survival, seedling survival and vegetative growth to mature plant survivorship.

**2** Demonstrating an effect of herbivory on plant performance does not constitute evidence that herbivory is important in plant population dynamics.

**3** Only by well replicated, manipulative field experiments can it be demonstrated that herbivore feeding influences plant population density.

**4** Existing evidence from herbivore exclusion experiments is biased in favour of vertebrates. Fencing experiments almost always exhibit pronounced effects of herbivore exclusion from the first year onwards. Insect herbivore exclusion using pesticides has only recently been attempted in natural vegetation, and although effects on plant populations are sometimes observed, they are usually less pronounced than those caused by vertebrates.

**5** The most extensive data on the role of herbivore feeding on plant population dynamics come from weed biocontrol, where alien plants have been reduced by release of imported herbivorous insects, but very little information is available on the dynamics of these interactions in their native setting.

**6** Herbivores may be of central importance in the dynamics of some plant populations, but as yet, very few data are available which give a convincing demonstration of the role of the herbivore as a key factor, let alone as a regulating factor. Before progress will be made towards understanding the role of herbivory in plant population dynamics, a great many more studies are needed on the relative importance of such processes as microsite limitation and seed limitation in regulating plant recruitment.

## INTRODUCTION

The impact of herbivory on the population dynamics of plants is poorly understood, despite decades of applied research in pest control and pasture management, and a recent surge of interest by theoreticians and experimentalists (Harper 1977; Crawley 1987a). One of the major aims of studies on plant–herbivore interactions is to understand how the animals influence the distribution and abundance of the plants. This is an ambitious aim, because of the large number of processes involved, and it is worth stressing at the outset that herbivores will never be the only factor affecting plant population dynamics. Indeed, in many cases they will not be the key factor influencing population change from year to year, or from one habitat to another. Sometimes, herbivory *may* be the regulating factor in plant dynamics, acting in a density-dependent manner that, in concert with the density-independent factors, determines the equilibrium population density (Oksanen 1983). At other times, herbivory may act in a minor, density-independent manner, playing no role at all in population regulation (Hairston, Smith & Slobodkin 1960). In non-equilibrium systems, abundant, polyphagous herbivores may prevent the establishment of certain plant species altogether.

Before any statement can be made about the role of herbivory in plant population dynamics, it is essential to establish the factor or factors limiting plant recruitment (microsites, intraspecific competition, rainfall, fungal diseases, and so on). We must distinguish clearly between the impact of herbivory on plant performance, and its effect on population dynamics. For example, it is one thing to show that herbivore feeding reduces seed production, but quite another to show that reduced seed production leads to reduced plant recruitment in the next generation. Similarly, it must be established whether or not different mortality factors are compensatory. Heavy seedling mortality caused by molluscs may have no effect on population dynamics if it simply reduces the mortality suffered later, as a result of self-thinning. This will be a recurrent theme throughout the chapter; until the limits of plant recruitment are known, the importance of herbivory remains difficult to assess.

Recent successes in biological weed control emphasize the potential impact of herbivores on plant population dynamics. The South American weevil *Cyrtobagous salviniae* has been used to clear the aquatic fern *Salvinia molesta* from infested rivers and lakes in Australia and New Guinea (Thomas & Room 1986), and a seed-head feeding weevil, *Rhinocyllus conicus*, introduced from Europe to the rangelands of Canada, has reduced impenetrable stands of the nodding thistle *Carduus nutans* to densities where they no longer interfere with grazing (Zwolfer & Harris 1984).

It is not clear, however, how much can be read into these biocontrol  examples, when considering the role of herbivores in the population dynamics of plants in more natural settings. Biological weed control is unusual in two important respects. First, both the plants and the animals used to control them are introduced, non-native species. Second, the herbivore is released from any constraints imposed by natural enemies attacking it in its native environment. Thus, both plant–plant and herbivore–enemy links in the trophic structure are novel, and the resulting plant–herbivore dynamics will almost certainly be different than they would be in a more closely coevolved, native system. Further, since successful weed control is often followed by improved management practice, intended to reduce the probability of resurgence of the weed, any resulting stable, low density weed population can not be attributed unequivocally to plant–herbivore dynamics (e.g. while the agent reduced the weed, plant competition may keep the weed scarce once control has been achieved; see Crawley 1986). Proof that the herbivore maintains the weed at low densities requires experimental removal of the control agent (using insecticides for example) followed by resurgence of the weed on the sprayed areas. Needless to say, this kind of test has never been carried out.

## DIRECT EFFECTS OF HERBIVORY

Native herbivores have the potential to affect the rates of development, fecundity and mortality of the plants upon which they feed. This section reviews the recent literature on these direct effects. The following section draws attention to the kinds of indirect influences that can occur within and between different trophic levels, when interactions between one plant species and another, and between herbivores and their natural enemies are added to the equation.

### Plant growth

Leaf-eating and sap-sucking species can affect both growth (Karban 1985) and form (Whitham & Mopper 1985). For leaf-feeding species, an important question concerns the nature of the relationship between feeding and plant performance. Is there, for example, direct proportionality between feeding and growth, or is there evidence for threshold levels of feeding, below which damage is negligible? Again, does plant regrowth tend to be exactly-, under- or over-compensating?

For sap-sucking herbivores, the relationship between feeding and plant growth is complicated by a number of interacting factors. Many sucking species inject saliva into the phloem, and this can affect plant growth (Dixon 1986). A great many plant diseases are transmitted by sucking insects (Menges & Loucks 1984). Phloem-feeders typically excrete copious honeydew that can coat the upper surfaces of leaves and form a substrate for a range of fungal species that may have direct or indirect effects on photosynthetic efficiency (Choudhury 1985).

Herbivore feeding may have a variety of effects on tissue longevity, and on the allocation of resources between different tissues. For example, severe defoliation causes root growth to cease almost immediately, and this exacerbates any nutrient or water limitation suffered by the plant (Richards 1984; Seastedt 1985). Similarly, root-pruning invertebrate herbivores cause reductions in shoot growth-rate (Ingham & Detling 1984; Roberts & Morton 1985). Reductions in root or shoot growth-rate can allow neighbouring plants, relatively less affected by herbivory, to occupy the soil or canopy volume (Boryslawski & Bentley 1985; Cottam, Whittaker & Malloch 1986). Indeed, changes in plant size and shape due to herbivory may be amongst the most potent effects in altering the competitive relationships between individuals in communities where competition between mature plants is intense (Harper 1977).

Failure to detect differences in plant growth or form between grazed and ungrazed individuals should not be taken as evidence that herbivory is irrelevant in plant dynamics. After 6 years of insect exclusion, for example, there was no significant difference in total shoot production between sprayed and unsprayed oaks (*Quercus robur*), but there were dramatic differences in seed production over the same period (Crawley 1985). There is enormous variation in growth between individual plants in any single year, and, for perennials, between the same individual plants from one year to the next. Furthermore, the same individual plants tend to be consistently heavily or lightly attacked by herbivores year after year (Crawley 1987b). This exceptionally high variability is exacerbated by within-plant variation (Whitham, Williams & Robinson 1984), and serves to emphasize the need for high replication, careful randomization and well-planned controls in designing experiments on field populations.

Finally, there may be a trade-off between growth and defoliation, as when plants invest limited resources in expensive anti-herbivore traits of various kinds, rather than in dry matter increase. For example, a number of studies have reported positive correlations between defoliation and plant growth, in which plants with the highest rates of defoliation grew faster. The correlation may come about as a result of interspecific differences in leaf chemistry (leaves that are constrained by other factors to be long-lived, need to be more heavily defended, and therefore tend to suffer lower rates of defoliation, Coley 1980), or intraspecific differences in phenology (plants that have earlier bud burst suffer higher rates of attack by defoliating caterpillars, but grow faster than late flushing trees in years when caterpillars are scarce, Crawley 1983).

### Death rate of mature plants

It is unusual for otherwise healthy plants to be killed by herbivores. Nonetheless, both insects and vertebrates are capable of inflicting substantial plant mortality under certain circumstances. When the population densities of large, mobile, polyphagous vertebrate herbivores have been allowed to increase to unusually high levels (usually through the intervention of humans), they can cause devastating mortality (e.g. elephants and trees, Wijngaarden 1985; domestic livestock in arid regions, Newsome & Corbett 1977). Insects, too, can reach population densities at which mass plant mortality is observed (e.g. defoliating species like spruce budworm and gypsy moth in north-eastern North America, Faeth 1987; or cambium-feeding species like bark beetles in the Rocky Mountains, Berryman, Stenseth & Isaev 1986). Given the large numbers of species, and very high potential rates of population growth of tree-feeding insects, it is noteworthy that so few species ever reach the densities at which mature plants are killed (e.g. out of the many thousands of native, foliage-feeding insects found on trees in the United States, only thirteen are recorded as major outbreaking pests, Faeth 1987). Even when mass mortality occurs, it is not observed in all plant species. For example, 3 consecutive years of defoliation by gypsy moth caused 31% mortality amongst *Quercus rubra* but less than 5% mortality in other *Quercus* species in the same New Jersey forests (Stalter & Serrao 1983). Heavy herbivore-induced mortality is particularly likely in plants that have only one terminal meristem. For example, tropical forest palms suffer about twice the rate of mortality of other woody plants in the same habitat (De Steven & Putz 1985), and long-lived, sub-alpine monocarpic rosette plants are especially vulnerable to vertebrate herbivory (Young 1985).

Herbivore-induced mortality of established plants is normally associated with some other, exacerbating circumstance. Most commonly perhaps, mass mortality is observed in plant populations that show even-aged, over-mature age structures. In such cases, it could be argued that plant deaths were inevitable from one cause or another, and no great significance can be attached to the fact that foliage-feeding insects, rather than bark-feeders, fungi, viral disease or wind-throw, were the cause. In other cases, herbivores reach unusually damaging levels on plants that have been subject to drought (White 1984), waterlogging (Kozlowski 1984), exposure to storms (Sprugel 1976), air pollution (Flueckiger & Braun 1986), or unusually high or low levels of soil nutrients (Batzli 1986). In such cases of 'stress-related' herbivore mortality, the cause of the increase in herbivore damage might come about as a result of increases in the food quality of the stressed plants (e.g. increased availability of amino acids) or as a result of impaired ability on the part of

Smith 1984). There are too few data to calculate meaningful average rates of predispersal seed predation for a single species, let alone to compare different plant species.

Postdispersal seed predation by rodents, birds and ants is an important source of mortality, particularly in desert environments (e.g. Brown, Reichman & Davidson (1979) record granivory-rates as high as 75% of all seeds produced). In temperate environments, extremely high rates of postdispersal seed predation have been recorded. In many experiments where seeds were placed on the ground at different densities, and at different distances from parent plants, losses of 100% are frequently recorded. There is a substantial recent literature on postdispersal predation (Davidson, Inouye & Brown 1984; Andersen & Ashton 1985; Janzen 1985; Webb & Willson 1985), but few of these studies report the effect of seed mortality on plant recruitment. The size of the seed crop can affect the death-rate experienced subsequently by the seeds; very large (mast) crops may satiate the seed predators (Ballardie & Whelan 1986; Nilsson & Wastljung 1987).

A major methodological difficulty in studies of postdispersal mortality, is that many of the seed-feeders are also the principal agents of seed dispersal (De Steven & Putz 1984; Stork 1984; Courtney & Manzur 1985; Kjellson 1985; Davidar & Morton 1986; Jensen & Nielsen 1986). Thus, while it is easy to measure the rates of seed removal, it is by no means clear how many of the removed seeds will eventually die through being eaten. The occasional acorn taken 25 km by a jay, may be the only seed to grow to reproductive age from a cohort of tens of thousands of seeds. This kind of event is virtually impossible to observe. Even when seed removal is more local, seeds are frequently cached by birds and small mammals. While undiscovered caches sometimes form the basis for seedling recruitment, it is by no means clear that survival of cache-germinating seedlings is high (West 1968), or that caches do go undiscovered by the animals that made them (Smith 1975).

### Germination and seedling establishment

Those seeds that escape the ravages of pre- and postdispersal seed-predators may find their way to suitable germination microsites. Others may enter a bank of dormant seeds within the soil, or succumb to other mortality agents (fungal disease, over-heating, desiccation, etc.). Of those that germinate successfully, herbivores may take a substantial toll at the seedling stage (Louda 1984; Fye 1984; O'Dowd & Gill 1984; Clark & Clark 1985; Pigott 1985; Mills 1986). Herbivores like molluscs, that may have little effect at other stages in the plant's life-cycle, can devastate populations of seedlings (Duggan 1988). The effect of insect herbivory on seedling mortality varies with the kind of insect. Large, mobile, polyphagous insects like grasshoppers and some sawflies can cause heavy seedling mortality (Parker 1985). Sessile, monophagous insects that feed on seedlings tend to be so small that seedling death-rates are increased little if at all (Mills 1984).

Vertebrate herbivores are capable of eradicating certain plant species by selective removal of seedlings, so long as the seedlings are relatively scarce, and the preference exhibited by the herbivores is sufficiently high. In such cases, the herbivore exhibits a linear functional response to the density of the seedlings; there is no refuge for the seedlings from these mobile herbivores, and no saturating asymptote to the functional response that might allow the seedlings to escape through predator satiation. This kind of exclusion at the seedling stage is poorly documented for natural plant communities, but is observed when newly sown, mixed pastures are selectively grazed by sheep, or seedlings germinating from the seed bank are selectively grazed by rabbits (Crawley, unpublished data). Nevertheless, this does represent one of the most potent means by which herbivory might reduce plant species richness (for a natural example, see Parker & Root 1981).

## INDIRECT EFFECTS OF HERBIVORY

Plants do not generally respond passively to attack by herbivores. They react in a variety of ways, including the production of noxious chemicals (Haukioja & Neuvonen 1985; Rhoades 1985; Edwards, Wratten & Greenwood 1986; and references in Tallamy 1985), or improved physical defences (McNaughton *et al.* 1985) that may reduce subsequent attack. They repair damaged tissues by callus formation, they cut off partially damaged tissues by inducing premature abscission of leaves, fruits or twigs (Faeth 1986; Whitham 1987), they produce new, regrowth shoots (Fox & Bryant 1984; Chapin, Bryant & Fox 1985), and they alter their internal physiological processes of production and distribution of photosynthate (Reinartz 1984; Hunt & Nicholls 1986). A good deal of recent work, and a certain amount of controversy, centres on the degree to which these processes can make good the losses induced by herbivory. At one extreme, is a body of opinion suggesting that some plants over-compensate for herbivory, and therefore have higher fitness when they are grazed than when they are not. At the other extreme, some argue that herbivory is always deleterious to plant fitness. The controversy is summarized by Crawley (1987a).

### *Regrowth and compensation*

When shoots or inflorescences are removed by herbivores, many plants are able to initiate previously dormant meristems which develop into new shoots or inflorescences. Typically, there is a delay in such regrowth, so that young regrowth shoots or inflorescences are exposed to different conditions of herbivory or pollinator availablity than were experienced by their parent tissues (Crawley 1983).

Vegetative regrowth shoots are often juvenile in growth form. This means that their morphology and chemistry are different from shoots of the same age on other parts of the plant (e.g. higher up, out of range of the tallest terrestrial herbivores). There is little general pattern to emerge from studies of subsequent herbivory on regrowth shoots. For vertebrates the regrowth tissues are sometimes less palatable (e.g. snowshow hares fed willow regrowth, Bryant, Wieland & Clausen 1985) and sometimes more palatable (e.g. moose feeding on browsed birches, Danell, Huss-Danell & Bergstrom 1985). For insects, regrowth tissues tend to be more attractive for feeding larvae (Meijden *et al.* 1984), gall-forming cynipids (Washburn & Cornell 1981), or ovipositing adult moths (Crawley & Nachapong 1984). Only rather infrequently, are regrowth tissues less attractive than primary tissues of the same age to specialist, adapted insects (Haukioja, Niemela & Siren 1985).

The contribution of regrowth tissues to plant fitness has been studied in rather few cases. Vegetative regrowth in grasses and shrubs almost certainly increases their competitive ability, compared with plants that do not regrow (McNaughton & Chapin 1985; Belsky 1986a). Even when regrowth of reproductive shoots allows recovery from 100% loss of the initial crop of flowers, regrowth seeds are almost always produced in smaller numbers and/ or are smaller in size than primary seeds (Elmqvist *et al.* 1987). Regrowth seeds of ragwort, for example, are lighter than primary seeds and produce seedlings that are less competitive when grown with pasture grasses (Islam & Crawley 1983; Crawley & Nachapong 1985). Recent work on scarlet gilia, *Ipomopsis aggregata*, in Arizona, shows that plants grazed by ungulates produce 2.4 times as many seeds as ungrazed plants, because the plant supports four flowering regrowth shoots, but only one primary shoot (Paige & Whitham 1987). It is not clear, however, why the regrowth shoots escape herbivory, nor why ungrazed plants would not benefit from branching from the base (see Crawley 1987a).

### Interspecific plant competition

One of the most profound community-level effects of herbivory acts through its role in modifying the relative competitive abilities of the component plant species (Harper 1977). Some plant species may be so intolerant of grazing that they are eliminated directly from the community by selective herbivory. More usually, however, different plant species suffer herbivory at different rates, or have their competitive abilities affected to differing degrees by the same level of herbivory (Grant *et al.* 1985).

Harper (1977) lists a number of cases where plant species that would dominate a plant community in the absence of herbivory, are disadvantaged by herbivore feeding to the point where other, less preferred or more grazing-tolerant plants attain dominance or codominance. This kind of frequency-dependent herbivore feeding can be a potent force in increasing plant-species richness, allowing the recruitment of grazing-tolerant, but otherwise inferior competitors to the plant community. Such changes in community composition are typically reversed when herbivore grazing is relaxed, as in exclusion experiments (Bakker *et al.* 1983; Anderson & Jonasson 1985; Lodge & Whalley 1985; Bazely & Jefferies 1986).

Irreversible community changes can result from herbivory occurring at moderate grazing intensities, when selective herbivory allows an advantage to a plant that is both relatively unpalatable and yet intolerant of defoliation. The classic example is the grass *Nardus stricta* in upland pastures in northern Britain. At low grazing intensities, this unpalatable grass spreads, increasing in abundance at the expense of more palatable grasses like *Agrostis capillaris* that are preferentially taken by sheep. When sheep are excluded, however, the grassland does not revert to its former abundance of *Agrostis*; the *Nardus* is capable of resisting invasion even though the factor that initially gave it its competitive edge, is removed (Grant *et al.* 1985; Welch 1986). Westoby, Noy Meir & Walker (1988) give other examples of apparently irreversible changes under grazing.

### Apparent competition

Manipulative field experiments are currently seen as the most powerful method available for investigating cause and effect in population dynamic systems (Bender, Case & Gilpin 1984). When one species of plant is experimentally reduced in abundance and a second species increases as a result, conventional wisdom would interpret this as evidence that the two species were competitors. The result could, however, be due to 'apparent competition', caused by the feeding responses of shared herbivores (a notion developed initially by Holt & Pickering (1985) in the context of shared predators and diseases). The experimentally reduced plant density may reduce the abundance or alter the feeding behaviour of a herbivore, and it is this change in herbivory, rather than a reduction in interspecific plant competition, that causes the numerical response observed in the second plant species.

A fine example of this is provided by Rice (1987). When previously over-

grazed, semi-arid rangeland in California is protected from cattle grazing by
fencing, a number of previously abundant, alien herbs like *Erodium* spp. are
greatly reduced in abundance, while perennial grasses grow back to achieve
their former dominance. A clear case, it might be supposed, of interspecific
competition, with the grasses shading out the rosette-forming herbs. Rice,
however, erected cages inside the cattle exclosure to exclude voles, and the
*Erodium* returned in the vole exclosures. Evidently, the increase in grass
cover afforded increased protection for the voles from their avian predators,
and this enabled the rodents to forage so effectively beneath the grass cover,
that they were able to prevent *Erodium* establishment. Thus, selective
herbivory, rather than interspecific competition, was the cause of the decline
of *Erodium* following exclusion of the cattle.

### Beneficial effects of herbivores

A number of seed-feeding herbivores probably have a net beneficial effect on
plant fitness, when the benefits obtained through seed dispersal outweigh the
costs of the seeds that are eaten and killed. Less obvious beneficial effects of
herbivory may occur when one mortality factor reduces the intensity of
another, as when, for example, weevil attack on seeds reduces the probability
that the seed will be eaten by small mammals. If the probability of surviving
weevil attack is greater than the probability of surviving small mammal
attack, then weevil attack may increase plant fitness (Crawley 1987a).

Other, more conventional benefits of herbivores involve the creation of
establishment microsites. A great many herbivorous animals create such
microsites in the course of their digging activities (rabbits in temperate
grasslands, porcupines in deserts) or in places where they trample, scratch,
defecate or urinate. Seedling recruitment may be more or less confined to
these local disturbances caused by herbivores (Rabinowitz, Rapp & Dixon
1984; Hobbs & Mooney 1985; Rissing 1986).

The notion that defoliation itself may benefit individual plants has been
debated vigorously in recent years (references in Owen & Weigert 1984). The
current consensus appears to be that while herbivory could, in principle,
increase plant fitness via over-compensatory regrowth (McNaughton 1986),
the evidence overwhelmingly points to the fact that it virtually never does
(Belsky 1986b). As discussed earlier, the increased seed production by grazed
*Ipomopsis* plants raised more questions than it answered (Paige & Whitham
1987), and, while this is the best evidence to date of herbivory increasing
plant performance, it is an isolated case, from which it is difficult to draw
any general conclusions.

*Herbivores in plant succession*

Grazers have traditionally been seen as 'deflecting' successions, rather than as integral parts of them (Edwards & Gillman 1987). Despite the dramatic effects on vegetation of excluding vertebrate herbivores by fencing, remarkably little detailed work has been carried out on the population dynamics of plants during succession, and even less on the role of herbivores in this process. From the small number of experimental studies published to date it is clear that vertebrate herbivores have a greater impact on secondary successions than invertebrates (Brown 1982). If facilitation is a major process (as it may be in many primary successions, Connell & Slatyer 1977), then herbivore feeding should slow down or reverse succession, because it will slow down the rate of environmental amelioration by the dominant plants (Westhoff & Sykora 1979; Bakker 1985; Joenje 1985). Herbivores might halt successions simply by preventing the establishment of tree cover, as with rabbits in chalk grassland. The animals are maintained at relatively high densities by an abundance of grassland plants, and are able to eliminate potential woody invaders by selective grazing on their seeds and seedlings. If inhibition is the main factor affecting the rate of vegetational change, then herbivores should speed succession by weakening the dominant plants and allowing the invasion and establishment of later successional dominants (Toorn & Mook 1982; McBrien, Harmsen & Crowder 1983).

As with so many other aspects of plant–herbivore biology, it is too soon to suggest any general role for herbivory in succession, save to point out that the traditional model of grazing as acting to oppose successional trends in an elastic, reversible way, does not hold for a great many systems (Welch 1986; Westoby, Noy Meir & Walker 1988).


## PLANT POPULATION DYNAMICS

The net change in abundance resulting from births and deaths of plants, and from immigration and emigration of seed, produces characteristic patterns of population dynamics. Plotted over time, plant population density fluctuates to a greater or lesser degree. It may show long-term upward or downward trends, or it may exhibit some form of cyclic behaviour. The object here is to determine how herbivore feeding affects each of these components of dynamics: the existence of an equilibrium, the magnitude of fluctuations, and the direction of long-term trends. Because these are separate issues, relating to regulating factors, key factors, and stability properties, it is worth considering them one at a time.

# M. J. CRAWLEY

## Equilibrium and non-equilibrium dynamics

A persistent controversy centres on the question of whether or not populations are regulated. At one extreme, there are those who believe that most systems are non-equilibrium. They argue that the frequency of disturbance is so great, and spatial heterogeneity so pervasive, that systems have no time to come to equilibrium (Andrewartha & Birch 1984; Strong *et al.* 1984). At the other extreme is a curious alliance between adherents of the 'balance of nature' school of field ecology, and hard-line theoreticians. For completely different reasons, they consider it helpful to view plant communities as structured largely by equilibrium dynamics (May 1973, Mayr 1982).

The truth lies somewhere between these extremes. There are certainly some systems that are driven by immigration, and which suffer periodic extinction of a number of their species. Just as obviously, there are other plant communities that are persistent, and appear to be highly resistant to invasion by immigrating species (Crawley 1986). The fundamental point is that one does not find equilibrium or non-equilibrium systems, but that within any one system, some plant species may exhibit non-equilibrium dynamics, while others may show highly stable, equilibrium interactions with their herbivores.

For example, in the sheep-grazed, upland *Nardus* grasslands discussed earlier, both equilibrium and non-equilibrium interactions occur within the same community. Grazing tips the competitive balance between the grasses in favour of *Nardus* because the sheep feed almost exclusively on the more rapidly-growing *Agrostis*. The dynamics of this interaction are understood most clearly as an equilibrium, with the grazing intensity and the initial abundances of the two grasses determining the outcome. On the other hand, palatable grasses are sufficiently abundant that sheep are maintained at a relatively high density, so that any rare, palatable plant species will have non-equilibrium dynamics. All the invading individuals are discovered by the sheep and eradicated. At any one time, sampling might detect individuals of these plant species, but their dynamics are driven entirely through immigration, and all established individuals eventually succumb to the sheep. Upland tree species like *Quercus petraea*, *Betula pubescens* and *Sorbus aucuparia* fall into this category. Remove the sheep, of course, and succession to woodland would begin. A new equilibrium would be reached in which the *Nardus* was excluded following canopy closure of the trees.

## Equilibrium densities

Lotka-Volterra models of plant herbivore interactions are based on the assumption that the equilibrium plant population is determined entirely by the ratio between the herbivore's death-rate and its feeding conversion efficiency. The higher the animals' death-rate, or the lower their conversion efficiency, the higher the plant abundance. Neither the demographic attributes of the plants themselves (like their growth-rate), nor their carrying capacity have any effect on the size of the equilibrium plant population (see Crawley 1983, for details). But the dynamics need not be like this. For example, the plant may be microsite limited, with the herbivores exerting no depressive effect on abundance, perhaps because of an ungrazeable reserve coupled with regrowth after herbivory. In this case, year to year fluctuations in microsite availability would be both the key *and* the regulating factor. Competition between seeds for limited microsites represents a classic form of intraspecific contest competition.

No matter what the nature of the regulating factors, the existence of an equilibrium density depends absolutely upon the existence of density dependence in one or more of the factors. The precise density at which equilibrium occurs is determined by the *interaction* between density-dependent and density-independent factors (see Watkinson 1986).

## Stable, unstable and cyclic dynamics

Theoretical analysis has provided a detailed understanding of the kinds of factors than can promote stability and instability in plant herbivore dynamics, and what kinds of processes can cause cyclical dynamics (see May 1973; Crawley 1983). Most unfortunately, there has been little attempt by biologists studying dynamics in the field to discriminate between the various possibilities, in order to determine what actually produces the dynamics observed in the field. The best example of this problem concerns cyclic populations of herbivores, such as those of snowshoe hares, lemmings or autumnal moth. Despite decades of research we are still no closer to understanding what causes the cycles. We know what *could* cause them (disease, induced changes in plant quality, refuges, predators, time-lags or combinations of these) but we don't know what *does* cause them (Fox & Bryant 1984; Haukioja, Suomela & Neuvonen 1985; Jonasson *et al.* 1986; Krebs *et al.* 1986).

Discerning the causes of observed dynamics is by no means easy, but it is one of the fundamental challenges facing population biologists. It requires that we have a firm grasp of the basic principles of key and regulating factors, and of the nature of equilibrium and non-equilibrium dynamics. Herbivores could, in principle, regulate plant abundance at stable, equilibrium densities, well below the carrying capacity set by the availability of space and by the supply rates of various essential resources. Some cases of biological weed control may exhibit this kind of dynamics. Unstable plant herbivore dynamics, dominated by positive feedbacks, have been documented in certain bark-beetles, where, once a threshold beetle density is passed, more and more of the trees become susceptible to attack, with the end result that all susceptible trees are eventually killed, and beetle numbers crash (Berryman, Stenseth & Isaev 1985). Non-equilibrium plant–herbivore dynamics occur whenever plant-population density is maintained by seed immigration (e.g. Keddy 1981).

There are no general patterns of plant–herbivore dynamics either within or between communities. Within one community we may find some plant–herbivore interactions that are stable, others that are unstable; some interactions that are equilibrium, others non-equilibrium. While we know a certain amount about the dynamics of hypothetical, model communities, so few studies have been carried out, and the difficulties of detecting regulation are so profound, that predictions concerning the dynamic structure of real communities may well be impossible for many years to come.

### Coexistence and plant species richness

The effect of herbivory on plant species richness has long been debated. Darwin (1859), experimenting with his lawn, observed:

'If turf which has long been mown, . . . be let to grow, the more vigorous plants gradually kill the less vigorous, though fully grown, plants: thus out of twenty species growing on a little plot of turf (three feet by four) nine species perished from the other species being allowed to grow up freely.'

This example highlights a fundamental misconception about plant species richness. What Darwin actually observed was that without mowing, the individual plants grew bigger. Fewer large plants can fit inside a quadrat of 12 square feet, so it is statistically inevitable that fewer species will be found per quadrat when mowing is stopped. The important question is whether any plant species were lost from the lawn as a whole. Of course there may have been extinctions, but the fixed area quadrat technique was not appropriate for determining this.

The most common effect of herbivory is to alter the relative abundance of plant species. The former dominant becomes rarer, some rare, grazing tolerant plants become commoner, some rare plants become even rarer, and a few may be eaten to extinction. Grazing may bring about a net increase in species richness by allowing the incursion of entirely new, grazing tolerant species, formerly excluded by plant competition. Such changes are widely documented in the literature on over-grazed rangelands (Harper 1977; Crawley 1983).

The text book recipe whereby herbivory increases plant species richness is by providing frequency-dependent attack, to the disadvantage of common plants and the advantage of scarce ones. This has rarely been documented in natural communities (Dirzo 1984; Cottam 1985; Molgaard 1986). More often, changes in species richness result from herbivores eating preferred species to extinction, or opening up the community to invasion by grazing tolerant or toxic species (Crawley 1983).

## THE IMPACT OF DIFFERENT KINDS OF HERBIVORES

There are serveral well-documented cases of insect herbivores affecting recruitment in natural plant populations (Louda 1983; Davidson, Samson & Inouye 1985; Meijden *et al.* 1985), but these are greatly outnumbered by studies showing the profound impact of vertebrates (references in Crawley 1983; Huntly 1987). The relative importance of vertebrate and invertebrate herbivores can be judged by the simple experiment of comparing the effects of erecting a fence with the effects of spraying an area with insecticides. In almost all cases, the fence produces a much more pronounced change in plant populations (Wood & Andersen 1988; Crawley & Brown 1988). The principal cause of this difference may lie in the relative importance of natural enemies to the two groups; insects may be kept at low densities by their natural enemies (Hassell 1978), whereas vertebrate herbivores are more frequently food-limited (Sinclair, Dublin & Borner 1985). Within the insects, large, mobile, polyphagous species like grasshoppers and locusts can be thought of as honorary vertebrates, exerting relatively important effects on plant dynamics. More sedentary, specialist insects like lepidoptera and aphids are less likely to have a profound effect on plant population dynamics (Crawley 1983).

Where complex factorial experiments have been carried out to determine the effects of interactions between several different kinds of herbivores, the results have been impossible to generalize. For example, Duggan (1988) excluded insects, molluscs and rabbits singly and severally from populations of the crucifer, *Cardamine pratensis*. The insects were mainly pod-feeding moths and beetles, accounting for about 50% seed mortality in normal conditions. Seed production was only increased markedly following insecticide application when rabbits and slugs were also excluded, because these other animals took the bulk of the immature fruits. Rabbit and slug feeding were more or less compensatory; if more pods survived because of rabbit fencing, more were taken by slugs and vice versa.

The main difficulty with experiments of this kind is their very small scale. In order for one person to carry out the necessary field work, each plot can only contain a few plants if replication is to be satisfactory. We need many more such experiments, and we need them to be funded at a level that allows sufficient pairs of hands to sample large numbers of reasonably large plots.

## HERBIVORES AND THE DYNAMICS OF PLANTS WITH DIFFERENT LIFE-HISTORIES

Just as different herbivores have broadly different patterns of impact on plants, so different life-histories of plants may be influenced by herbivory in different ways. There is no doubt, for example, that recruitment of trees from seed is markedly affected by both invertebrate and vertebrate seed-predators in both temperate and tropical ecosystems (Crawley 1985; Janzen 1985). Similarly, it is clear that herbivory is relatively unimportant in the population dynamics of most of the annual plants characteristic of ruderal communities (Brown 1982). More subtle effects of herbivory on the fecundity of plants, and on the outcome of competition between plants and their neighbours, have tremendous potential influence, but, as yet, we have too few experimental studies to say whether plants of different life-histories are affected in characteristically different ways by this kind of sub-lethal herbivory. For example, while it is tempting to think of plants like grasses as being 'grazing-adapted', individual grass species differ widely from one another in their tolerance of defoliation (Richards 1984). Other kinds of plants, such as trees, exhibit just as wide a range of responses to herbivore feeding (Coley 1980; Crawley 1987b).

## CONCLUSION

Of all the factors that might influence the population dynamics of plants (microsite limitation, intraspecific competition, competition with other plants, diseases, disturbance, occurrence of suitable weather conditions for recruitment, and so on), herbivory retains an enigmatic position. Conventional wisdom has it that herbivores are usually not important in plant population dynamics (Hairston, Smith and Slobodkin 1960). Plants are assumed to be regulated by competition, and herbivores by natural enemies.

While these models may have an element of truth in them, it is not clear that there are any sensible generalizations to be made about population dynamics within entire trophic levels. On the limited evidence available, a rich variety of interactions between plants and herbivores within a single community is indicated. Thus, in any one habitat, some plants may be maintained at low density by the action of seed-feeding insects, others may have their seed production limited by vertebrate folivores, still others may be unaffected by herbivory but suffer microsite limitation, and so forth. All plants are likely to be influenced by herbivores at some time or another; some plants may be limited by herbivores most of the time. Some plant herbivore interactions will be well defined by equilibrium dynamics; others will be stochastic, non-equilibrium interactions characterized by spatial and temporal variation, immigration and periodic local extinction. It is difficult to generalize about how commonly each of these types of dynamics occurs because there simply aren't enough detailed case studies. In several cases, the outcome of herbivory is known; the challenge now is to understand the processes which bring these patterns about.

## REFERENCES

**Andersen, A. N. & Ashton, D. H. (1985).** Rates of seed removal by ants at heath and woodland sites in southeastern Australia. *Australian Journal of Ecology*, **10**, 381–90.

**Anderson, M. & Jonasson, S. (1985).** Rodent cycles in relation to food resources on an alpine heath. *Oikos*, **46**, 93–106.

**Andrewartha, H. G. & Birch, L. C. (1984).** *The Ecological Web. More on the Distribution and Abundance of Animals.* University of Chicago Press, London.

**Bakker, J. P. (1985).** The impact of grazing on plant communities, plant populations and soil conditions on salt marshes. *Vegetatio*, **62**, 391–98.

**Bakker, J. P., Bie, S. de, Dallinga, J. H., Tjaden, P. & Vries, Y. de (1983).** Sheep-grazing as a management tool for heathland conservation and regeneration in the Netherlands. *Journal of Applied Ecology*, **20**, 541–60.

**Ballardie, R. T. & Whelan, R. J. (1986).** Masting, seed dispersal and seed predation in the cycad *Macrozamia communis*. *Oecologia*, **70**, 100–05.

**Batzli, G. O. (1986).** Nutritional ecology of the California vole: effects of food quality on reproduction. *Ecology*, **67**, 406–12.

**Bazely, D. R. & Jefferies, R. L. (1986).** Changes in the composition and standing crop of salt marsh communities in response to the removal of a grazer. *Journal of Ecology*, **74**, 693–706.

**Belsky, A. J. (1986a).** Population and community processes in a mosaic grassland in the Serengeti, Tanzania. *Journal of Ecology*, **74**, 841–56.

**Belsky, A. J. (1986b).** Does herbivory benefit plants? A review of the evidence. *American Naturalist*, **127**, 870–92.

**Bender, E. A., Case, T. J. & Gilpin, M. E. (1984).** Perturbation experiments in community ecology: theory and practice. *Ecology*, **65**, 1–13.

**Benkman, C. W., Balda, R. P. & Smith, C. C. (1984).** Adaptations for seed dispersal and the compromises due to seed predation in limber pine. *Ecology*, **65**, 632–42.

**Bernays, E. A. & Lewis, A. C. (1986).** The effect of wilting on palatability of plants to *Schistocerca gregaria*, the desert locust. *Oecologia*, **70**, 132–35.

**Berryman, A. A., Stenseth, N. C. & Isaev, A. S. (1987).** Natural regulation of herbivorous forest insect populations. *Oecologia*, **71**, 174–84.

Boryslawski, Z. & Bentley, B. L. (1985). The effect of nitrogen and clipping on interference between C3 and C4 grasses. *Journal of Ecology*, **73**, 113–21.

Brown, J. H., Reichman, O. J. & Davidson, D. W. (1979). Granivory in desert ecosystems. *Advances in Ecological Research*, **10**, 201–27.

Brown, V. K. (1982). The phytophagous insect community and its impact on early successional habitats. *Proceedings of the 5th International Symposium on Insect-Plant Relationships* (Ed. by J. H. Visser & A. K. Minks), pp. 205–13. Pudoc, Wageningen.

Bryant, J. P., Wieland, G. D. & Clausen, T. (1985). Interactions of snowshoe hare and feltleaf willow in Alaska. *Ecology*, **66**, 1564–73.

Chapin, F. S., Bryant, J. P. & Fox, J. F. (1985). Lack of induced chemical defense in juvenile Alaskan woody plants in response to simulated browsing. *Oecologia*, **67**, 457–59.

Choudhury, D. (1985). Aphid honeydew: a reappraisal of Owen and Wiegert's hypothesis. *Oikos*, **45**, 287–90.

Clark, D. B. & Clark, D. A. (1985). Seedling dynamics of a tropical tree: impacts of herbivory and meristem damage. *Ecology*, **66**, 1884–92.

Coley, P. D. (1980). Effects of leaf age and plant life history patterns on herbivory. *Nature*, **284**, 545–46.

Connell, J. H. & Slatyer, R. O. (1977). Mechanisms of succession in natural communities and their role in community stability and organization. *American Naturalist*, **111**, 1119–44.

Cottam, D. A. (1985). Frequency-dependent grazing by slugs and grasshoppers. *Journal of Ecology*, **73**, 925–33.

Cottam, D. A., Whittaker, J. B. & Malloch, A. J. C. (1986). The effects of chrysomelid beetle grazing and plant competition on the growth of *Rumex obtusifolius*. *Oecologia*, **70**, 452–56.

Coulson, R. N. (1979). Population dynamics of bark beetles. *Annual Review of Entomology*, **24**, 417–47.

Courtney, S. P. & Manzur, M. I. (1985). Fruiting and fitness in *Crataegus monogyna*: the effects of frugivores and seed predators. *Oikos*, **44**, 398–406.

Crawley, M. J. (1983). *Herbivory. The Dynamics of Animal-Plant Interactions*. Blackwell Scientific Publications, Oxford.

Crawley, M. J. (1985). Reduction of oak fecundity by low density herbivore populations. *Nature*, **314**, 163–64.

Crawley, M. J. (1986). The population biology of invaders. *Philosophical Transactions of the Royal Society of London*, **B314**, 711–31.

Crawley, M. J. (1987a). Benevolent herbivores? *Trends in Ecology and Evolution*, **2**, 167–68.

Crawley, M. J. (1987b). The effects of insect herbivores on the growth and reproductive performance of English oak. *Insects-Plant. Proceedings of the 6th International Symposium on Insect-Plant Relationships. Pau, France* (Ed. by V. Labeyrie, G. Fabres & D. Lachaise), pp. 307–11. Junk, Dordrecht.

Crawley, M. J. & Brown, R. A. (1988). The effect of grazing by rabbits on the yield of winter wheat. *Journal of Applied Ecology*, in press.

Crawley, M. J. & Nachapong, M. (1984). Facultative defences and specialist herbivores? Cinnabar moth (*Tyria jacobaeae*) on the regrowth foliage of ragwort (*Senecio jacobaea*). *Ecological Entomology*, **9**, 389–93.

Crawley, M. J. & Nachapong, M. (1985). The establishment of seedlings from primary and regrowth seeds of ragwort (*Senecio jacobaea*). *Journal of Ecology*, **73**, 255–61.

Danell, K., Huss-Danell, K. & Bergstrom, R. (1985). Interactions between browsing moose and two species of birch in Sweden. *Ecology*, **66**, 1867–78.

Darwin, C. (1859). *The Origin of Species*. John Murray, London.

Davidar, P. D. & Morton, E. S. (1986). The relationship between fruit crop sizes and fruit removal rates by birds. *Ecology*, **67**, 262–65.

Davidson, D. W., Inouye, R. S. & Brown, J. H. (1984). Granivory in a desert ecosystem: experimental evidence for indirect facilitation of ants by rodents. *Ecology*, 65, 1780–86.

Davidson, D. W., Samson, D. A. & Inouye, R. S. (1985). Granivory in the Chihuahuan Desert: interactions within and between trophic levels. *Ecology*, 66, 486–502.

De Steven, D. & Putz, F. E. (1984). Impact of mammals on early recruitment of a tropical canopy tree, *Dipteryx panamensis*, in Panama. *Oikos*, 43, 207–16.

De Steven, D. & Putz, F. E. (1985). Mortality rates of some rain forest palms in Panama. *Principes*, 29, 162–65.

Dirzo, R. (1984). Herbivory: a phytocentric overview. *Perspectives on Plant Population Ecology* (Ed. by R. Dirzo & J. Sarukhan), pp. 141–65. Sinauer, Massachusetts.

Dixon, A. F. G. (1986). The distribution of aphid infestation in relation to the demography of plant parts: a reply. *American Naturalist*, 127, 410–13.

Duggan, A. (1988). *Population ecology of* Cardamine pratensis L. *and* Anthocharis cardamines L. Unpublished PhD thesis, University of London.

Edwards, P. J. & Gillman, M. P. (1988). Herbivores and plant succession. *Colonization, Succession and Stability* (Ed. by A. J. Gray, M. J. Crawley & P. J. Edwards), pp. 295–314. Blackwell Scientific Publications. Oxford.

Edwards, P. J., Wratten, S. D. & Greenwood, S. (1986). Palatability of British trees to insects: constitutive and induced defences. *Oecologia*, 69, 316–19.

Elmqvist, T., Ericson, L., Danell, K. & Salomonson, A. (1987). Flowering shoot production and vole bark herbivory in a boreal willow. *Ecology*, 68, 1623–29.

Faeth, S. H. (1986). Indirect interactions between temporally separated herbivores mediated by the host plant. *Ecology*, 67, 479–94.

Faeth, S. H. (1987). Community structure and folivorous insect outbreaks: the roles of vertical and horizontal interactions. *Insect Outbreaks* (Ed. by P. Borbosat & J. C. Shultz). Academic Press, New York.

Flueckiger, W. & Braun, S. (1986). Effect of air pollutants on insects and host plant/insect relationships. *How are the Effects of Air Pollutants on Agricultural Crops influenced by the Interaction with other Limiting Factors*, pp. 79–91. Commission of the European Communities, Brussels.

Foster, W. A. (1984). The distribution of the sea lavender aphid *Staticobium staticis* on maritime saltmarsh and its effects on host plant fitness. *Oikos*, 42, 97–104.

Fox, J. F. & Bryant, J. P. (1984). Instability of the snowshoe hare and woody plant interaction. *Oecologia*, 63, 128–35.

Fye, R. E. (1984). Damage to vegetable and forage seedlings by overwintering *Lygus hesperus* (Heteroptera: Miridae) adults. *Journal of Economic Entomology*, 77, 1141–43.

Grant, S. A., Suckling, D. E., Smith, H. K., Torvell, L., Forbes, T. D. A. & Hodgson J. (1985). Comparative studies of diet selection by sheep and cattle: the hill grasslands. *Journal of Ecology*, 73, 987–1004.

Hairston, N. G., Smith, F. E. & Slobodkin, L. B. (1960). Community structure, population control and competition. *American Naturalist*, 94, 421–25.

Halse, S. A. & Trevenen, H. J. (1985). Damage to mesic pastures by skylarks in north-western Iraq. *Journal of Applied Ecology*, 22, 337–46.

Harper, J. L. (1977). *Population Biology of Plants*. Academic Press, London.

Harris, P. (1981). Stress as a strategy in the biological control of weeds. *Biological Control in Crop Protection* (Ed. by G. C. Papavizas), pp. 333–40. Allanheld, Osmun, Toronto.

Hassell, M. P. (1978). *The Dynamics of Arthropod Predator-Prey Systems*. Princeton University Press, New Jersey.

Haukioja, E. & Neuvonen, S. (1985). Induced long-term resistance of birch foliage against defoliators: defensive or accidental? *Ecology*, 66, 1303–08.

Haukioja, E., Niemela, P. & Siren, S. (1985). Foliage phenols and nitrogen in relation to growth,

O'Dowd, D. J. & Gill, A. M. (1984). Predator satiation and site alteration following fire: mass reproduction of alpine ash (*Eucalyptus delegatensis*) in southeastern Australia. *Ecology*, 65, 1052–66.

Oksanen, L. (1983). Trophic exploitation and arctic phytomass patterns. *American Naturalist*, 122, 45–52.

Owen, D. F. & Wiegert, R. G. (1984). Aphids and plant fitness: 1984. *Oikos*, 43, 403.

Paige, K. N. & Whitham, T. G. (1987). Overcompensation in response to mammalian herbivory: the advantage of being eaten. *American Naturalist*, 129, 407–16.

Parker, M. A. (1985). Size-dependent herbivore attack and the demography of an arid grassland shrub. *Ecology*, 66, 850–60.

Parker, M. A. & Root, R. B. (1981). Insect herbivores limit habitat distribution of a native composite *Machaeranthera canescens*. *Ecology*, 62, 1390–92.

Pigott, C. D. (1985). Selective damage to tree-seedlings by bank voles (*Clethrionomys glareolus*). *Oecologia*, 67, 367–71.

Rabinowitz, D., Rapp, J. K. & Dixon, P. M. (1984). Competitive abilities of sparse grass species: means of persistence or cause of abundance. *Ecology*, 65, 1144–54.

Rausher, M. D. & Feeny, P. (1980). Herbivory, plant density, and reproductive success: the effect of *Battus philenor* on *Aristolochia reticulata*. *Ecology*, 61, 905–17.

Reinartz, J. A. (1984). Life history variation of common mullein (*Verbascum thapsus*). I. Latitudinal differences in population dynamics and timing of reproduction. II. Plant size, biomass partitioning and morphology. III. Differences among sequential cohorts. *Journal of Ecology*, 72, 897–912, 913–25 & 927–36.

Reynolds, D. N. (1984). Populational dynamics of three annual species of alpine plants in the Rocky Mountains. *Oecologia*, 62, 250–55.

Rhoades, D. F. (1985). Offensive-defensive interactions between herbivores and plants: their relevance in herbivore population dynamics and ecological theory. *American Naturalist*, 125, 205–38.

Rice, K. J. (1987). Interaction of disturbance, patch size and herbivory in *Erodium* colonization. *Ecology*, 68, 1113–15.

Richards, J. H. (1984). Root growth response to defoliation in two *Agropyron* bunchgrasses: field observations with an improved root periscope. *Oecologia*, 64, 21–25.

Rissing, S. W. (1968). Indirect effects of granivory by harvester ants: plant species composition and reproductive increase near ant nests. *Oecologia*, 68, 231–34.

Roberts, R. J. & Morton, R. (1985). Biomass of larval Scarabaeidae (Coleoptera) in relation to grazing pressures in temperate, sown pastures. *Journal of Applied Ecology*, 22, 863–74.

Schroeder, D. & Goeden, R. G. (1986). The search for arthropod natural enemies of introduced weeds for biocontrol—in theory and practice. *Biocontrol News and Information*, 7, 147–55.

Seastedt, T. R. (1985). Maximization of primary and secondary productivity by grazers. *American Naturalist*, 126, 559–64.

Sheppard, A. (1987). *Insect herbivore competition and the population dynamics of* Heracleum sphondyllium L. (*Umbelliferae*). Unpublished PhD thesis. University of London.

Sinclair, A. R. E., Dublin, H. & Borner, M. (1985). Population regulation of Serengeti wildebeest: a test of the food hypothesis. *Oecologia*, 65, 266–68.

Smith, C. C. (1975). The coevolution of plants and seed predators. *Coevolution of Animals and Plants* (Ed. by L. E. Gilbert & P. H. Raven), pp. 53–77. University of Texas Press, Austin.

Sprugel, D. G. (1976). Dynamic structure of wave-generated *Abies balsamea* forests in the northwestern United States. *Journal of Ecology*, 64, 889–911.

Stalter, R. & Serrao, J. (1983). The impact of defoliation by gypsy moths on the oak forest at Greenbrook Sanctuary, New Jersey. *Bulletin of the Torrey Botanical Club*, 110, 526–29.

Stamp, N. E. (1984). Effect of defoliation by checkerspot caterpillars (*Euphydryas phaeton*) and sawfly larvae (*Macrophya nigra* and *Tenthredo grandis*) on their host plants (*Chelone* spp.). *Oecologia*, **63**, 275–80.

Stephenson, A. G. (1984). The regulation of maternal investment in an indeterminate flowering plant (*Lotus corniculatus*). *Ecology*, **65**, 113–21.

Stork, V. L. (1984). Examination of seed dispersal and survival in red oak, *Quercus rubra* (Fagaceae), using metal-tagged acorns. *Ecology*, **65**, 129–22.

Strong, D. R., Simberloff, D., Abele, L. G. & Thistle, A. B. (1984). *Ecological Communities: conceptual Issues and Evidence*. Princeton University Press, New Jersey.

Tallamy, D. W. (1985). Squash beetle feeding behaviour: an adaptation against induced cucurbit defences. *Ecology*, **66**, 1574–79.

Thomas, P. A. & Room, P. M. (1986). Taxonomy and control of *Salvinia molesta*. *Nature*, **320**, 581–84.

Thompson, J. N. (1985). Postdispersal seed predation in *Lomatium* spp. (Umbelliferae): variation among individuals and species. *Ecology*, **66**, 1608–16.

Toorn, J. van der, & Mook, J. H. (1982). The influence of environmental factors and management on stands of *Phragmites australis*. I. Effects of burning, frost and insect damage on shoot density and shoot size. *Journal of Applied Ecology*, **19**, 477–99.

Waloff, N. & Richards, O. W. (1977). The effect of insect fauna on growth, mortality and natality of broom, *Sarothamnus scoparius*. *Journal of Applied Ecology*, **14**, 787–98.

Washburn, J. O. & Cornell, H. V. (1981). Parasitoids, patches, and phenology: their possible role in the local extinction of a cynipid gall wasp population. *Ecology*, **62**, 1597–607.

Watkinson, A. R. (1978). The demography of a sand dune annual: *Vulpia fasciculata*. II. The dynamics of seed populations. *Journal of Ecology*, **66**, 35–44.

Watkinson, A. R. (1986). Plant population dynamics. *Plant Ecology* (Ed. by M. J. Crawley), pp. 137–84. Blackwell Scientific Publications, Oxford.

Webb, S. L. & Willson, M. F. (1985). Spatial heterogeneity in post-dispersal predation on *Prunus* and *Uvularia* seeds. *Oecologia*, **67**, 150–53.

Welch, D. (1986). Studies in the grazing of heather moorland in north-east Scotland. V. Trends in *Nardus stricta* and other unpalatable graminoids. *Journal of Applied Ecology*, **23**, 1047–58.

West, N. E. (1968). Rodent-influenced establishment of ponderosa pine and bitterbrush seedlings in central Oregon. *Ecology*, **49**, 1009–11.

Westhoff, V. & Sykora, K. V. (1979). A study of the influence of desalination on the Juncetum gerardii. *Acta Botanica Neerlandica*, **28**, 505–12.

Westoby, M., Noy Meir, I. & Walker, B. H. (1988). Irreversible rangeland successions. *Journal of Arid Lands Research*, in press.

White, T. C. R. (1984). The abundance of insect herbivores in relation to the availability of nitrogen in stressed food plants. *Oecologia*, **63**, 90–105.

Whitham, T. G. (1987). Evolution of territoriality by herbivores in response to host plant defences. *American Zoologist*, **27**, 359–69.

Whitham, T. G. & Mopper, S. (1985). Chronic herbivory: impacts on architecture and sex expression of pinyon pine. *Science*, **228**, 1089–91.

Whitham, T. G., Williams, A. G. & Robinson, A. M. (1984). The variation principle: individual plants as temporal and spatial mosaics of resistance to rapidly evolving pests. *A New Ecology: Novel Approaches to Interactive Systems* (Ed. by P. W. Price, C. N. Slobodchikoff & W. S. Gaud), pp. 15–51. John Wiley, New York.

Wijngaarden, W. van (1985). Elephants-trees-grass-grazers: relationships between climate, soil, vegetation and large herbivores in a semi-arid savanna ecosystem (Tsavo, Kenya). ITC Publication, **4**, 1–159. Wageningen.

Wood, D. M. & Andersen, M. C. (1988). Predispersal seed predation in *Aster ledophyllus*: does it matter? Experimental evidence from Mount St. Helens. *Ecology*, in press.

**Young, T. P. (1985).** *Lobelia telekii* herbivory, mortality, and size at reproduction: variation with growth rate. *Ecology*, **66**, 1879–83.

**Zwolfer, H. & Harris, P. (1984).** Biology and host specificity of *Rhinocyllus conicus* (Froel.) (Col., Curculionidae), a successful agent for biocontrol of the thistle, *Carduus nutans L. Zeitschrift für angewandte Entomologie,* **97**, 36–62.

# 19. PLANT PARASITISM: THE POPULATION DYNAMICS OF PARASITIC PLANTS AND THEIR EFFECTS UPON PLANT COMMUNITY STRUCTURE

## A. R. WATKINSON[1] AND C. C. GIBSON[2]

[1]*School of Biological Sciences, University of East Anglia, Norwich NR4 7TJ, UK and*
[2]*Nature Conservancy Council, All Saints House, High Street, Colchester, Essex CO1 1UG, UK*

## SUMMARY

**1** The range of plant parasites is large and includes viruses, bacteria, fungi, gall-forming insects and other plants. About 3000 species of higher plant in fifteen families are known to be parasitic on other plants. These range from holoparasites to hemiparasites and from root parasites to aerial parasites.

**2** The nature of parasitism, in particular amongst the parasitic Scrophulariaceae, is reviewed. The population dynamics of four hemiparasitic species (*Euphrasia stricta, Pedicularis palustris, Rhinanthus angustifolius* and *R. minor*) are then examined in detail. The basic reproductive rate is calculated for each species and it is shown how a complex of density-dependent and density-independent factors interact with the host environment to determine abundance.

**3** The number of host species recorded for members of the hemiparasitic Scrophulariaceae has been widely documented and ranges from four to seventy-nine per species. Far less information is available on host selectivity, but certain families including the Gramineae and Leguminosae would appear to be preferred hosts.

**4** It is demonstrated that differential host selectivity by *Rhinanthus minor* may alter the competitive balance between pairs of species under controlled conditions.

**5** Differential host attack provides a mechanism by which parasitic plants in general may be expected to affect the structure of plant communities. In natural plant communities infested with *R. minor* it is shown that the effect of the hemiparasite is generally to decrease species diversity through parasitism of the competitively non-dominant species. Finally, the influence of plant parasites on community structure is compared with that of herbivores and predators.

A. R. WATKINSON AND C. C. GIBSON

## INTRODUCTION

organisms, including viruses, bacteria, fungi and other plants, live as parasites of plants. They obtain their nutrients from one or only a few host individuals, normally causing harm but not immediate death of the host. Thus, whilst the parasite benefits from the parasite–host interaction, the host plant suffers a decrease in fitness through either a decrease in survival, reproduction and/or growth. This $+/-$ interaction between the two interacting organisms is similar to that between herbivores and plants. It might be expected, therefore, that the effects of parasites on both the community structure and population dynamics of plants would be rather similar to those of herbivores.

A useful distinction among parasites can be made between the microparasites and the macroparasites (Anderson & May 1979). The microparasites, which include plant viruses and bacterial diseases, multiply directly within the host. In contrast, macroparasites grow within their host and multiply by producing infective stages that are released from the host to infect new hosts. Plant macroparasites include higher fungi, gall-forming and mining insects and a number of flowering plants. This chapter concentrates on the flowering plants. First, we examine the nature of parasitism in parasitic vascular plants with particular reference to the Scrophulariaceae. Second, we calculate the basic reproductive rate for a range of hemiparasitic species and examine the factors that determine their abundance. Third, we investigate host selectivity and its effects on the competitive relationships between plants. Finally we consider the effects of parasitism on plant community structure.


## PARASITISM IN PLANTS

About 3000 species of higher plant in fifteen families are known to be parasitic on other plants. The percentage of parasitic members in each family ranges from less than 1% in the Lauraceae to 100% in several families including the Loranthaceae, Santalaceae, Rafflesiaceae and Orobanchaceae (Ozenda 1965; Atsatt 1973). Parasitic plants are, therefore, not a homogeneous group in terms of their taxonomic affinities, morphology, mode of parasitism or life-history, and it has been suggested (e.g. Kuijt 1969) that parasitism in higher plants has evolved independently at least nine times.

Holoparasites (e.g. Orobanchaceae, Rafflesiaceae and Balanophoraceae) have no chlorophyll and are totally dependent upon their hosts for water, mineral nutrients and photosynthetic products. They are thus obligate parasites. Hemiparasites, in contrast, such as many members of the

Santalaceae, Loranthaceae and Scrophulariaceae, retain at least a degree of photosynthetic ability, and generally obtain only water and mineral nutrients from their hosts (Smith, Muscatine & Lewis 1969). This division is, however, not absolute. Certain of the hemiparasitic Scrophulariaceae (e.g. *Tozzia*) and all of the Cuscutaceae are considered to be virtual holoparasites as they have only a very small capacity for photosynthesis (Kuijt 1969). The majority of hemiparasites are obligate parasites in that they must have a host in order to complete their life-cycle, although a few parasitic members of the Scrophulariaceae have been grown autotrophically, at least under laboratory conditions (e.g. Hodgson 1973).

A second major division within parasitic plants is in the physical position they occupy on their hosts. Aerial parasites are found in only three families (Cuscutaceae, Lauraceae and Loranthaceae). They make parasitic connections with the above-ground structures of their hosts, whilst the species in all the remaining families are root parasites, attaching on to roots or other underground organs. Aerial parasites and root holoparasites are clearly distinctive and their parasitic mode of existence has been recognized for many centuries (Kuijt 1969). There is, however, a certain amount of confusion between aerial parasites and epiphytes (e.g. MacDougal 1911), and between root holoparasites and saprophytes (Kuijt 1969). On the other hand, root hemiparasites are very 'ordinary' plants in that they are not readily distinguished from autotrophs. As a consequence their parasitic existence has been recognized only since the mid-19th century (Mitten 1847; Decaisne 1847), primarily as a result of difficulties in the cultivation of such species. The only known parasitic gymnosperm, *Podocarpus ustus*, was not recognized until as late as 1959 (De Laubenfels 1959), and it is certainly possible that other species will be found to be parasitic.

Irrespective of the type of parasitic plant involved, the method by which parasitism is effected is basically the same. Contact with the host is through a structure termed the haustorium, which forms a continuity between the vascular systems of the host and the parasite. In holoparasites, the haustoria contain both xylem and phloem elements, although the phloem connection is not necessarily continuous and movement of photosynthate may occur via the apoplast (Raven 1983). The haustoria of hemiparasites generally contain only a xylem connection (Musselman & Dickison 1973; Weber 1977; Tsivion 1978), although in the haustoria of certain species (e.g. *Castilleja*) structures resembling phloem elements have been reported (Dobbins & Kuijt 1973). In order that the parasite can abstract vascular fluids from the host it is clear that there must be a pressure gradient across the haustoria. Several mechanisms to establish this have been suggested, including the presence of unusually large numbers of stomata on the leaves of the parasite to create the

dependent processes observed in populations of *R. minor* were also sufficient
to account for the observed range of densities found in natural field
populations. In addition, stochastic models showed the expected range of
variation in population size that might be expected (Gibson 1986).

Can it be concluded then that the host populations have no effect on the
population dynamics of *R. minor*? No, it is rather that the host populations
in this case can be treated as constant, in much the same way as human host
populations can be treated as constant in the study of many micro- and
macroparasitic infections of humans. A clear demonstration of the effect of
host populations on the population dynamics of another *Rhinanthus* species,
*R. angustifolius*, has been provided by De Hullu (1984). In comparing the
performance of *R. angustifolius* in three phases of a grassland succession she
showed that the structure of the host vegetation, the phenology of the host
species and host quality all had an effect on the population dynamics of *R.
angustifolius*. The three phases of the grassland succession corresponded to
*Rhinanthus* populations that were either increasing, at maximum size or
decreasing. Interestingly the growth (Fig. 19.2) and fecundity of *R.
angustifolius* was greatest in the plants from phase I of the succession, and yet
the highest densities of plants occurred in phase II where host quality was
lower and the number of seeds produced per plant lower. Similarly Keddy

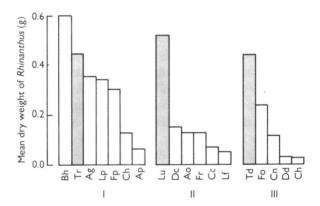

FIG. 19.2.   The dry weight production of *Rhinanthus angustifolius* when grown with a range of
host species after 11 weeks. The hosts were selected from phases I, II and III of a successional
series in which the *Rhinanthus* populations were respectively expanding, maximal and
contracting. The shaded columns indicate members of the Leguminosae. Bh: *Bromus
hordeaceus*, Tr: *Trifolium repens*, Ag: *Alopecurus geniculatus*, Lp: *Lolium perenne*, Fp: *Festuca
pratensis*, Ch: *Carex hirta*, Ap: *Alopecurus pratensis*, Lu: *Lotus uliginosus*, Dc: *Deschampsia
cespitosa*, Ao: *Anthoxanthum odoratum*, Fr: *Festuca rubra*, Cc: *Cynosurus cristata*, Lf: *Lychnis
flos-cuculi*, Td: *Trifolium dubium*, Fo: *Festuca ovina*, Cn: *Carex nigra*, Dd: *Danthonia decumbens*,
Ch: *Carex hostiana* (data from De Hullu 1984).

(1982) found that the maximum biomass production per plant in *Cakile edentula* did not correspond with those areas where the density of the plant was at a maximum. In both cases it is quite clear that any inferences about the physiological tolerances of plants based on observations of their abundance along a gradient would be quite misleading. The reasons for the lower densities of *R. angustifolius* in phase I of the succession, despite its higher fecundity, are various (De Hullu 1985) but it is clear that difficulties of establishment in the denser vegetation of phase I are a key factor.

Figure 19.2 shows that *R. angustifolius* performs particularly well when parasitizing members of the Leguminosae, no matter what the stage of the succession. Indeed the Leguminosae are generally considered to be good hosts for hemiparasites, although there are a number of exceptions (Snogerup 1982). One might predict, therefore, that if a hemiparasite were to have a significant effect on the population dynamics of any of its hosts then this would be on members of the Leguminosae. Unfortunately there is no specific information on the population dynamics of host populations in the presence and absence of hemiparasites. Nevertheless there are indications from the experiments of Ter Borg & Bastiaans (1973) and Rabotnov (1959) that the abundance of legumes decreases in the presence of hemiparasites. Although they did not measure genet or module number directly they showed that the yield per unit area of the Leguminosae and the proportion of the Leguminosae in the sward generally decreased in the presence of *Rhinanthus angustifolius* (Fig. 19.3). The effects on other dicotyledonous plants and grasses varied. Moreover, the total yield on each of the plots generally decreased in the presence of *R. angustifolius*, perhaps because of the strong suppression of the Leguminosae (Ter Borg & Bastiaans 1973) and a consequent decrease in the nitrogen availability on each plot.

## SELECTIVITY AND THE MEDIATION OF COMPETITION BY PARASITES

The extent to which a parasite influences the population dynamics of its hosts will depend only in part upon the direct effects of the parasite on growth, survival and reproduction. In addition, parasites may commonly be expected to have indirect effects on host populations through their effects on between-species interactions such as competition (Price *et al.* 1986). This section examines the host range of a number of hemiparasitic Scrophulariaceae, the evidence for selectivity between hosts and the effects of such selective host utilization on plant competition.

Considerable effort has gone into cataloguing the species that can act as hosts for parasites, under both natural and artificial conditions. For members

Table 19.1(a).   The effect of removing *Rhinanthus* on species diversity at four sites in eastern England (data from Gibson 1986).

| Site | *Rhinanthus* | Simpson's index | Shannon–Wiener index |
|---|---|---|---|
| 1 | Present | 0.138 | 2.33 |
|   | Removed | 0.193 | 2.10 |
| 2 | Present | 0.259 | 1.62 |
|   | Removed | 0.221 | 1.87 |
| 3 | Present | 0.205 | 1.86 |
|   | Removed | 0.193 | 1.93 |
| 4 | Present | 0.187 | 2.06 |
|   | Removed | 0.184 | 2.22 |

Table 19.1(b).   Species responding to the removal of *Rhinanthus*. Preferred hosts recorded at that site (\*\*) or another site (\*), avoided hosts at that site (††) or another site (†). No data on host preference were available for site 4 (data from Gibson 1986).

| Site | Increased abundance | Decreased abundance |
|---|---|---|
| 1 | *Koeleria macrantha*\*\* | *Festuca rubra*† |
|   |   | *Carex arenaria*† |
|   |   | *Elymus farctus*†† |
|   |   | *Bromus hordeaceus* |
| 2 | *Honkenya peploides*\*\* | *Elymus farctus*†† |
|   | *Plantago lanceolata*\* | *Carex arenaria*†† |
|   | *Leontodon autumnalis* | *Festuca rubra*†† |
| 3 | *Ononis repens*\*\* | *Elymus farctus*† |
| 4 | *Holcus lanatus* | *Carex flacca* |
|   | *Filipendula ulmaria* | *Carex hostiana* |

It would appear that the usual effect of the presence of *R. minor* on a community is to decrease species diversity. It is particularly informative then to examine the one exception that was observed to this apparent trend. On one of the dune sites the removal of *R. minor* resulted in a decrease in diversity rather than an increase, primarily as a result of an increase in the abundance of *Koeleria macrantha*. This is known to be a preferred host (Gibson 1986) and the inference from this experiment is that it is also a competitive dominant. A more detailed analysis of the species that responded positively and negatively to the removal of *R. minor* (Table 19.1b) on the various sites shows that those which responded positively had previously been shown by examination of their haustorial connections to be preferred hosts whilst those which responded negatively were avoided hosts. Presumably on most sites, however, the preferred hosts were not competitively dominant.

An understanding of host preferences is, therefore, a key issue in understanding the likely effects of parasites on plant species diversity. Do parasites, for example, respond specifically to exudates from the host roots, or do potential hosts possess defences that prevent their utilization? Evidence for the former comes from hemiparasites with small seeds, which have been shown to respond to root exudates in both their germination and growth (Kuijt 1969). Hemiparasites with larger seeds and thus a longer time to attach to suitable hosts, in contrast, generally appear to show no response to root exudates. Rather it is likely that their selectivity, in part at least, reflects mechanisms that result in the differential susceptibility of hosts to parasites. For example, the degree of corkiness in bark affects the attachment of mistletoes (Loranthaceae), whilst sclerenchymatous barriers halt the haustorial development of dodders (Cuscutaceae). Such negative selectivity is, however, an unlikely explanation for the selectivity of *Pedicularis* spp. for deep rooting woody hosts in arid environments (Sprague 1962), or for the high frequency of leguminous hosts associated with many hemiparasites such as *Rhinanthus*. Instead some form of positive host selection is implicated. However, the only evidence which suggests that exudates may affect *Rhinanthus* is from De Hullu (1985), who showed that the stratification of seeds of *R. angustifolius* amongst the roots of certain hosts delayed germination.

## CONCLUDING REMARKS

Parasitism, herbivory and predation are all interactions that involve an increase in fitness of one organism at the expense of another. One might, therefore, expect many similarities in the population dynamics of such organisms and in their effects upon community structure. Similarly, because of differences in selectivity and the amount of host damage caused by parasites, one might expect as much variation in the dynamical relationship amongst hosts and parasites as is found amongst herbivores and plants, and predators and prey. For example, some have argued that herbivores may regulate the abundance of plants (e.g. Brues 1946) and that herbivores in turn are food-limited. Others, in contrast, have argued (e.g. Hairston, Smith & Slobodkin 1960) that herbivore abundance is determined by predators and parasites and that herbivores are typically too scarce in relation to their food supply to have any impact on plant abundance.

Some parasites can also be expected to regulate the abundance of their hosts whilst others will probably be too rare to affect host abundance at all. But as Crawley (1983) has pointed out for plant–herbivore systems, it is more likely that there is a continuum between these two extremes, with the bulk of examples lying somewhere in the middle. This is certainly true for *Rhinanthus*, the parasitic plant for which we have most demographic information. As the plant is not host specific it is extremely unlikely that the abundance of host roots will regulate the abundance of the hemiparasite. Rather it can be expected that a complex of density-independent and density-dependent factors will determine the number of individuals. Fecundity, seed predation and seedling establishment are all potential regulators of the numbers of *Rhinanthus* (Gibson 1986). This is not of course to say that the host environment has no impact on the population dynamics of the hemiparasite. The presence of legumes in the sward has a clear impact on hemiparasite fecundity (De Hullu 1984), and the structure of the vegetation, in particular its openness, is a major determinant of the number of suitable safe microsites for seedling establishment (De Hullu 1985). Unfortunately, far less information is available on how the hemiparasite affects the population dynamics of its hosts. We nevertheless have a clear indication from observations on legumes (e.g. Rabotnov 1959) and a number of other plants, such as *Koeleria macrantha* (Gibson 1986), that the abundance of some host species is strongly affected by the presence or absence of *Rhinanthus*.

The complex dynamical behaviour between a parasitic plant and its host species still remains to be explored. This will no doubt be as difficult to study for a non-specific hemiparasite and its hosts as it will be, for example, for the rabbit and the plants that it grazes. Fortunately, to understand the effects of parasites and herbivores on plant community structure does not require a full understanding of this dynamical behaviour. Crawley (1983) has claimed a central role for herbivores in determining the structure and dynamics of plant communities. Certainly there is ample evidence that they affect the physical, three-dimensional structure of communities, species richness and the relative abundance of species within the community. It would be absurd to claim such an important role in the determination of community structure for parasitic flowering plants, except in a few communities. (Such would not, however, be the case for parasites as a whole.) The mechanisms by which herbivores and parasites affect diversity, however, are remarkably similar. We have, in particular, demonstrated here that selective parasitism may affect species diversity. For herbivores it has long been argued (see Harper 1977; Crawley 1983) that selective grazing on what would otherwise be the dominant plant species will inevitably lead to an increase in species diversity by modifying the competitive relationships between the component plant species. Conversely, selective feeding on already uncompetitive plants can be expected to reduce plant diversity. In the case of *Rhinanthus* (Gibson 1986), both effects have been demonstrated, but it would appear that it is usually the less competitive plants that are parasitized. As a consequence the effect of the presence of the hemiparasite in vegetation is usually to decrease diversity. The reason for this, and the question of whether or not it is true of all hemiparasites, remain to be explored.

## REFERENCES

Anderson, R. M. & May, R. M. (1979). Population biology of infectious diseases. *Nature*, **280**, 361–67 & 455–61.

Atsatt, P. R. (1970). Biochemical bridges between vascular plants. *Biochemical Evolution* (Ed. by K. L. Chambers), pp. 53–68. Proceedings of the 29th Annual Biology Colloquium. Oregon State University Press.

Atsatt, P. R. (1973). Parasitic flowering plants: how did they evolve? *American Naturalist*, **107**, 502–10.

Bentley, S. J. B. & Whittaker, J. B. (1979). Effects of grazing by a chrysomelid beetle, *Gastrophysa viridula*, on competition between *Rumex obtusifolius* and *Rumex crispus. Journal of Ecology*, **67**, 79–90.

Björkmann, E. (1960). *Monotropa hypopitys* L.—an epiparasite on tree roots. *Physiologia Plantarum*, **13**, 308–25.

Bormann, F. H. (1966). The structure, function and ecological significance of root grafts in *Pinus strobus* L. *Ecological Monographs*, **36**, 1–26.

Botha, P. J. (1948). The parasitism of *Alectra vogelii* Benth, with special reference to the germination of its seeds. *Journal of South African Botany*, **14**, 63–80.

Brown, R. (1965). The germination of angiospermous parasite seeds. *Encyclopaedia of Plant Physiology*, **15**, 925–32.

Brues, C. T. (1946). *Insect Dietary*. Harvard University Press, Cambridge, Massachusetts.

Burdon, J. J. & Chilvers, G. A. (1977). The effect of barley mildew on barley and wheat competition in mixtures. *Australian Journal of Botany*, **25**, 59–65.

Burdon, J. J., Groves, R. H., Kaye, P. E. & Speer, S. S. (1984). Competition in mixtures of susceptible and resistant genotypes of *Chondrilla juncea* differentially infected with rust. *Oecologia*, **64**, 199–203.

Campion-Bourget, F. (1983). La germination des graines des espèces françaises de *Rhinanthus* L. *Revue de Cytologie et de Biologie Végétales—le Botaniste*, **6**, 15–94.

Crawley, M. J. (1983). *Herbivory: The Dynamics of Animal–Plant Interactions*. Blackwell Scientific Publications, Oxford.

Curtis, E. J. C. & Cantlon, J. E. (1965). Studies of the germination process in *Melampyrum lineare. American Journal of Botany*, **52**, 552–55.

De Hullu, E. (1984). The distribution of *Rhinanthus angustifolius* in relation to host plant species. *Proceedings of the Third International Symposium on Parasitic Weeds* (Ed. by C. Parker, L. J. M. Musselman, R. M. Polhill & A. K. Wilson), pp. 43–53. Icarda, Aleppo, Syria.

De Hullu, E. (1985). Population dynamics of *Rhinanthus angustifolius* in a succession series. PhD thesis, University of Groningen.

De Hullu, E., Brouwer, T. & Ter Borg, S. J. (1985). Analysis of the demography of *Rhinanthus angustifolius* populations. *Acta Botanica Neerlandica*, **34**, 23–32.

De Laubenfels, D. J. (1959). Parasitic conifer found in New Caledonia. *Science*, **130**, 97.

Decaisne, M. J. (1847). Sur le parasitisme de Rhinanthacées. *Annales des Sciences Naturelles, Série 3,* 8, 5–9.

Dobbins, D. R. & Kuijt, J. (1973). Studies on the haustorium of *Castilleja*. II. The endophyte. *Canadian Journal of Botany,* 51, 923–31.

During, H. J., Schenkeveld, A. J., Verkaar, H. J. & Willems, J. H. (1985). Demography of short-lived forbs in chalk grassland in relation to vegetation structure. *The Population Structure of Vegetation* (Ed. by J. White), pp. 341–70. Junk, Dordrecht.

Fisher, J. T., Reid, C. P. P. & Hawksworth, F. G. (1979). Water relations of dwarf mistletoes (*Arceuthobium* spp.). *Proceedings of the Second Symposium on Parasitic Weeds* (Ed. by L. J. Musselman, A. D. Worsham & R. E. Eplee), pp. 34–44. North Carolina State University, Raleigh, North Carolina.

Fitter, A. H. (1977). Influence of mycorrhizal infection on competition for phosphorus and potassium by two grasses. *New Phytologist,* 79, 119–25.

Francis, R. & Read, D. J. (1984). Direct transfer of carbon between plants connected by vesicular-arbuscular mycorrhizal mycelium. *Nature,* 307, 53–56.

Fresco, L. (1980). Ecological response curves of *Rhinanthus serotinus*; a synecological study. *Acta Botanica Neerlandica,* 29, 533–39.

Galil, J. (1984). *Ficus religiosa* L.—the tree splitter. *Botanical Journal of the Linnean Society,* 88, 185–203.

Gibson, C. C. (1986). *The population and community biology* of Rhinanthus minor L. Ph.D. thesis, University of East Anglia.

Govier, R. N. (1966). *The water relationships of the hemiparasites and their hosts, with special reference to* Odontites verna (Bell.) Dum. Ph.D. thesis, University of Wales.

Govier, R. N. & Harper, J. L. (1965). Hemiparasitic weeds. *Proceedings of the 7th British Weed Control Conference,* pp. 577–82. British Crop Protection Council, Brighton.

Grubb, P. J., Kelly, D. & Mitchley, J. (1982). The control of abundance in communities of herbaceous plants. *The Plant Community as a Working Mechanism* (Ed. by E. I. Newman), pp. 79–97. British Ecological Society Special Publication number 1. Blackwell Scientific Publications, Oxford.

Hairston, N. G., Smith, F. E. & Slobodkin, L. B. (1960). Community structure, population control, and competition. *American Naturalist,* 94, 421–25.

Harper, J. L. (1977). *Population Biology of Plants.* Academic Press, London.

Hodgson, J. F. (1973). *Aspects of the Carbon Nutrition of Angiospermous Parasites.* Ph.D. thesis, University of Sheffield.

Karlsson, T. (1974). Recurrent ecotypic variation in Rhinanthaceae and Gentianaceae in relation to hemiparasitism and mycotrophy. *Botaniska Notiser,* 127, 527–39.

Keddy, P. A. (1982). Population ecology along an environmental gradient: *Cakile edentula* on a sand dune. *Oecologia,* 52, 348–55.

Klaren, C. H. (1973). Compounds of benzoic acid in hemiparasitic Scrophulariaceae. *Acta Botanica Neerlandica,* 22, 452–55.

Kuijt, J. (1969). *The Biology of Parasitic Flowering Plants.* University of California Press, Berkeley.

MacDougal, D. T. (1911). An attempted analysis of parasitism. *Botanical Gazette,* 52, 249–66.

Mitten, W. (1847). On the economy of the roots of *Thesium linophyllum* Hook. *London Journal of Botany,* 6, 146–48.

Musselman, L. J. & Dickison, W. C. (1973). The structure and development of the haustorium in parasitic Scrophulariaceae. *Botanical Journal of the Linnaean Society,* 70, 183–212.

Musselman, L. J. & Mann, W. F. (1977). Host plants of some Rhinanthoideae (Scrophulariaceae) of eastern North America. *Plant Systematics and Evolution,* 127, 45–53.

Ozenda, P. (1965). Recherches sur les phanerogames parasites. I. Revue des travaux recents. *Phytomorphology,* 15, 311–38.

Piehl, M. A. (1962). The parasitic behaviour of *Dasistoma macrophylla*. *Rhodora*, 64, 331–36.

Piehl, M. A. (1963). Mode of attachment, haustorium structure and hosts of *Pedicularis canadensis*. *American Journal of Botany*, 50, 978–85.

Piehl, M. A. (1966). The root parasites of *Cordylanthus* and some of its ecological implications. *American Journal of Botany*, 53, 622.

Price, P. W., Westoby, M., Rice, B., Atsatt, P. R., Thompson, J. N. & Mobley, K. (1986). Parasite mediation in ecological interactions. *Annual Review of Ecology and Systematics*, 17, 487–505.

Rabotnov, T. A. (1959). The effect of *Rhinanthus major* Ehrh. upon the crops and the composition of the flood land herbage. *Byulleten' M. Obschchestva Ispyt. Priroda*, 64, 105–07.

Raven, J. A. (1983). Phytophages of xylem and phloem: a comparison of animal and plant sap-feeders. *Advances in Ecological Research*, 13, 135–234.

Saunders, A. R. (1933). Studies in phanerogamic parasitism, with particular reference to *Striga lutea* Lour. *Union of South Africa Department of Agriculture Science Bulletin, Number 128*.

Schmidt, K. P. & Levin, D. A. (1985). The comparative demography of reciprocally sown populations of *Phlox drummondii* Hook. I. Survivorship, fecundities, and finite rates of increase. *Evolution*, 39, 396–404.

Schulze, E. O. & Ehleringer, J. R. (1984). The effect of nitrogen supply on growth and water-use efficiency in xylem-tapping mistletoes. *Planta*, 162, 268–75.

Senn, G. (1913). Der osmotische Druk einiger Epiphyten und Parasiten. *Verhandlungen der Naturforschenden Gesellschaft in Basel*, 24, 179–83.

Smith, D., Muscatine, L. & Lewis, D. (1969). Carbohydrate movement from autotrophs to heterotrophs in parasitic and mutualistic symbiosis. *Biological Reviews*, 44, 17–90.

Snogerup, B. (1982). Host influence on northwest European taxa of *Odontites* (Scrophulariaceae). *Annales Botanici Fennici*, 19, 17–30.

Sprague, E. F. (1962). Parasitism in *Pedicularis*. *Madroño*, 16, 192–200.

Stevens, R. A. & Eplee, R. E. (1979). *Striga* germination stimulants. *Proceedings of the Second Symposium on Parasitic Weeds* (Ed. by L. J. Musselman, A. D. Worsham & R. E. Eplee), pp. 211–18. North Carolina State University, Raleigh, North Carolina.

Sunderland, N. (1960). Germination of the seeds of angiospermous root parasites. *The Biology of Weeds* (Ed. by J. L. Harper), pp. 83–93. Blackwell Scientific Publications, Oxford.

Ter Borg, S. J. (1985). Population biology and habitat relations of some hemiparasitic Scrophulariaceae. *The Population Structure of Vegetation* (Ed. by J. White), pp. 463–87. Junk, Dordrecht.

Ter Borg, S. J. & Bastiaans, J. C. (1973). Host parasite relations in *Rhinanthus serotinus*. I. The effect of growth conditions and host; a preliminary review. *Proceedings of the European Weed Research Council Symposium on Parasitic Weeds*, pp. 236–46. Malta University Press.

Tsivion, Y. (1978). Physiological concepts of the association between parasitic angiosperms and their hosts—a review. *Israel Journal of Botany*, 27, 103–27.

Visser, J. (1975). Germination stimulants of *Alectra vogelii* Benth. seed. *Zeitschrift für Pflanzenphysiologie*, 74, 464–69.

Watkinson, A. R. (1985). On the abundance of plants along an environmental gradient. *Journal of Ecology*, 73, 569–78.

Watkinson, A. R. & Davy, A. J. (1985). Population biology of salt marsh and sand dune annuals. *Vegetatio*, 62, 487–97.

Watkinson, A. R. & Harper, J. L. (1978). The demography of a sand dune annual: *Vulpia fasciculata*. I. The natural regulation of populations. *Journal of Ecology*, 66, 15–33.

Weber, H. C. (1976). Ueber Wirtspflanzen und Parasitismus einiger mitteleuropäischer Rhinanthoideae (Scrophulariaceae). *Plant Systematics and Evolution*, 125, 97–107.

Weber, H. C. (1977). Anatomische studien an den Haustorien (Kontaktorganen) von *Thesium*—Arten (Santalaceae). *Berichte der Deutschen Botanischen Gesellschaft*, 90, 439–58.

natural undisturbed communities, while not infrequent, rarely caused extensive mortality (Browning 1974; Dinus 1974; Schmidt 1978; Segal *et al.* 1980, 1982); epidemics arose only when the community was disturbed, e.g. by accidental or deliberate introduction of a novel pathogen, or by a reduction in structural or species diversity. Concurrently and independently of pathologists' views, theoreticians hypothesized that pathogens (or pests in general) play a significant role in a variety of ecological and evolutionary phenomena in natural systems, including: (1) the control of population size and dynamics, (2) the maintenance of genetic variation in populations (Clarke 1976), (3) the evolution and maintenance of sexual reproduction (Levin 1975, Hamilton 1980, Rice 1983), (4) the determination of large-scale and micro-scale distributions of species (Janzen 1970), and (5) the enhancement of species diversity (Holt & Pickering 1985). This rather imposing list appears contradictory to the above observations; despite apparently low levels of disease, theoreticians assign to pathogens a considerable effect on plant populations.

In the past decade the number of field and laboratory studies on pathogens in natural plant populations has expanded considerably (Burdon 1987a; Alexander 1988). The aim of this chapter is to summarize recent empirical data on plant disease in natural populations and to evaluate this information in light of the above theoretical developments. The studies included deal almost exclusively with fungal pathogens. Major unanswered questions concerning pathogens in natural populations are also identified.

# IMPACT ON PLANT POPULATION DYNAMICS

## *Theoretical background*

Plant disease will cause a decline in host population size if it affects the demographic parameters of a population, i.e. its birth-rate and/or death-rate (Harper 1977; Burdon & Chilvers 1982). Because the effect of disease on plants can be size/age dependent, it may have different consequences for population size depending on the size/age category of the plants that the disease strikes (Burdon & Shattock 1980). By killing seeds and seedlings, it lowers recruitment into the population. This causes an immediate reduction in the current population size and a rapid shift in its size/age structure. In addition, disease can decrease the population's birth-rate by reducing the seed output of adult plants, either directly by attacking reproductive parts or indirectly by reducing the plant's competitive ability, resource acquisition and growth. This reduction in birth-rate may affect the potential population size of future generations.

The impact of pathogens on plant populations can be either density-dependent or density-independent. If disease incidence increases with added host density, then pathogens should be important in the regulation of population size (Antonovics & Levin 1980; Burdon & Chilvers 1982). Density-dependent disease mortality causes negative feedback regulating pathogen populations as well, such that eventually plant population size should reach a dynamic equilibrium (Burdon & Shattock 1980; Burdon & Chilvers 1982). Density-dependent feedback, while directly related to plant density *per se*, may be influenced by the pattern of dispersion of the plants (Burdon & Chilvers 1976) and age structure (Schmidt 1978). When the impact of disease is density-independent and dependent on environmental conditions, e.g. weather, then the impact of disease on plant populations can be expected to be unpredictable and to result in population fluctuations.

### Empirical evidence

Disease is known to increase the death-rate of plants in natural populations. Pathogens kill at all life-stages: germinating seeds (Kirkpatrick & Bazzaz 1979; Neher, Augspurger, & Wilkinson 1988), seedlings (Hermann & Chilcote 1965; Augspurger 1983a, 1984), juveniles (Alexander & Burdon 1984; Parker 1986; Paul & Ayres 1986a), and reproductive adults (Newhook & Podger 1972). No data are available on the extent of death due to disease of dormant seeds in the seed bank and no study documents relative disease effects at all life-stages of any one species.

The fecundity of plants has also been shown to decline due to disease. The reproductive output of diseased plants decreases because pathogens directly attack reproductive structures such as flowers or inflorescence stalks (Lee 1981; Alexander 1987b; de Nooij & van der AA 1987), developing fruits (Alexander & Burdon 1984), or developing seeds (Kitajima & Augspurger, unpublished data). In addition, pathogens indirectly lower seed production by affecting the plant's development (Paul & Ayres 1987a), depressing its growth rate (Ben-Kalio & Clark 1979; Parker 1986; Paul & Ayres 1986a, b, c), or by increasing the plant's susceptibility to other abiotic and biotic stresses (Ayres 1984; Paul & Ayres 1987b). Size hierarchies and differential fecundity are exaggerated as a result of both direct mortality and growth suppression of diseased individuals (Paul & Ayres 1986a). Interestingly, some diseases lower flower production while enhancing survival and competitive ability (Clay 1984).

Studies reveal no clear pattern of density-dependent regulation by disease in natural populations. Disease incidence can be related positively to seedling density (Augspurger & Kelly 1984), though only up to a density threshold

(Alexander 1984), and unrelated to adult density (Pielou & Foster 1962). In a review of sixty-nine examples of primarily managed populations, Burdon & Chilvers (1982) found 57% and 35% of the examples to show a positive and negative effect, respectively, of host density on disease incidence. The majority of fungal diseases show a positive correlation with density. Disease levels can be affected by additional factors that interact with density. For example, disease incidence can be lower in both small and isolated patches (Jennersten, Nilsson & Watljung, 1983). The population's age structure, by affecting the temporal coincidence of plant susceptibility and pathogen activity, dictates the effective plant density and alters damping-off epidemics in experimental seedling populations (Neher, Augspurger, & Wilkinson 1988).

The impact of pathogens on population size by density-independent means can be inferred from observations that disease levels are correlated with weather variables. They are, for example, highest for mildew on *Phlox* populations in years with the coolest temperatures (Jarosz & Levy 1988). The influence of weather on the phenology of both host and pathogen determines the amount of their interaction. If, for example, host plants mature early in a cool season, they may escape the most deleterious effects of a pathogen whose activity is enhanced by warmer temperatures (Dinoor 1970). Geographical distributions of disease resistance are also correlated with climatic variables (Weltzien 1972). Less selection favouring a diversity of pathogen races arises when the climate is less suitable for disease and the interaction of host and pathogen is ephemeral and intermittent. Isolation of pathogen races on *Avena* spp. is lowest in years with drought (Oates, Burdon & Brouwer 1983). Presence of resistant *Avena sterilis* plants in Israel is lowest in arid regions with low humidity (Wahl 1970).

Knowledge of the mortality and fecundity of healthy and infected individuals, and of disease transmission levels, allows population projection models to be constructed. Using such data in computer simulation models of *Silena alba*, infected by the anther smut, *Ustilago violaceae*, Alexander & Antonovics (1988) conclude that three population fates (loss of pathogen, coexistence, loss of plant and pathogen) can arise, depending on the rates of plant recruitment and disease spread. Also, they conclude that host and pathogen can coexist without invoking genetic variation in resistance as a requirement for coexistence (see below).

No long-term study of natural, undisturbed populations demonstrates that negative feedback imposed by disease leads to an equilibrium in population size. Examples of pathogens reducing population size arise from studies of managed systems. The rust, *Phragmidium violaceum*, was introduced to reduce the population size of *Rubus constrictus* in Chile (Oehrens 1977).

Crown rust reduces the yield of ryegrass by 84% relative to fungicide-treated plants in a mixed pasture system (Latch & Lancashire 1970).

### Unanswered questions

Despite accumulating evidence that disease affects birth- and death-rates of natural populations, fundamental questions remain largely unaddressed. First, is the population size reduced to a size lower than if the pathogen were absent? That is, do other density-dependent mortality agents compensate in the absence of disease, thus reducing the population to a comparable size? The types of controlled experiments needed to examine this issue are similar to those used in managed systems, viz. the effect of the exclusion of pathogens, e.g. via fungicide application, or the addition of the pathogen to a disease-free population. Second, what is the relative impact of disease at different life-stages on population dynamics and size? Third, is the role of pathogens more important in an expanding phase than a stable phase of population growth (Harper 1977)? These two questions can be addressed first by long-term observations of field populations. In addition, models using field-derived data can be used to estimate long-term projections of population size and whether the pathogen and host are predicted to coexist or go extinct locally. Finally, on what spatial and temporal scales does density-dependent disease mortality operate to reduce population size? Experiments are needed covering a range of densities, clumping patterns, and age structures.

## MAINTENANCE OF GENETIC VARIATION

### Theoretical background

If genetic variation in resistance to pathogens exists and pathogens act as a major selective role causing differential fitness, then pathogens will affect the genetic structure for resistance in the population (Haldane 1949; Harlan 1976; Burdon 1985). Although pathogens have the potential to cause a superior resistant host genotype to dominate the population, theoreticians generally conclude that pathogens are responsible for the maintenance of genetic variation in resistance, especially via frequency-dependent selection favouring the rare genotype (Clarke 1976, 1979). Acquisition of resistance genes is hypothesized to come at a cost of general fitness in the absence of the pathogen (Harlan 1976). Several models have demonstrated how selection affects the frequency of genes for resistance and virulence in host–pathogen systems (Leonard 1977; Barrett 1978; Leonard & Czochor 1980; Levin 1982; Burdon, Oates & Marshall 1983; May & Anderson 1983). But there is a dispute as to whether host–pathogen interactions lead to a static, balanced polymorphism of resistant and susceptible individuals (Clarke 1976, 1979; Leonard & Czochor 1980) or dynamic cyclic polymorphisms (Hamilton 1982).

*Empirical evidence*

Evidence from natural populations is sought to support three theoretical issues: (1) phenotypic and genetic variation occurs in host resistance; (2) resistance increases host fitness in the presence of the pathogen and has a cost in the absence of the pathogen; and (3) frequency-dependent selection occurs favouring the rare genotype. Burdon (1985) points out the inherent difficulties in assessing the selective effect of pathogens. Resistant and susceptible phenotypes are hard to identify directly in the field. Complications arise from changes in the phenotypic expression of resistance genes and from variability in the environment. Antonovics and Alexander (1989) discuss the concept of fitness and emphasize the appropriate methods to measure it in both plant and pathogen populations.

Several studies have screened for phenotypic variation in resistance in natural populations (Wahl 1970; Zimmer & Rehder 1976; Wahl *et al.* 1978; Burdon & Marshall 1981a). They reveal that a large amount of phenotypic variation in resistance exists both within and between populations and that such resistance is pathogen race-specific (Segal *et al.* 1980; Miles & Lenné 1984; Harry & Clarke 1986). Spatial patterns of resistance occur on a geographic scale (Dinoor 1970; Burdon, Oates & Marshall 1983; Jarosz 1984; Dinoor & Eshed 1987), the frequency of resistance phenotypes being greater in areas where environmental conditions favour pathogen growth. Very complex spatial patterns of resistance exist on local population and neighbourhood scales (Snaydon & Davies 1972; Dinoor 1977; Burdon 1980; Parker 1985; Clarke, Bevan & Crute 1986; Harry & Clarke 1986). Small-scale variation in virulence is also known to exist in pathogen populations (Oates, Burdon & Brouwer 1983).

Tests for genetic variation in resistance are not widely available. One study infers genetic differences among clones of trees because their marked phenotypic variation in disease levels suggests strong genetic control (Copony & Barnes 1974). Another study (Dinoor 1977) estimates the number of resistance genes in a population based on phenotypic patterns of host resistance obtained in response to different pathogen races. Quantitative genetics tests, using offspring from either half-sibling or full-sibling host crosses transplanted into the field, demonstrate significant genetic variation among families in resistance (Barber 1966; Kinloch & Stonecypher 1969;

Alexander, Antonovics, & Rausher 1984; Miles & Lenné 1984; Parker 1985). The host's resistance depends, however, specifically on the source population of the pathogen (Parker 1985). Classical breeding studies demonstrate most directly the underlying genetic basis of the observed phenotypic resistance. The first study to document fully the genetic basis of race-specific resistance in whole populations uncovered one to three single dominant genes for resistance in each host line and estimates of ten to twelve resistance genes or alleles in a given population of *Glycine canescens* (Burdon 1987b). In *Amphicarpaea bracteata*, a disease resistance polymorphism occurs in the population; such resistance is governed by a single locus with resistance exhibiting nearly complete dominance (Parker 1988).

Parker (1986) also used this annual plant to demonstrate that the individual's lifetime fecundity is correlated negatively with its intensity of infection. This type of study that documents differential fecundity due to disease incidence provides only circumstantial evidence of selection by pathogens. It is unknown whether surviving phenotypes with higher fecundity are those with genetic resistance. While such resistance may enhance fitness in the presence of the disease, no study, to date, tests whether there is a fitness penalty for carrying resistance genes in the absence of disease.

Evidence that pathogens maintain genetic variation in populations is inconclusive. Wild relatives of crop species often have much variation for resistance to pathogens (Segal *et al.* 1982), and the greatest variation is associated with the centre of distribution of the plant, presumably where pathogen selective pressure is most persistent (Leppik 1970; Harlan 1976). Negative evidence comes from two field studies. In the clonal perennial, *Podophyllum peltatum*, pathogens differentially kill small ramets and may prevent the entry of new genotypes, thus maintaining the genetic uniformity of the 'colony' (M. A. Parker, unpublished data). Jarosz & Levy (1988) undertook the first field study of a natural population to investigate the effect of genetic variation on disease. They found that disease spread in experimental *Phlox* populations is not buffered by the presence of resistant plants, and suggest that in low density populations, natural selection strongly favours homogeneous levels of high resistance. In a broad survey of *Phlox* populations, Jarosz (1984) found that in shaded habitats that favour pathogen growth, populations have high or uniform resistance.

### Unanswered questions

Several studies now point to the existence and pattern of phenotypic and genetic variation for resistance in natural populations. Over what spatial scale have host and pathogen coevolved? To answer this requires a study

**2** Oscillations in genotype frequency occur during extended epidemics (Levin 1982).

**3** Asexual species live in habitats where the incidence of disease is less compared to habitats of sexual species from which they are derived (Levin 1982).

**4** In species with both modes of reproduction, high host density induces a switch towards sexual reproduction, assuming greater disease results from such crowding (Levin 1982).

**5** Populations that colonize a new area lacking in their pathogens minimize or eliminate sexual reproduction (Jaenike 1978).


## DETERMINANT OF SPECIES' DISTRIBUTIONS

### *Theoretical background*

Given the demonstrated ability of pathogens to kill large numbers of individuals, it is reasonable to postulate that pathogens may control the distribution of a species. Such control may affect the species' distribution on a regional, geographical scale (Weltzien 1972) or it may dictate habitat specificity on a local scale, restricting the host to 'refuge' areas in the midst of areas favourable for pathogen activity (Burdon & Shattock 1980). Host and pathogen may have different environmental conditions required to complete their life-cycle (Weir 1918). If the pathogen tolerates a wider range of conditions and has a broader range than the host, then it may appear that the pathogen is preventing host range expansion, while, in fact, environmental conditions restrict host range. Alternatively, the pathogen may actively restrict the host range when environmental conditions greatly favour the growth of the pathogen over the growth of the host. Then environmental stress on the host may be sufficient to make it more vulnerable (Ayres 1984), and the pathogen gains easy entry into the host and kills it. Coexistence of host and pathogen may require more comparable environmental requirements.

The ease of the host acquiring resistance may differ through the host range. Studies of crop progenitors demonstrate that resistance genes are most likely found in the centre of the host range, where pathogen and host have overlapped for a long period of time (Leppik 1970: Harlan 1976). At the margins of a host's range where conditions are sub-optimal, the loss of fitness imposed by a high level of resistance may restrict the host species geographically (Harlan 1976).

## Empirical evidence

Most evidence for this hypothesis is inferential and draws on observations of host patterns that may or may not be explained by differential pathogen activity. Evidence of cause and effect, as demonstrated via experimental addition or elimination of the pathogen, is lacking for the most part.

Wide-scale elimination of a species during epidemics, e.g. *Castanea dentata* due to the introduced pathogen, *Endothia parasitica* (Roame, Griffin & Rush 1986), documents the far-reaching effects a pathogen can have on a species' distribution. Use of a pathogen as a biological control agent, e.g. rust on *Chondrilla juncea* in Australia, suggests the potential of pathogens in restricting host range (Burdon, Groves & Cullen 1981), although generally such introductions are expected to lower density while not eliminating the host from areas (Cullen & Groves 1977). The problem with these examples is that the introduced pathogens lack their natural selective forces in the new environment and it is difficult to generalize that native pathogens control host populations to a comparable extent. Evidence is lacking from any natural system that pathogen effects are so extreme that the host cannot complete its life-cycle in an area.

On a geographical scale, Weir (1918) and Rochow (1970) noted that a species with a broad altitudinal range is attacked more severely at the lowest elevations. Weir (1918) suggested that the difficulty in cultivating *Larix* in lowland Germany is due to the increased destructiveness of its pathogen which inflicts minor damage in natural populations at higher elevations. Soil and water-logging conditions promoting damping-off disease in seedlings have been implicated as restricting the range of some *Pinus* species (Wilde & White 1939).

On a local population scale, pathogen activity has been associated with soil nitrogen levels and the height of the vegetation, which presumably enhances humidity (Snaydon & Davies 1972). Ridges with low soil moisture may allow some *Eucalyptus* species to escape from *Phytophthora cinnamomi* in Australia (Pratt, Heather & Shepherd 1973). Light-gaps in tropical forests serve as refuge areas for seedlings to escape from damping-off that causes extreme mortality in shaded conditions (Augspurger 1984). The high density of seedlings adjacent to parent trees is associated with damping-off epidemics that eliminate almost all seedlings within a restricted range near the parent tree (Augspurger 1983a). The net result of the above examples is that the species' distribution becomes non-continuous. A small-scale mosaic of host presence and absence arises depending on the spatial pattern of the above variables.

While pathogens accelerate extinction in local areas, they may not alter the overall spatial pattern of the host from the pattern that would arise in the absence of the pathogen. For example, damping-off may kill all seedlings in the shade, but in some species, because of their photosynthetic limitations, none would survive to adulthood in the shade, even in the absence of pathogens. Spatial patterns affected initially by pathogens thus shift subsequently to new patterns, within the constraints imposed by the early seedling mortality (Augspurger 1983b).

Experimental demonstration of those factors that control the species' distribution is limited. The promotion of disease in shade has been demonstrated experimentally for damping-off in tropical tree seedlings (Augspurger & Kelly 1984), for mildew of *Phlox* (Jarosz & Levy 1988), and for mildew of *Hordeum* (Dinoor & Eshed 1987). Augspurger & Kelly (1984) showed experimentally that plant density affects damping-off but that the incidence of disease is confounded by an interaction with distance from the parent tree. This result suggests the possibility of very local-scale patterns of pathogen resistance. Experimental transplants demonstrated that seedlings incur lower levels of damping-off when distributed around non-parent than parent trees (Augspurger & Kelly 1984). Lowered infection of cloned genotypes of *Amphicarpea bracteata* in areas away from their origin also supports the idea of local adaptation of pathogens to specific portions of the population (Parker 1985). The net result of this local differentiation is an alteration in the spatial pattern of particular genotypes in the population.

### Unanswered questions

Are host distributions explained by differences in environmental conditions governing the growth of host and pathogen or by changes in resistance patterns over space? The latter mechanism depends on the spatial scale over which pathogen and host have coevolved and whether acquisition of resistance brings larger fitness costs at the margins than at the centre of the host's distribution.

Seed dispersal results occasionally in a new discontinuous distribution of the host. Is such range expansion promoted because the host population escapes its former pathogens? Is host resistance evolved against pathogens in the old environment effective against novel pathogens in the new area? Because this type of dispersal is rare and unpredictable in its occurrence, it is difficult to study. Experimental manipulation of host dispersal on local or long-distance scales may be required to study these questions.

# INTERSPECIFIC INTERACTIONS AND SPECIES COEXISTENCE

## *Theoretical background*

Assuming disease lowers population size below the habitat's carrying capacity, then openings are created in the community for other species to fill. In this way species diversity is enhanced (Harper 1969). Janzen (1970) formulated a similar idea in a model in which natural enemies act to increase spacing among conspecifics, through disproportionately high attack on offspring near adults, thus allowing other species to occupy the intervening space. If natural enemies, such as pathogens, lower plant competitive ability, then exclusion of rarer species by dominant species is prevented and species coexistence is enhanced (Connell 1971; Chilvers & Brittain 1972).

Mathematical models have explored this set of hypotheses. An early model showed that two competing species can be maintained in equilibrium assuming density-dependent negative feedback from respective host-specific pathogens (Chilvers & Brittain 1972; Burdon & Chilvers 1974). Other assumptions include resource limitation and the dominant species being the most susceptible to the disease. In an extension of the model Gates *et al.* (1986) found that the stability of coexistence is not an inevitable outcome, but depends on growth-rates and relative competitive abilities of the plant species, and the transmission efficiency of the pathogen and its effect on individual host plants.

The amount of species diversification imposed by pathogen activity is expected to depend on the degree of host specificity (Dinoor & Eshed 1984). When species share a pathogen, the dominant species can exclude the alternate host (Holt & Pickering 1985). Additionally, the probability of these expected species' interactions depends on the community's spatial pattern, i.e. whether it is a random or non-random clumped mix of species (Burdon & Chilvers 1976).

## *Empirical tests*

In greenhouse (Burdon & Chilvers 1977) tests of interspecific competition using de Wit replacement series, barley loses its dominance over wheat when barley's pathogen is present. Similar results arise in greenhouse studies of intraspecific competition between different morphological forms of *Chondrilla juncea* (Burdon, Groves & Cullen 1981; Burdon *et al.* 1984). Both studies show that the advantage of the dominant, susceptible plant is lost when it is attacked by the pathogen and the resistant plants persist. In field tests diseased individuals are weakened in competition with healthy individuals of *Senecio vulgaris* (Paul & Ayres 1986c) and abiotic stress exaggerates the pathogen effect on the mixture (Paul & Ayres 1987b). In contrast, disease enhances the interspecific competitive ability of infected individuals of *Danthonia spicata*, possibly by increasing plant resistance to insect herbivory (Clay 1984).

In a review of studies of tropical tree species, testing aspects of Janzen (1970) and Connell's (1971) hypothesis, Clark & Clark (1984) concluded that most of the twenty-four studies show evidence of either density-dependent or distant-dependent mortality and that no individuals may survive within a minimum distance of the parent, thus promoting spacing among conspecifics and preventing dominance by one species. In only one of the twenty-four studies is disease demonstrated as the mortality agent affecting such spacing (Augspurger 1983a).

More direct evidence of the effect of pathogens on species diversity of natural communities can be found by observing the community before and after a disease outbreak. The following four host–pathogen systems illustrate this; the examples differ in whether the pathogen is indigenous or introduced by humans:

1   Harper (1969) points out an unintended experiment using *Hevea brasiliensis*, the rubber tree, that in nature grows interspersed in a very diverse community and suffers minimal disease. In monocultures in its native Brazil, epidemics of the local disease flourish. The inference is that pathogens promote floristic diversity in the natural communities where rubber trees grow.

2   When the pathogen, *Endothia parasitica*, was introduced accidentally from Asia into North American forests, populations of *Castanea dentata*, which comprised up to 85% of canopy trees, were effectively obliterated (Roame, Griffin & Elkins 1986). Changes in vegetation patterns several decades after the death of the dominant tree show the occasional entry of new species and a major shift toward a more equitable distribution among previously coexisting species (Woods & Shanks 1959; Day & Monk 1974; Stephenson 1986). As rare species become less rare they are less prone to species extinction and thus species diversity is maintained.

3   The pathogen, *Phytophthora cinnamomi*, has exploded in diverse communities of Australia. Its origin is disputed (Newhook & Podger 1972). It is unmatched in the variety of hosts it attacks (444 species, 131 genera, forty-eight families in Australia and New Zealand) and the range of communities it affects (Newhook & Podger 1972). Within a 5-year period after pathogen invasion the vegetation can change from a shrub woodland to an open sedge

woodland; common species decline in plant density and some face extinction, some species are eliminated, and the relative proportion of existing species changes (Weste 1981). In general, the data suggest that species diversity is lowered, at least temporarily, by the devastation of this very invasive pathogen. The pathogen is unique in its wide breadth of hosts and thus these diverse communities behave more like a monoculture for this pathogen (Dinoor & Eshed 1984). This example illustrates that when species share a common pathogen, some species are excluded from the community, as predicted by Holt & Pickering (1985).

4 A slow-developing epidemic of root-rot disease, caused by *Phellinus weirii*, an indigenous pathogen of coniferous forests in north-western USA, appears to control local community structure. The pathogen acts as a focal disturbance and creates circular areas devoid of trees at the spreading margin. Successional trees invade the void left by the disease. The successional vegetation has more species, higher species diversity, and greater heterogeneity of sizes/ ages than vegetation not yet attacked (McCauley & Cook 1980; Copsey 1985). Cook (1982) concludes that no equilibrium in forest composition has occurred after two-and-a-half oscillations of host and pathogens during a 350 year period. Likewise, Copsey (1985) finds no convergence in species composition between infected and non-infected areas. The persistence of this non-converging community is apparently due to an interacting feedback: (a) the most abundant species in uninfected forest is most susceptible to fungal attack, and (b) rare species in uninfected forests are killed more slowly, thus contributing more to regeneration after fungal attack (Copsey 1985).

## Unanswered questions

Pathogens are not exclusive in their effects on the community. Herbivores and physical disturbances caused by fire or wind can provide similar effects. Therefore, how common and pervasive are disturbances caused by pathogens, relative to the generation time of host species, and what is the relative role of pathogens in diversifying the community?

The pathogens discussed above differ in several aspects. The pathogen can be a specialist or generalist, the plant community can be highly diverse or simple, and the pathogen can cause a rampant or slow-developing epidemic. Given this diversity of situations, what generalities about pathogen effects on the community will arise with further study? Because all of our evidence comes from forest communities for which experimentation is not feasible, an alternative to waiting for more examples to arise naturally is to develop models, the components of which can be field-tested in simpler, short-lived, but natural communities.

## CONCLUSION

The studies reviewed here support the idea that while pathogens are not always conspicuous and frequently don't appear to cause much damage, they exert a significant effect on the population biology of their hosts. Growing numbers of ecologists are studying pathogens and providing at least strong corroborative evidence that they affect the host's population dynamics, genetic variation, spatial distribution, and interspecific interactions. Accidental or deliberate introduction of pathogens provides the most compelling experimental evidence of the impact of pathogens on host populations. These examples underscore the benefits that would accrue from a more deliberate experimental approach to the field study of natural populations, e.g. eliminating or adding plant species, manipulating age structures and spatial patterns, or excluding or adding the pathogen. Theoretical developments will continue to provide a stimulus and direction for empirical field studies. They also offer a convenient framework which integrates observational and experimental evidence, identifies those major ideas not yet adequately tested, and helps to structure new and more sophisticated questions that arise from initial tests of hypotheses.

## ACKNOWLEDGMENTS

The constructive comments of H. M. Alexander, C. A. Kelly and D. Thiede are gratefully acknowledged.

## REFERENCES

Alexander, H. M. (1984). Spatial patterns of disease induced by *Fusarium moniliforme* var. *subglutinans* in a population of *Plantago lanceolata*. *Oecologia*, 62, 141–43.

Alexander, H. M. (1987). Pollination limitation in a population of *Silene alba* infected by the anther-smut fungus, *Ustilago violacea*. *Journal of Ecology*, 75, 771–80.

Alexander, H. M. (1988). Heterogeneity and disease in natural populations. *Spatial Components of Plant Disease Epidemics* (Ed. by M. J. Jeger), Prentice-Hall, New York, in press.

Alexander, H. M. & Antonovics, J. (1988). Disease spread and population dynamics of anther-smut infection of *Silene alba* caused by the fungus *Ustilago violacea*. *Journal of Ecology*, 76, 91–104.

Alexander, H. M., Antonovics, J. & Rausher, M. D. (1984). Relationship of phenotypic and genetic variation in *Plantago lanceolata* to disease caused by *Fusarium moniliforme* var. *subglutinans*. *Oecologia*, 65, 89–93.

Alexander, H. M. & Burdon, J. J. (1984). The effect of disease induced by *Albugo candida* (white rust) and *Peronospora parasitica* (downy mildew) on the survival and reproduction of *Capsella bursa-pastoris* (shepherd's purse). *Oecologia*, 64, 314–18.

Antonovics, J. & Alexander, H. M. (1989). The concept of fitness in plant-fungal pathogen systems. *Plant Disease Epidemiology*, Volume II (Ed. by K. J. Leonard & W. Fry), McGraw-Hill Book Co., New York, in press.

Antonovics, J. & Levin, D. A. (1980). The ecological and genetic consequences of density-dependent regulation in plants. *Annual Review of Ecology and Systematics*, 11, 411–52.

Augspurger, C. K. (1983a). Seed dispersal by the tropical tree, *Platypodium elegans*, and the escape of its seedlings from fungal pathogens. *Journal of Ecology*, 71, 759–71.

Augspurger, C. K. (1983b). Offspring recruitment around tropical trees: changes in cohort distance with time. *Oikos*, 40, 189–96.

Augspurger, C. K. (1984). Seedling survival of tropical tree species: interactions of dispersal distance, light-gaps, and pathogens. *Ecology*, 65, 1705–12.

Augspurger, C. K. & Kelly, C. K. (1984). Pathogen mortality of tropical tree seedlings: experimental studies of the effects of dispersal distance, seedling density, and light conditions. *Oecologia*, 61, 211–17.

Ayres, P. G. (1984). Interactions between environmental stress injury and biotic disease physiology. *Annual Review of Phytopathology*, 22, 53–75.

Barber, J. C. (1966). Variation among half-sib families from the loblolly pine stands in Georgia. *Georgia Forestry Research Paper*, 37, 1–5.

Barrett, J. A. (1978). A model of epidemic development in variety mixtures. *Plant Disease Epidemiology* (Ed. by P. R. Scott & A. Bainbridge), pp. 129–37. Blackwell Scientific Publications, Oxford.

Ben-Kalio, V. D. & Clarke, D. D. (1979). Studies on tolerance in wild plants: effects of *Erysiphe fischeri* on the growth and development of *Senecio vulgaris*. *Physiological Plant Pathology*, 14, 203–11.

Bradshaw, A. D. (1959). Population differentiation in *Agrostis tenuis*. II. The incidence and significance of infection by *Epichloe typhina*. *New Phytologist*, 58, 310–15.

Bremermann, J. H. (1980). Sex and polymorphism as strategies in host-pathogen interactions. *Journal of Theoretical Biology*, 87, 671–702.

Browning, J. A. (1974). Relevance of knowledge about natural ecosystems to development of pest management programs for agro-ecosystems. *Proceedings of the American Phytopathology Society*, 1, 191–99.

Burdon, J. J. (1980). Variation in disease resistance within a population of *Trifolium repens*. *Journal of Ecology*, 68, 737–44.

Burdon, J. J. (1985). Pathogens and the genetic structure of plant populations. *Studies on Plant Demography* (Ed. by J. White), pp. 313–25. Academic Press, New York.

Burdon, J. J. (1987a). *Diseases and Plant Population Biology*. Cambridge University Press, Cambridge.

Burdon, J. J. (1987b). Phenotypic and genotypic patterns of resistance to the pathogen *Phakopsora pachyrhizi* in populations of *Glycine canescens*. *Oecologia*, 73, 257–67.

Burdon, J. J. & Chilvers, G. A. (1974). Fungal and insect parasites contributing to niche differentiation in mixed species stands of Eucalypt saplings. *Australian Journal of Botany*, 22, 103–14.

Burdon, J. J. & Chilvers, G. A. (1976). The effect of clumped planting patterns on epidemics of damping-off disease in cress seedlings. *Oecologia*, 23, 17–29.

Burdon, J. J. & Chilvers, G. A. (1977). The effect of barley mildew on barley and wheat competition in mixtures. *Australian Journal of Botany*, 25, 59–65.

Burdon, J. J. & Chilvers, J. A. (1982). Host density as a factor in plant disease ecology. *Annual Review of Phytopathology*, 20, 143–66.

Burdon, J. J., Groves, R. H. & Cullen, J. M. (1981). The impact of biological control on the distribution and abundance of *Chondrilla juncea* in southeastern Australia. *Journal of Applied Ecology*, **18**, 957–66.

Burdon, J. J., Groves, R. H., Kaye, P. E. & Speer, S. S. (1984). Competition in mixtures of susceptible and resistant genotypes of *Chondrilla juncea* differentially infected with rust. *Oecologia*, **64**, 199–203.

Burdon, J. J. & Marshall, D. R. (1981a). Inter- and intra-specific diversity in the disease-response of *Glycine* species to the leaf-rust fungus *Phakopsora pachyrhizi*. *Journal of Ecology*, **69**, 381–90.

Burdon, J. J. & Marshall, D. R. (1981b). Biological control and the reproductive mode of weeds. *Journal of Applied Ecology*, **18**, 649–58.

Burdon, J. J., Oates, J. D. & Marshall, D. R. (1983). Interactions between *Avena* and *Puccinia* species. I. The wild hosts: *Avena barbata* Pott ex Link, *A. fatua* L., *A. ludoviciana* Durieu. *Journal of Applied Ecology*, **20**, 571–84.

Burdon, J. J. & Shattock, R. C. (1980). Disease in plant communities. *Applied Biology*, **5**, 145–219.

Chilvers, G. A. & Brittain, E. G. (1972). Plant competition mediated by host-specific parasites—a simple model. *Australian Journal of Biological Science*, **25**, 749–56.

Clark, D. A. and Clark, D. B. (1984). Spacing dynamics of a tropical rain forest tree: evaluation of the Janzen-Connell model. *American Naturalist*, **124**, 769–88.

Clarke, B. (1976). The ecological genetics of host-parasite relationships. *Genetic Aspects of Host-parasite Relationships* (Ed. by A. E. R. Taylor & R. Muller), pp. 87–103. Blackwell Scientific Publications, Oxford.

Clarke, B. (1979). The evolution of genetic diversity. *Proceedings of the Royal Society of London B*, **205**, 453–74.

Clarke, D. D., Bevan, J. R. & Crute, I. R. (1986). Genetic interactions between wild plants and their parasites. *Genetics and Plant Pathogenesis* (Ed. by P. R. Day & G. J. Jellis), pp. 195–206. Blackwell Scientific Publications, Oxford.

Clay, K. (1984). The effect of the fungus *Atkinsonella hypoxylon* (Clavicipitaceae) on the reproductive system and demography of the grass *Danthonia spicata*. *New Phytologist*, **98**, 165–75.

Clay, K. (1986). Induced vivipary in the sedge *Cyperus virens* and the transmission of the fungus *Balansia cyperi* (Clavicipitaceae). *Canadian Journal of Botany*, **64**, 2984–88.

Connell, J. H. (1971). On the role of natural enemies in preventing competitive exclusion in some marine animals and in rain forest trees. *Dynamics of Populations* (Ed. by P. J. den Boer & G. R. Gradwell), pp. 298–312. Centre for Agricultural Publishing and Documentation, Wageningen.

Cook, S. A. (1982). Stand development in the presence of a pathogen, *Phellinus weirii*. *Forest Succession and Stand Development Research in the Northwest* (Ed. by J. Means), pp. 159–63. Forest Research Laboratory, U.S. Forest Service, Corvallis, Oregon.

Copony, J. A. & Barnes, B. V. (1974). Clonal variation in the incidence of Hypoxylon canker on trembling aspen. *Canadian Journal of Botany*, **52**, 1475–81.

Copsey, A. D. (1985). *Long-term effects of a native forest pathogen*, Phellinus weirii: *changes in species diversity, stand structure, and reproductive success in a* Tsuga mertensiana *forest in the central Oregon high Cascades*. Ph.D. dissertation, University of Oregon.

Cullen, J. M. & Groves, R. H. (1977). The population biology of *Chondrilla juncea* L. in Australia. *Proceedings of the Ecological Society of Australia*, **10**, 121–34.

Day, F. P. & Monk, C. D. (1974). Vegetation patterns on a southern Appalachian watershed. *Ecology*, **55**, 1064–74.

de Nooij, M. P. & van der AA, H. A. (1987). *Phomopsis subordinaria* and associated stalk disease in natural populations of *Plantago lanceolata* L. *Canadian Journal of Botany*, **65**, 2318–25.

Dinoor, A. (1970). Sources of oat crown rust resistance in hexaploid and tetraploid wild oats in Israel. *Canadian Journal of Botany*, **48**, 153–61.

**Dinoor, A. (1977).** Oat crown rust resistance in Israel. *Annals of the New York Academy of Science,* **287**, 357–66.

**Dinoor, A. & Eshed, N. (1984).** The role and importance of pathogens in natural plant communities. *Annual Review of Phytopathology,* **22**, 443–66.

**Dinoor, A. & Eshed, N. (1987).** The analysis of host and pathogen populations in natural ecosystems. *Populations of Plant Pathogens: their Dynamics and Genetics* (Ed. by M. S. Wolfe & C. E. Caten), pp. 75–88. Blackwell Scientific Publications, Oxford.

**Dinus, R. J. (1974).** Knowledge about natural ecosystems as a guide to disease control in managed forests. *Proceedings of the American Phytopathology Society,* **1**, 184–90.

**Gates, D. J., Westcott, M., Burdon, J. J. & Alexander, H. M. (1986).** Competition and stability in plant mixtures in the presence of disease. *Oecologia,* **68**, 559–66.

**Haldane, J. B. S. (1949).** Disease and evolution. *La Ricerca Science Supplement,* **19**, 68–76.

**Hamilton, W. D. (1980).** Sex versus non-sex versus parasite. *Oikos,* **35**, 282–90.

**Hamilton, W. D. (1982).** Pathogens as causes of genetic diversity in their host populations. *Population Biology of Infectious Diseases* (Ed. by R. M. Anderson & R. M. May), pp. 269–96. Springer-Verlag, New York.

**Harlan, J. R. (1976).** Disease as a factor in plant evolution. *Annual Review of Phytopathology,* **14**, 31–51.

**Harper, J. L. (1969).** The role of predation in vegetational diversity. *Brookhaven Symposium in Biology,* **22**, 48–62.

**Harper, J. L. (1977).** *Population Biology of Plants.* Academic Press, London.

**Harry, I. B. & Clarke, D. D. (1986).** Race-specific resistance in groundsel (*Senecio vulgaris*) to the powdery mildew *Erysiphe fischeri. New Phytologist,* **103**, 167–75.

**Hermann, R. K. & Chilcote, W. W. (1965).** Effect of seed beds on germination and survival of Douglas fir. Research Paper (Forest Management Research) *Oregon Forestry Research Laboratory,* **4**, 1–28.

**Holt, R. D. & Pickering, J. (1985).** Infectious diseases and species coexistence: a model of Lotka-Volterra form. *American Naturalist,* **126**, 196–211.

**Jaenike, J. (1978).** An hypothesis to account for the maintenance of sex within populations. *Evolutionary Theory,* **3**, 191–94.

**Janzen, D. H. (1970).** Herbivores and the number of tree species in tropical forests. *American Naturalist,* **104**, 501–28.

**Jarosz, A. M. (1984).** *Ecological and evolutionary dynamics of* Phlox-Erysiphe cichoracearum *interactions.* Ph.D. dissertation, Purdue University.

**Jarosz, A. M. & Levy, M. (1988).** Effects of habitat and population structure on powdery mildew epidemics in experimental *Phlox* populations. *Phytopathology,* **78**, 358–62.

**Jennersten, O., Nilsson, S. G. & Wastljung, U. (1983).** Local plant populations as ecological islands: the infection of *Viscaria vulgaris* by the fungus *Ustilago violacea. Oikos,* **41**, 391–95.

**Kinloch, B. B., Jr. & Stonecypher, R. W. (1969).** Genetic variation in susceptibility to fusiform rust in seedlings from a wild population of loblolly pine. *Phytopathology,* **59**, 1246–55.

**Kirkpatrick, B. L. & Bazzaz, F. A. (1979).** Influence of certain fungi on seed germination and seedling survival of four colonizing annuals. *Journal of Applied Ecology,* **16**, 515–27.

**Latch, G. C. M. & Lancashire, J. A. (1970).** The importance of some effects of fungal diseases on pasture yield and composition. *XI International Grassland Congress,* 688–91.

**Lee, J. A. (1981).** Variation in the infection of *Silene dioica* by *Ustilago violacea* in north west England. *New Phytologist,* **87**, 81–89.

**Leonard, K. J. (1977).** Selection pressures and plant pathogens. *Annals of the New York Academy of Science,* **287**, 207–22.

**Leonard, K. J. & Czochor, R. J. (1980).** Theory of genetic interactions among populations of plants and their pathogens. *Annual Review of Phytopathology,* **18**, 237–58.

**Leppik, E. E. (1970).** Gene centers of plants as sources of disease resistance. *Annual Review of Phytopathology*, **8**, 323–44.

**Levin, B. R. (1982).** Evolution of parasites and hosts. *Biology of Infectious Disease* (Ed. by R. M. Anderson & R. M. May), pp. 213–43. Springer-Verlag, New York.

**Levin, D. A. (1975).** Pest pressure and recombination systems in plants. *American Naturalist*, **109**, 437–51.

**Levin, S. A. (1982).** Some approaches to the modelling of coevolutionary interactions. *Coevolution* (Ed. by M. H. Nitecki), pp. 21–65. University of Chicago Press, Chicago.

**May, R. M. & Anderson, R. M. (1983).** Parasite–host coevolution. *Coevolution* (Ed. by D. J. Futuyma & M. Slatkin), pp. 186–206. Sinauer Associates, Sunderland, Massachusetts.

**McCauley, K. J. & Cook, S. A. (1980).** *Phellinus weirii* infestation of two mountain hemlock forests in the Oregon Cascades. *Forest Science*, **26**, 23–29.

**Miles, J. W. & Lenné, J. M. (1984).** Genetic variation within a natural *Stylosanthes guianensis*, *Colletotrichum gloeosporioides* host-pathogen population. *Australian Journal of Agricultural Research*, **35**, 211–18.

**Neher, D. A., Augspurger, C. K. & Wilkinson, H. T. (1988).** Influence of age structure of plant populations on damping-off epidemics. *Oecologia*, **74**, 419–24.

**Newhook, F. J. & Podger, F. D. (1972).** The role of *Phytophthora cinnamomi* in Australian and New Zealand forests. *Annual Review of Phytopathology*, **10**, 299–326.

**Oates, J. D., Burdon, J. J. & Brouwer, J. B. (1983).** Interactions between *Avena* and *Puccinia* species. II. The pathogens: *Puccinia coronata* Cda and *P. graminis* Pers. f. sp. *avenae* Eriks & Henn. *Journal of Applied Ecology*, **20**, 585–96.

**Oehrens, E. (1977).** Biological control of the blackberry through the introduction of rust, *Phragmidium violaceum*, in Chile. *F.A.O. Plant Protection Bulletin*, **25**, 26–28.

**Parker, M. A. (1985).** Local population differentiation for compatibility in an annual legume and its host-specific fungal pathogen. *Evolution*, **39**, 713–23.

**Parker, M. A. (1986).** Individual variation in pathogen attack and differential reproductive success in the annual legume, *Amphicarpaea bracteata*. *Oecologia*, **69**, 253–59.

**Parker, M. A. (1987).** Pathogen impact on sexual vs. asexual reproductive success in *Arisaema triphyllum*. *American Journal of Botany*, **74**, 1758–63.

**Parker, M. A. (1988).** Polymorphism for disease resistance in the annual legume *Amphicarpaea bracteata*. *Heredity*, **60**, 27–31.

**Paul, N. D. & Ayres, P. G. (1986a).** The impact of a pathogen (*Puccinia lagenophorae*) on population of groundsel (*Senecio vulgaris*) overwintering in the field. I. Mortality, vegetative growth and the development of size hierarchies. *Journal of Ecology*, **74**, 1069–84.

**Paul, N. D. & Ayres, P. G. (1986b).** The impact of a pathogen (*Puccinia lagenophorae*) on populations of groundsel (*Senecio vulgaris*) overwintering in the field. II. Reproduction. *Journal of Ecology*, **74**, 1085–94.

**Paul, N. D. & Ayres, P. B. (1986c).** Interference between healthy and rusted groundsel (*Senecio vulgaris* L.) within mixed populations of different densities and proportions. *New Phytologist*, **104**, 257–69.

**Paul, N. D. & Ayres, P. G. (1987a).** Survival, growth and reproduction of groundsel (*Senecio vulgaris*) infected by rust (*Puccinia lagenophorae*) in the field during summer. *Journal of Ecology*, **75**, 61–71.

**Paul, N. D. & Ayres, P. G. (1987b).** Water stress modifies intraspecific interference between rust (*Puccinia lagenophorae*) infected and healthy groundsel *Senecio vulgaris*. *New Phytologist*, **106**, 555–66.

**Pielou, E. C. & Foster, R. E. (1962).** A test to compare the incidence of disease in isolated and crowded trees. *Canadian Journal of Botany*, **40**, 1176–79.

**Pratt, B. H., Heather, W. A. & Shepherd, C. J. (1973).** Recovery of *Phytophthora cinnamomi* from native vegetation in a remote area of New South Wales. *Transactions of the British Mycological Society*, **60**, 197–204.

Rice, W. R. (1983). Parent-offspring pathogen transmission: a selective agent promoting sexual reproduction. *American Naturalist*, **121**, 187–203.

Roame, M. K., Griffin, G. J. & Rush, J. R. (1986). Chestnut blight, other Endothia diseases, and the genus Endothia. *American Phytopathology Society*, St. Paul, Minnesota.

Rochow, T. F. (1970). Ecological investigations of *Thlaspi alpestre* L. along an elevational gradient in the Central Rocky Mountains. *Ecology*, **51**, 649–56.

Schmidt, R. A. (1978). Disease in forest ecosystems: the importance of functional diversity. *Plant Disease. Vol. II. How Disease Develops in Populations.* (Ed. by J. G. Horsfall & E. B. Cowling), pp. 287–315. Academic Press, New York.

Segal, A., Manisterski, J., Fischbeck, G. & Wahl, I. (1980). How plants defend themselves in natural ecosystems. *Plant Disease: An Advanced Treatise* (Ed. by J. G. Horsfall & E. B. Cowling), pp. 75–102. Academic Press, New York.

Segal, A., Manisterski, J., Browning, J. A., Fischbeck, G. & Wahl, I. (1982). Balance in indigenous plant populations. *Resistance to Diseases and Pests in Forest Trees* (ed. by H. M. Heybroek, Z. R. Stephan & K. von Weissenbery), pp. 361–70. Center for Agricultural Publishing and Documentation, Wageningen.

Snaydon, R. W. & Davies, M. S. (1972). Rapid population differentiation in a mosaic environment. II. Morphological variation in *Anthoxanthum odoratum*. *Evolution*, **26**, 390–405.

Stephenson, S. L. (1986). Changes in a former chestnut-dominated forest after a half century of succession. *American Midland Naturalist*, **116**, 173–79.

Wahl, I. (1970). Prevalence and geographic distribution of resistance to crown rust in *Avena sterilis*. *Phytopathology*, **60**, 746–49.

Wahl, I., Eshed, N., Segal, A. & Sobel, Z. (1978). Significance of wild relatives of small grains and other wild grasses in cereal powdery mildews. *The Powdery Mildews* (Ed. by D. M. Spencer), pp. 83–100. Academic Press, New York.

Weir, J. R. (1918). Notes on the altitudinal range of forest fungi. *Mycologia*, **10**, 4–14.

Weltzien, H. C. (1972). Geophytopathology. *Annual Review of Phytopathology*, **10**, 277–98.

Weste, G. (1981). Changes in the vegetation of sclerophyll shrubby woodland associated with invasion by *Phytophthora cinnamomi*. *Australian Journal of Botany*, **29**, 261–76.

Wilde, S. A. & White, D. P. (1939). Damping-off as a factor in the natural distribution of pine species. *Phytopathology*, **29**, 367–69.

Woods, F. W. & Shanks, R. E. (1959). Natural replacement of chestnut by other species in the Great Smoky Mountains National Park. *Ecology*, **40**, 349–61.

Zimmer, D. E. & Rehder, D. (1976). Rust resistance of wild *Helianthus* species of the North Central United States. *Phytopathology*, **66**, 208–11.

# 21. AN APOPHASIS OF PLANT POPULATION BIOLOGY

JOHN L. HARPER

*Unit of Plant Population Biology, University College of North Wales, Bangor, Gwynedd LL57 2UW, UK*

Apophasis was a process used by theologians to describe the nature of God by defining the extent of their ignorance. This chapter attempts an apophasis of plant population biology. This may be worthwhile if it is true that the beginning of knowledge is the awareness of ignorance.

A population biologist seeks to explain and then hopes to predict the abundance of organisms. There are clearly two quite different types of explanation—proximal and ultimate (final). Proximal explanations are made by showing how the present properties of organisms place limits on the range of environments that they discover and exploit and the speed with which they do so. Ultimate explanations invoke evolutionary forces (e.g. natural selection) to explain how the organisms came to have those present properties that impose those present limits on their behaviour, abundance and distribution. It is very easy to confuse these types of explanation, especially when the same word 'adaptation' is used to describe *both* the present match between organisms and their environments *and* the process by which the match is supposed to have evolved.

There is of course no reason why we should suppose that the ultimate forces that led to the evolution of the present properties of organisms bear any close relationship to the present consequences of those properties. Unfortunately, most of the evidence for the nature of forces that have governed the direction of evolution (and so the source of ultimate explanations) comes from observations of proximal events. We make the assumption that the forces that shaped the past were essentially the same as those we can observe today.

Sex is an example of a property that has certain clear present consequences in the behaviour of individuals, populations and species. It is hard to resist the temptation to explain the evolution of sex in terms of its present consequences. However it may be just one of many properties that, having been acquired by populations during odd episodes of history, are irreversible and hang like millstones round the necks of evolving populations. Proximal explanations of the consequences of sex (and other properties) may have

435

little to do with ultimate explanations of its evolution (see e.g. Gouyon, Gliddon & Couvet, 1988). An apophasis of plant population biology needs to stress that we are ignorant of how far we can obtain ultimate explanations of biological phenomena by examining their proximal consequences. True ultimate explanations might need to show how the present behaviour of organisms has become limited within narrow specializations rather than how they have become perfected fits to their present environments.

All known organisms are specialists, capable of growth and reproduction in only a limited range of conditions. Environments are patchy in space and time. The abundance of any particular type of organism depends on the abundance and distribution of patches in which its special requirements are met and whether it disperses to and is able to colonize these patches. The result is that only tiny subsets of the world's flora are present in any area we choose to study. The overwhelming characteristic of the flora of a region is the absence of all species but for a few egregious oddities. The specializations that distinguish taxa are responsible for excluding them from most of the world's habitats.

Darwin's evolutionary optimism saw the process of natural selection as a perfecting force, creating '. . . from the war of nature, from famine and death, the most exalted object which we are capable of conceiving . . .'. This attitude finds itself expressed in Sewall Wright's adaptive landscapes in which populations are envisaged as moving, under natural selection, across contours of genetic variation to the tops of hills—adaptive peaks (Wright 1931, 1932, 1935). An alternative view, an evolutionary pessimism, is that of Huffaker (1964) who saw instead specialization as 'a deepening rut in evolution'. The appropriate evolutionary landscape would then be of pits and troughs, with populations more or less deeply embedded in them; some indeed descending deep on their way to extinction! It may be that sex itself is a deep rut in the evolutionary process, from which it is very difficult to escape. We may be paying too much attention to trying to find the forces that lead to the evolution of sex—the more important question may often be what are the forces that constrain populations from losing the sexual habit. The answers may be quite different in higher plants (in which apomixis offers the opportunity for easy escape from outbreeding, without inbreeding depression) and in animals in which the formation of a zygote so often seems to be an obligatory condition for the start of a new life-cycle.

In this volume, Antonovics reexamines Haldane's dictum that the population size of a species will only be affected by changes in genotypes if the resultant phenotypes respond differently to density-dependent factors. He concludes that this may be true only under very specialized 'perfect' conditions. If the evolution of specializations, forced by fitness differentials, may have the effect of reducing population size, an adaptive landscape of troughs and ruts may be the most realistic picture of the real world. However Antonovics points out that '. . . we have no data or developed methodology that can provide even a tentative answer to the question of whether the "genetic component" does or does not influence population size.' This is one example of a part of the science of population biology that is defined at present by the extent of our ignorance.

## *Plasticity*

The genotype of a plant acts on its development by determining how it interacts with the environment. Many specializations represent narrow limits within which phenotypes may vary—this is especially true of the form of insect–pollinated flowers. Other organs may vary widely in form in different parts of the same plant (e.g. leaves of aquatic *Ranunculus* spp., juvenile and adult leaves of many *Eucalyptus* spp., *Acacia* spp., *Valeriana dioica*) or between plants of the same species or clone. This phenomenon of plasticity is especially well known in higher plants (see e.g. Bradshaw, 1965) but is probably characteristic of all modular organisms (Harper, Rosen & White 1986).

All organisms that grow by the branching iteration of their modular parts have an inherent plasticity in the number of these parts. The size of an oak tree, a strawberry clone or a coral is determined by the number of its modules. The number usually increases with age, but is profoundly influenced by environmental conditions of irradiance, nutrient and water supply and by the effect of neighbours on the availability of these resources. As a result the size (and hence usually the extent and sometimes even the timing of reproduction) of modular organisms becomes phenotypically highly plastic. Weiner (1988) reviews the literature on plant to plant variation in size that occurs within populations and results in hierarchies of reproductive output. Gottlieb (1977) has shown that much of this variation in fitness between individuals may have no genetic base and may indeed blur and slow the effects of selective forces in shaping evolutionary change.

At the level of the module however, it is not at all clear that phenotypic plasticity is more strongly developed among plants than among animals. The shapes and sizes of leaves, flowers, fruits and seeds may not, on average, be phenotypically more plastic than body size in *Drosophila*, rabbits or snails. In higher plants it is as if the tight control of development has been shifted from the genetic individual to its parts. Nevertheless there remain intriguing variations in the plasticity of modular parts. The flowers of annual cornfield poppies (*Papaver rhoeas* and *P. dubium*) are highly plastic, the length of petals and the number of stamens per flower varying at least five-fold. However in *Ranunculus* spp., e.g. *R. bulbosus*, petal size scarcely varies and in other species of Ranunculaceae, e.g. *Aquilegia* spp., stamen number has become rigidly canalized.

Levin 1988 argues that the ability to respond to different environments by phenotypic plasticity lessens the forces of selection (and drift) that operate within populations and so slows the pace of evolutionary change and divergence. How then does natural selection act on plasticity itself? By slowing down the speed of evolutionary response to a given selective force the risk of becoming trapped in an evolutionary rut is reduced. Over the long term, populations of individuals with high phenotypic plasticity may be less likely to go extinct, though, over the short term, those individuals with a precise canalized response to their immediate environment may often be fitter. Studies of comparative plasticity between and within species and within individual plants are still extremely rare (an exception is Marshall, Levin & Fowler 1986). We are still in almost complete ignorance of the genetic control of plasticity and homeostasis of the modules of higher plants. The great interest stimulated among animal breeders by the work of Lerner (1958) on homeostasis seems to have had little impact in the plant sciences.

Plants grow in environments that are patchy in space and time. The theoreticians' vision of panmictic populations in homogeneous environments developing to equilibrial conditions was created to make it easier to do sums; not because it represents the real world. Hubbell (1979) and Hubbell & Foster (1986) have argued from studies of a Panamanian forest, that in such patchy environments the composition of the forest is determined by populations of species (or at least guilds of species) that vary randomly—the ecological equivalent of genetic drift. However, Fowler (1988) argues that there is a variety of scenarios in which competition—and frequency-dependent interactions—may occur in plant communities yet be very difficult to detect or measure in the field (although they are easy to demonstrate in deliberately constructed experimental communities).

It is extremely difficult to design the appropriate experiments that might reveal how much of the composition of natural communities is determined by forces such as density and frequency dependence and how much by unrecognized heterogeneity in the communities over space and time. The boundaries of our ignorance of even the most general principles are drawn by the sheer limitation in the number of appropriately designed studies. As Fowler points out, proving Hubbell's hypothesis is extremely difficult as it is a null hypothesis. If we fail to find pattern and order in systems it may be that we have simply failed to search in the right way or that there are quite simple processes that govern the population dynamics which in their operation give results indistinguishable from chaos (May 1981). It is not certain what is the appropriate scientific method that resolves this situation—presumably only continuing the search for causation and continuing to adopt random processes as a fall-back position until or unless we discover causation.

## Somatic mutation

A number of population biologists have become excited about the role of somatic mutation as a source of variation on which natural selection may act and as a source of variation within single clones (genetic mosaics) that might give protection against disease and other hazards. It is remarkable that this enthusiasm for naturally occurring somatic mutation is found among ecologists and not among plant breeders. Three sets of evidence are frequently cited for the importance of somatic change in natural systems. Durrant's work (Durrant 1958, 1962, 1971) established beyond dispute that heritable changes could be induced in flax (*Linum usitatissimum*). What is extraordinary in this case is that the phenomenon seems to be almost entirely restricted to a very few species. There is simply no evidence that the phenomenon is widespread in the plant kingdom. If the phenomenon were at all frequent it would have served as a powerful tool for plant breeders! This has not happened. The other two frequently cited papers are those of Whitham & Slobodchikoff (1981) and Gill (1986). Both papers give convincing evidence that different parts of the same genet behave differently. However in neither case is there evidence that the observed differences between plant parts involve changes in the genome.

There are many alternative explanations of differences in behaviour between plant parts. It is for example rather easy to ensure that different branches on the same tree bear their main fruit crops in different years by inhibiting fruiting for 1 year on selected branches. Asynchronous fruiting between the branches then continues for several years. It is a commercial procedure that may be used to prevent alternating years of high and low fruit production. A common source of what may look like somatic variation occurs as a result of virus infection producing quite different morphologies and physiologies on different parts of the same genet. The history of virus disease within potato varieties (which are single genets) is ample proof of this source of somatic variation. Differences in susceptibility to disease in parts of the same genet have been claimed to be the result of genetic mosaicism—but virus infection may have the same result. Indeed Gibbs (1980) showed that seedlings of *Kennedya rubicunda* infected with yellow mosaic virus suffered less from grazing by rabbits than healthy seedlings and that in palatability trials infected plant material was consumed at a significantly lower rate than healthy material. It is technically quite a laborious procedure to prove that viruses are not responsible for somatic variation within plants. Even the demonstration that the somatic change is transmitted through gametes to progeny does not rule out viruses as the cause because of the ease with which some viruses are pollen- or ovule-transmitted.

Another source of evidence that somatic mutation is an important element in plant evolution comes from the known origin of horticultural cultivars from vegetative 'sports'. Some of these are certainly viral infections. Indeed the first records of somatic 'mutation' that proved to be virus infection are the illustrations of broken-flowered tulips in the works of Dutch realist painters. The extent of viral infections of plants in nature is virtually unknown. Attention has been concentrated almost entirely on crop plants and the most recent survey of virus distributions in natural populations is that of MacClement & Richards (1956), which reports natural incidence of infection as high as 10% and in one genus, *Plantago*, above 64% of the individuals were infected in nine study sites in England. Mackenzie (1985) found virus infections in seven out of fourteen populations of *Primula vulgaris* sampled in North Wales. The effects of virus infections are likely to be reported as somatic mutations if only part of a genet is infected.

For most of the somatic cultural 'sports' in horticultural plants there is no evidence at all that they involve changes in the genome or that they contribute genes to the population gene pool. This is the crucial test of whether they are material on which natural selection acts.

Of course somatic variants may easily be obtained from plant material that has passed through tissue culture–somaclonal variation (e.g. Evans & Sharp 1983). The formation of callus appears to be a necessary step in eliciting this variation. The abnormalities are generated only during passage through the disorganized condition of callus culture and appear to be a product of the treatment: regeneration from wound callus has long been known as a procedure that can generate polyploids and other genetic changes in the karyotype. It may be that somatic mutation is extremely rare or absent in organized multicellular meristems, or that it is a common occurrence but that diplontic selection weeds out mutant cells. It is perhaps significant that some of the strongest evidence of somatic variation without an intervening callus stage is that of Klekowski, Kazarinova-Fukshansky & Mohr (1985) from studies of ferns. It is in such plants with single apical cells that the opportunity for diplontic selection would be expected to be least. If somatic mutation is a common feature within plants it should be especially common in mosses where the haploid tissues should immediately display even recessive mutations.

Antolin & Strobeck (1985) have modelled the ways in which somatic mutation might influence the variation that is exposed to natural selection—but the value of their model awaits the critical data that show whether the phenomenon is real or not (and if it is real, whether it is of significant magnitude). Schaal (1988) contributes some of the strongest evidence for somatic mutation as a real phenomenon in flowering plants. Her data were obtained from careful analysis of ribosomal DNA in *Solidago altissima* and *Brassica campestris*. Despite the strength of her evidence, she is extremely guarded in her conclusions: '. . . whether this somatic variation is inherited and enters into the gene pool remains to be determined . . . The significance of these processes in population biology is yet to be determined.' Perhaps the most important experiments needed in this area are not those that examine whether or not somatic mutation occurs, but whether it makes a significant contribution to the genetic variation on which selection may act. The crucial experiments are those that will give us measures of the relative contributions made to *genetic* variance by variation between and within genets.

The problem of the role of somatic mutation is a good illustration of apophasis. The extent of our ignorance is not easily mapped if exciting ideas are rapidly assimilated into a body of literature before they have become more than ideas or hypotheses. The risk is greater because authors are unwilling to report, and editors reluctant to publish, purely negative results from hypothesis testing. The boundaries of ignorance are set by knowledge—not by hypotheses.

## Life-cycles and growth form

The life-time activity of mobile unitary animals can be seen as a partitioning of time and resources between a variety of activities such as hunting for food and mates, competing with other individuals, escaping from enemies and rearing offspring. Arber (1950) pointed out that almost every activity of an animal that is related to its ability to move is achieved by plants through some aspect of growth or form. It is not surprising therefore that plant population biologists are increasingly concerned with the ways in which form is determined—both its canalization and its plasticity.

In the case of mobile animals it is often convenient to compare the 'costs' of various activities in terms of the ways in which some resource, usually energy, is partitioned. Energy spent in searching for food is energy that is not available for escaping from enemies or for some other activity. Natural selection may be seen as acting on the ways in which limited resources are allocated between the variety of activities, each of which brings rewards but also incurs costs. This same approach may be followed for plants, e.g. Harper

& Ogden (1970) (though currencies other than energy or carbon may then be appropriate).

An alternative approach for plants and other modular organisms is to compare the ways in which modules of structure are allocated to various activities such as reproduction, growth, defence, etc. Higher plants grow by the iteration of modular units of structure which are generated from meristems. Meristems may be allocated to growth, reproductive or other structures. The very form of the plant is therefore the consequence of the ways in which meristems have been allocated. The time schedules of growth and of reproduction are locked together because of the dependence of both on a limited resource of meristems. Thus the annual plant may be annual because it switches all its available meristems to reproduction and so puts an end to further growth. A perennial habit is possible only if some meristems remain in a vegetative condition.

The reproductive schedule of a tree is characterized by a more or less long period in which no meristems are allocated to reproduction (the juvenile phase) followed by a sudden or gradual slowing of growth as a proportion of meristems end their lives in a lethal act of reproduction. The timing of changes in the patterns of allocation of meristems largely defines the reproductive schedule of a plant and can quite readily respond to selection over generations or as phenotypic change in the life of an individual. The switch of meristems from continuing production of new modules and meristems to a lethal reproductive fate is often a response to photoperiodic stimulus as, for example, in annual cereals.

The responsiveness of the switch mechanism to selection has been dramatically illustrated by the selection of rapid-cycling populations of *Brassica* spp. The commercial brassica crops and their wild relatives normally require 6 months or more to complete their life-cycle from seed to seed. A period of innate seed dormancy is normally added to this so that the life-history is completed only once in a year. Williams & Hill (1986) selected and interpollinated early flowering forms from a world collection of brassicas and radish (*Raphanus*) and repeatedly selected for (1) minimum time from sowing to flowering, (2) rapid seed maturation, (3) absence of seed dormancy, (4) small plant size and (5) high female fertility. This gave populations of *B. campestris* that can flower 16 days after sowing, complete the whole life-cycle in 36 days and are capable of completing ten whole life-cycles in a year!

The placement of reproductive as opposed to vegetative meristems is a major determinant of form and accounts for much of the variation in the structure of plant canopies: plants with terminal floral and axillary vegetative meristems clearly produce a different canopy structure from those that bear axillary floral and terminal vegetative meristems. Such differences in the pattern of meristem allocation account for much of the variation in tree form that has been summarized in a classification of the shapes of tropical trees (Hallé 1986; Hallé & Oldemann 1970; Hallé, Oldemann & Tomlinson 1978). It would be of great interest to know the levels of heritability of factors controlling the allocation of meristems in wild species. The effects of both reproductive schedules and form on the population biology of plants are clearly potentially very profound. We need to know much more about the degree of plasticity and the genetic canalization of the processes in different species. In particular we need to know much more about the heritability of morphology and life-history properties in different species. Is it the case that in some species the properties are readily selected and in others they are resistant to microevolutionary change?

Form is, however, more than just the result of the allocation of meristems to different fates (to survive or die, to generate more vegatative meristems or reproductive organs); the lengths of internodes and angles of branching also make a major contribution to the way in which a canopy is displayed to light, flowers are displayed to pollinators and seeds are poised for dispersal. Clearly the resources used in making long internodes are costs to the plants—they might have been used in making other organs. One consequence of bearing short internodes is that the plant acquires a compact 'phalanx' growth form, with more or less tightly packed modules. As a consequence the modules may enter into each other's resource depletion zones (Harper 1985)—tightly packed leaves on a plant are more likely to shade each other and tightly packed roots on a single plant are more likely to deprive each other of access to water and nutrients. A compensating effect is that the territory of phalanx plants may be extremely resistant to invasion by other individuals of the same or different species. Typical phalanx growth forms among herbs are tussock grasses (e.g. *Deschampsia caespitosa* is an extreme example among temperate species), the dense clones formed by some clonal herbs (e.g. *Mercurialis perennis*) and shrubs (e.g. *Gaultheria shallon*) and the dense canopies of some trees (e.g. *Cupressus* spp, *Taxus baccata*, *Abies* and *Picea* spp.).

Plants with internodes that are long relative to the size of the leaves or shoots that they bear, (whether on aerial shoots or on rhizomes), develop a canopy with more widely dispersed leaves and shoots, that enter less severely into each others resource-depletion zones and so produce less intense competition for resources within the individual plant. However, as a consequence the canopies are more invadable and the shoots, though more effective as invaders of new resource zones, are more easily penetrated and invaded by the shoots of individuals of the same or of other species. The strawberry (*Fragaria vesca*), *Potentilla reptans* and rhizomatous grasses (e.g.

*Agropyron repens* and *Agrostis stolonifera*) are examples of extreme guerilla forms among herbs. However, guerilla growth forms can also be recognized in the canopy forms of trees. Species such as *Acer pseudoplatanus* and to a lesser extent *Quercus robur* and *Fagus sylvatica* are capable of extending long internodes (and so branches) into light gaps, giving them exploratory qualities that are denied to the more tightly canalized phalanx growth forms of, e.g. *Picea* and *Abies* spp. The branches of a sycamore wander and explore gaps in a canopy—the branches of a fir may die if they find themselves in shade but are under too tight a programme of development to grow out and 'wander' into light gaps.

### The physiology of individuals in populations

There is a clear contrast between species in which the form of the whole genet is under tight developmental control and those in which there is opportunistic local response to the environmental conditions experienced by the parts. At one extreme is the firmly canalized growth form of trees like *Picea* and *Abies* in which strong forces of correlative inhibition keep the expression of form of the genet within narrow limits. At the other extreme are species such as the aquatics *Lemna* spp., *Pistia stratiotes*, *Salvinia* spp., in which physical connection and all opportunity for developmental coordination of the modules is rapidly lost. In most plants, even though the integrity of physical connection is maintained, the individual parts have a high degree of physiological autonomy (Watson 1984; Watson & Casper 1984). Hutchings & Slade (1988) discuss the ways in which the balance between autonomy and regulation acts to control growth form and resource capture in the woodland herb *Glechoma hederacea*. Experiments of this sort are rare and they emphasise how ignorant we are of the physiology of plants in real environments.

The overwhelming majority of studies made by plant physiologists use isolated plants or plant parts (leaves, roots, etc.). Yet in the real world most plants grow with neighbours—and these determine much of the quality of the local environment. There is of course a well-developed science of the physiology of natural, agricultural and forest ecosystems—concerned with the carbon assimilation, water balance and nutrient cycling of areas of vegetation—but these studies almost inevitably neglect the behaviour of individual plants within the systems. The result is that we know a great deal about the physiology of individual plants outside their communities and a great deal about the physiology of communities without reference to the behaviour of individual plants. The physiology of *the individual in the population* is a wide open area of ignorance. Rozema *et al.* argue this case in the present volume and they also indicate the importance of competition between organisms for limiting resources as a major force in plant population biology.

In most studies of the reaction of individual plants to the presence of neighbours their behaviour is usually interpreted as *either* the withdrawal by one of the resources that could have allowed the faster growth of the other (competition); *or* the provision by one of the resources that allow the faster growth of one another (e.g. nitrogen fixation by legumes in a population of grasses); *or* the creation by one of physical conditions (e.g. humidity) that hinder the growth of another; *or* the production by one of chemicals that suppress the growth of another as in the still strictly unproved case of allelopathy.

There is another important way in which plants may respond to the presence of neighbours; this is the reaction to environmental cues from neighbours which change their behaviour and so minimize competition. The enforcement of seed dormancy by canopy-filtered radiation has the effect of preventing seeds from developing into seedlings in environments that would be likely to prove lethal. The demonstration by Deregibus *et al.* (1985) that the development of grass tillers may be inhibited by canopy-filtered radiation is another example of potential competition between plants or plant parts that is 'anticipated' by cues and then avoided. The growth and branching of white clover (*Trifolium repens*) appears to be regulated by canopy-filtered radiation in the same way so that the plants respond not just to reduced photosynthetically active radiation in shade but also to the nature of the shade cast specifically by leaves (Solangaarachchi & Harper 1987). Mechanisms like these may be as important in regulating the intensity of competition between plants as territorial behaviour is in animals—the processes have much in common.

## Patchiness

We know a great deal about the effects of the density of populations, of pure and mixed species, on the growth of plants in experimental models and in agronomic practice. We still know extraordinarily little about how these effects are produced. Moreover, most of what we know comes from experiments in which spacing and mixing are designed to give more or less homogeneous populations: edge effects are usually excluded in the protocol of experimental design. One of the great areas of ignorance is of the effects of patchiness in populations and what happens at edges. Most of the natural world is patchy and most of the exciting things probably happen at edges. It is unfortunate that the literature concerned with pattern (patchiness) in vegetation has been concerned to describe the patchiness of species distribution (e.g. Greig-Smith 1983) and not with the much more difficult description of the patchy distribution of individuals or genets. This makes it impossible to use the existing elegant data on pattern in plant communities for important exercises like the measurement of genetic areas or for purely demographic studies. The science of plant population biology needs different procedures if we are to gain an understanding of the effects of patchiness on population dynamics comparable to its known effects in some animals (see e.g. Hassell 1981).

The patchiness of natural populations makes for some obvious comparisons with the patchy occurrence of diseases. The older science of epidemiology may indeed have much to offer to plant population biology. Areas of land become 'infected' by colonizing plants and these areas may become foci for infection of neighbouring areas. Areas of land may develop vegetation that confers 'immunity' against further infection. Newly susceptible land areas become available for infection and the chance that this will occur depends on the proximity of these 'habitable areas' to those already infected. It seems likely that whole blocks of formal epidemiological theory might usefully be absorbed into plant population biology and the approach of Carter & Prince (1988) may help to discover the missing link between plant population dynamics and more classical studies of the dynamics of vegetation.

*Parasitism, predation and mutualism: major forces or minor complications in the biology of plant populations*

Much of the early development of plant population biology as an experimental science was concerned to establish relationships between plant density and performance in pure stands and in mixtures. Relationships such as the Law of Constant Final Yield and the $-3/2$ thinning law were described by Japanese workers (e.g. Kira, Ogawa & Shinozaki 1953; Yoda *et al.* 1963) mainly for single species stands, and de Wit (1960) developed replacement series models that allowed the behaviour of two species mixtures to be analysed and described. Such models, in which the ecological world is composed solely of higher plants interacting with each other in an otherwise wholly physical environment, have developed greatly in sophistication. But the real world is a great deal more complex as evidenced by the chapters in this volume by Augspurger on the role of plant pathogens, Crawley on herbivory, and Turkington on the microorganisms of the rhizosphere. Watkinson adds another dimension to the plant–plant interactions by discussing the parasitic interactions between higher plants.

We simply do not know whether it is the direct density- and frequency-dependent interactions between higher plants coupled with physical heterogeneity in the environment and mediated by dispersal phenomena that account for the greater part of the character of natural vegetation—in which case all these other phenomena of predation, parasitism etc. are just noise in the system. However it may be that, in the majority of natural communities, it is these parasites, predators and mutualists that are the real determinants of community structure. It may be that it is competitive interactions between higher plants that are the forces that confuse the picture.

It is a worry for those who expect science to produce simplifying generalizations of nature, that research seems continually to expose new elements that need to be taken into consideration in accounting for the character of the natural world. New complications keep emerging, rather than a simpler story. Agronomists are well aware that plant viruses, nematodes and possible virus–nematode interactions are often major determinants of plant performance in the field. These are two extra forces (with their potential further interactions!) to add to those that are already being studied by population biologists.

It is not just that population biologists keep discovering new forces that may need to be considered as possible explanations of population phenomena; it is perhaps even more worrying that so many studies reveal highly species-specific interactions. Indeed, in the case of pathogens, the interactions are commonly specific at the race, or even individual genotype level (see e.g. Burdon 1987). Is it the case that plant population biology is developing by the accumulation of ever more individualistic special cases. The problem is not of course peculiar to the study of plants—a recent book on human evolutionary ecology (Foley 1987) is, perhaps cynically, titled *Another Unique Species*! If this is how the science is developing, the boundaries of our ignorance are becoming defined by a cloud of unique spots of knowledge, rather than by broad sweeping lines of generalization.

Grime *et al.* (1987) describe an experiment in which they established microcosms of many species of higher plant with and without VA mycorrhiza. After a year, a mycorrhizal network of interspecies connections appeared to have established: the species diversity of the microcosms was quite different in the presence and absence of the mycorrhiza. Most notably the dominant species (*Festuca ovina*) was strongly suppressed in the mycorrhizal micro-cosms. It may be that in this case *Festuca* reacted to the fungus as a pathogen and the increased diversity of the mycorrhizal microcosms reflected the reduced competitive force from the dominant. However, whatever the mechanisms involved, it is no longer realistic to neglect the role of other organisms in the population biology of higher plants. It may be that it was natural selection from the animal (and microbial) kingdom that provides the ultimate explanation of the variety in the plant kingdom that we know today; it may be these same kingdoms that retain proximal responsibility for the present patterns of abundance among the diversity that they created.

## The dynamics of plant populations

Plant population biologists are concerned to explain patterns in the numbers of plants over space and time. Some of these patterns develop over intervals of time that are long compared to the life of humans. The life-span of genets of rhizomatous plants and of trees commonly exceed that of humans and it is not surprising that we have few records over long enough time-spans to contain the essential elements of their population dynamics. The dynamics of *Pinus strobus* populations in New England appears to be determined essentially by fires and hurricanes. A major fire occurred in forests in New Hampshire in 1665 and a series of hurricanes in 1635, 1898, 1909 and 1929 (Henry and Swan 1974). In this case it was possible to reconstruct some of the dynamics of the populations from data frozen in fallen tree trunks—but few research programmes would have included the experience of such crucial events within their time-span.

We depend for much of our information about natural population dynamics on those individuals who have set up permanent recording sites and made arrangements for repeated census after they stopped making them themselves. White (1985) has reviewed the history of long-term studies— they are relatively few and may have given us a biased view of the dynamics of natural vegetation. A partial substitute for long-term studies is possible when communities in which different stages in community development can be found as a mosaic in space at the same time. This allows the dynamics of the whole to be inferred from its parts. Much of what we know about the population dynamics of tropical rainforest comes from this type of analysis and is illustrated by Whitmore (1988).

Perhaps the most important developments in plant population biology and vegetation science in the past 20 years have stemmed from the belated appreciation of the work of Watt (1947). He emphasized that communities are mosaics in space and time—that the apparently stable mixtures of species in a community are really space–time mosaics of replacements, local extinctions and colonizations (the regeneration niche of Grubb 1977). Martínez-Ramos, Sarukhán & Piñero (1988) describe this process for Mexican forests and also describe ways in which the annual 'tree', *Kochia trichophylla*, may be used to explore intimate 'tree-to-tree' relationships in experimental patches. The emphasis on patchiness in the chapters of both

Whitmore & Martínez-Ramos, Sarukhán & Piñero again stresses that most of the exciting phenomena in plant population biology occur at edges—where interactions between plants, and the elements in population processes can be followed in the performance of individuals. The patchiness may in itself allow density- and frequency-dependent processes to be operating within the fine scale of a mosaic, yet not be detectable in the larger scale of the community (Fowler 1988).

It might be expected that, if generalizations are to appear in plant population biology, it will be the study of annual plants that contributes most. This is because it is clearly easy to follow whole life-cycles of several generations in relatively short research programmes and because they are especially suited to experimental population studies. The most extensive data set is that of Symonides (1988). Her data show that a whole variety of population dynamics may be found among annual plants—the only feature they seem to have in common is that the generation time is short. Perhaps the important point is that, 20 years ago, it would have been quite impossible to recognize similarities and differences in the population dynamics of different groups of plants. Data sets for the natural dynamics are now beginning to be assembled and it is at last possible to see commonalities beginning to emerge—though perhaps not across the groupings where they might have been expected.

Data sets of the quality of those obtained by Symonides are needed in much greater numbers if generalizations are to emerge. There are few technological advances that substitute for immensely laborious field census-compiling. There are few funding bodies prepared to invest in research programmes of appropriate length. The science of population biology depends on committed individuals, often working over long periods and often in their spare time, to produce the data sets that justify the activities of the experimentalists and theoreticians. At present the balance between those producing data about population processes in the real world and those attempting interpretation appears to be grossly unbalanced. Sadly, a century of phytogeography and descriptive community studies has assembled records of the world flora that are virtually useless to those who wish to interpret how populations behave and how vegetation comes to be what it is.

## REFERENCES

**Antolin, M. F. & Strobeck, C. (1985).** The population genetics of somatic mutation in plants. *American Naturalist,* **126,** 52–62.

**Arber, A. (1950).** *The Natural Philosophy of Plant Form.* Cambridge University Press, London.

**Bradshaw, A. D. (1965).** Evolutionary significance of phenotypic plasticity in plants. *Advances in Genetics,* **13,** 115–55.

**Burdon, J. J. (1987).** *Diseases and Plant Population Biology.* Cambridge University Press, London.

**Carter, R. N. & Prince, S. D. (1988).** Distribution limits from a demographic viewpoint. *Plant Population Ecology* (Ed. by A. J. Davy, M. J. Hutchings & A. R. Watkinson), pp. 165–84. Blackwell Scientific Publications, Oxford.

**Deregibus, V. A., Sanchez, R. A., Casal, J. J., Trlica, M. J. (1985).** Tillering responses to enrichment of red light beneath the canopy in a humid natural grassland. *Journal of Applied Ecology*, **22**, 199–206.

**Durrant, A. (1958).** Environmental conditioning of flax. *Nature (Lond.)*, **181**, 928–9.

**Durrant, A. (1962).** The induction of heritable changes in *Linum. Heredity*, **17**, 27–61.

**Durrant, A. (1971).** The induction and growth of flax genotrophs. *Heredity*, **27**, 277–98.

**Evans, D. A. & Sharp, W. (1983).** Single gene mutations in tomato plants regenerated from tissue culture. *Science*, **221**, 949–51.

**Foley, R. (1987).** *Another Unique Species: Patterns in Human Evolutionary Ecology.* Longman, London.

**Fowler, N. (1988).** The effects of environmental heterogeneity in space and time on the regulation of populations and communities. *Plant Population Ecology* (Ed. by A. J. Davy, M. J. Hutchings & A. R. Watkinson), pp. 249–69. Blackwell Scientific Publications, Oxford.

**Gibbs, A. (1980).** A plant virus that partially protects its wild legume host against herbivores. *Intervirology*, **13**, 42–7.

**Gill, D. E. (1986).** Individual plants as genetic mosaics: ecological organisms versus evolutionary individuals. In *Plant Ecology* (Ed. by M. J. Crawley), pp. 321–43. Blackwell Scientific Publications, Oxford.

**Gottlieb, L. D. (1977).** Genotypic similarity of large and small individuals in a natural population of the annual plant *Stephanomeria exigua* spp. *coronaria* (Compositae). *Journal of Ecology*, **65**, 127–34.

**Gouyon, P. H., Gliddon, C. J. & Couvet, D. (1988).** The evolution of reproductive systems: a hierarchy of causes. *Plant Population Ecology* (Ed. by A. J. Davy, M. J. Hutchings & A. R. Watkinson), pp. 23–33. Blackwell Scientific Publications, Oxford.

**Greig-Smith, P. (1983).** *Quantitative Plant Ecology.* Blackwell Scientific Publications, Oxford.

**Grime, J. P., Mackey, J. M. L., Hillier, S. H. & Read, D. J. (1987).** Floristic diversity in a model system using experimental microcosms. *Nature (Lond.)*, **328**, 420–2.

**Grubb, P. J. (1977).** The maintenance of species-richness in plant communities: the importance of the regeneration niche. *Biological Reviews*, **52**, 107–45.

**Hallé, F. (1986).** Modular growth in seed plants. *The Growth and Form of Modular Organisms* (Ed. by J. L. Harper, B. Rosen & J. White), pp. 77–88. The Royal Society, London.

**Hallé, F. & Oldemann, R. A. A. (1970).** *Essai sur l'architecture et dynamique de croissance des arbres tropicaux.* Masson, Paris.

**Hallé, F., Oldemann, R. A. A. & Tomlinson, P. B. (1978).** *Tropical Trees and Forests: an Architectural Analysis.* Springer-Verlag, Berlin.

**Harper, J. L. (1985).** Modules, branches and the capture of resources. *Population Biology and Evolution of Modular Organisms.* (Ed. by J. B. C. Jackson, L. W. Buss & R. E. Cook), pp. 1–34. Yale University Press, New Haven.

**Harper, J. L., Rosen, B. & White, J. (1986).** Preface. *The Growth and Form of Modular Organisms* (Ed. by J. L. Harper, B. Rosen & J. White), pp. 3–6. The Royal Society, London.

**Harper, J. L. & Ogden, J. (1970).** The reproductive strategy of higher plants. I. The concept of strategy with special reference to *Senecio vulgaris. Journal of Ecology*, **58**, 681–98.

**Hassell, M. P. (1978).** *The Dynamics of Arthropod Predator-Prey Systems.* Princeton University Press, Princeton, New Jersey.

**Hassell, M. P. (1981).** Arthropod predator-prey systems. *Theoretical Ecology. Principles and Applications.* (Ed. by R. M. May), pp. 105–31. Blackwell Scientific Publications, Oxford.

Henry, J. D. & Swan, J. M. A. (1974). Reconstructing forest history from live and dead plant material—an approach to the study of forest succession in southwestern New Hampshire. *Ecology*, **55**, 772–83.

Hubbell, S. P. (1979). Tree dispersion, abundance and diversity in a tropical dry forest. *Science*, **203**, 1299–309.

Hubbell, S. P. & Foster, R. B. (1986). Biology, chance and history, and the structure of tropical tree communities. *Community Ecology* (Ed. by J. M. Diamond & T. J. Case), pp. 314–29. Harper and Row, New York.

Huffaker, C. B. (1964). Fundamentals of Biological Weed Control. *Biological Control of Pests and Weeds*. (Ed. by P. DeBach & E. I. Schlinger), pp. 74–117. Chapman and Hall, London.

Hutchings, M. J. & Slade, A. J. (1988). Morphological plasticity, foraging and integration in clonal perennial herbs. *Plant Population Ecology* (Ed. by A. J. Davy, M. J. Hutchings & A. R. Watkinson), pp. 83–109. Blackwell Scientific Publications, Oxford.

Kira, T., Ogawa, H. & Sakazaki, K. (1953). Intraspecific competition among higher plants. I. Competition-density-yield inter-relationships in regularly dispersed populations. *Journal of the Institute of Polytechnics, Osaka City University*, **D4**, 1–16.

Klekowski, E. J., Kazarinova-Fukshansky, N. & Mohr, H. (1985). Shoot apical meristems and mutation: stratified meristems and angiosperm evolution. *American Journal of Botany*, **72**, 1788–800.

Lerner, I. M. (1958). *The Genetic Basis of Selection*. John Wiley, New York.

Levin, D. A. (1988). Plasticity, canalization and evolutionary stasis in plants. *Plant Population Ecology* (Ed. by A. J. Davy, M. J. Hutchings & A. R. Watkinson), pp. 35–45. Blackwell Scientific Publications, Oxford.

MacClement, W. D. & Richards, M. G. (1956). Virus in wild plants. *Canadian Journal of Botany*, **34**, 793–9.

Mackenzie, S. (1985). Reciprocal transplantation to study local specialisation and the measurement of components of fitness. Ph.D. thesis, University of Wales.

Marshall, D. L., Levin, D. A. & Fowler, N. L. (1986). Plasticity of yield components in response to stress in *Sesbania macrocarpa* and *Sesbania vesicaria* (Leguminosae). *American Naturalist*, **127**, 508–21.

Martinez-Ramos, M., Sarukhán, J. & Piñero, D. (1988). The demography of tropical trees: the case of *Astrocaryum mexicanum* at Los Tuxtlas tropical rainforest. *Plant Population Ecology* (Ed. by A. J. Davy, M. J. Hutchings & A. R. Watkinson), pp. 293–313. Blackwell Scientific Publications, Oxford.

May, R. M. (1981). Models for single populations. *Theoretical Ecology: Principles and Applications*. 2nd Edn. (Ed. by R. M. May), pp. 5–29. Blackwell Scientific Publications, Oxford.

Rozema, J., Scholten, M. C. T., Blaauw, P. A. & Van Diggelen, J. (1988). Distribution limits and physiological tolerances with particular reference to the salt marsh environment. *Plant Population Ecology* (Ed. by A. J. Davy, M. J. Hutchings & A. R. Watkinson), pp. 137–64. Blackwell Scientific Publications, Oxford.

Schaal, B. A. (1988). Somatic variation and genetic structure in plant populations. *Plant Population Ecology* (Ed. by A. J. Davy, M. J. Hutchings & A. R. Watkinson), pp. 47–58. Blackwell Scientific Publications, Oxford.

Solangaarachchi, S. M. & Harper, J. L. (1987). The effect of canopy filtered light on the growth of white clover *Trifolium repens*. *Oecologia*, **72**, 372–6.

Symonides, E. (1988). Population dynamics of annual plants. *Plant Population Ecology* (Ed. by A. J. Davy, M. J. Hutchings & A. R. Watkinson), pp. 221–48. Blackwell Scientific Publications, Oxford.

Watson, M. A. (1984). Developmental constraints: effects on population growth and patterns of resource allocation in a clonal plant. *American Naturalist*, **123**, 411–26.

**Watson, M. A. & Casper, B. B. (1984).** Morphogenetic constraints on patterns of carbon distribution in plants. *Annual Review of Ecology and Systematics*, **15**, 233–58.

**Watt, A. S. (1947).** Pattern and process in the plant community. *Journal of Ecology*, **35**, 1–22.

**Weiner, J. (1988).** Variation in the performance of individuals in plant populations. *Plant Population Ecology* (Ed. by A. J. Davy, M. J. Hutchings & A. R. Watkinson), pp. 59–81. Blackwell Scientific Publications, Oxford.

**White, J. (1985).** The census of plants in vegetation. *The Population Structure of Vegetation. Handbook of Vegetation Science, Part III.* (Ed. by J. White), pp. 33–88. Junk, Dordrecht.

**Whitham, T. G. & Slobodchikoff, C. N. (1981).** Evolution by individuals, plant herbivore interactions, and mosaics of genetic variability: the adaptive significance of somatic mutations in plants. *Oecologia*, **49**, 287–92.

**Whitmore, T. C. (1988).** The influence of tree population dynamics on forest species composition. *Plant Population Ecology* (Ed. by A. J. Davy, M. J. Hutchings & A. R. Watkinson), pp. 271–91. Blackwell Scientific Publications, Oxford.

**Williams, P. H. & Hill, C. B. (1986).** Rapid-cycling populations of *Brassica. Science*, **232**, 1385–9.

**Wit, C. T. de (1960).** On competition. *Verslagen van Landbouwkundige Onderzoekingen*, **66**, 1–82.

**Wright, S. (1931).** Evolution in Mendelian populations. *Genetics*, **16**, 97–159.

**Wright, S. (1932).** The role of mutation, inbreeding, crossbreeding, and selection in evolution. *Proceedings of the 6th Congress of Genetics*, **1**, 356–66.

**Wright, S. (1935).** Evolution in populations in approximate equilibrium. *Journal of Genetics*, **30**, 257–66.

**Yoda, K., Kira, T., Ogawa, H. & Hozumi, K. (1963).** Self-thinning in overcrowded pure stands under cultivated and natural conditions. *Journal of Biology*, Osaka City University, **14**, 107–29.

# AUTHOR INDEX

Page numbers in *italic* indicate the main entry

# SUBJECT INDEX

468